Environmental Chemistry

A Modular Approach

Environmental Chemistry

A Modular Approach

Ian Williams

University of Central Lancashire, Preston, UK

JOHN WILEY & SONS, LTD

Chichester • New York • Weinheim • Brisbane • Singapore • Toronto

Other Wiley Editorial Offices

John Wiley & Sons, Inc., 605 Third Avenue,
New York, NY 10158-0012, USA

Wiley-VCH Verlag GmbH, Pappelallee 3,
D-69469 Weinheim, Germany

John Wiley & Sons Australia Ltd, 33 Park Road, Milton,
Queensland 4064, Australia

John Wiley & Sons (Asia) Pte Ltd, 2 Clementi Loop #02-01,
Jin Xing Distripark, Singapore 0512

John Wiley & Sons (Canada) Ltd, 22 Worcester Road,
Rexdale, Ontario M9W 1L1, Canada

British Library Cataloguing in Publication Data

A catalogue record for this book is available from the British Library

ISBN 0 471 48941 7 (ppc)   ISBN 0 471 48942 5 (pbk)   ac ✓

Typeset in 10.5/12.5pt Palatino by Laser Words, Chennai, India
Printed and bound in Great Britain by Bookcraft (Bath) Ltd
This book is printed on acid-free paper responsibly manufactured from sustainable
forestry, in which at least two trees are planted for each one used for paper production.

This book is dedicated to the memory of
Dr Elizabeth Mayer-Abraham.

Contents

4 The Structure and Composition of the Earth

Contents

Preface

As a child, one of my ambitions was to play rugby union for Llanelli RFC and Wales. Not surprisingly, given my lack of talent for the game, the best that I managed was to be a regular substitute for my school team. However, as an adult, one of my ambitions was to write a textbook that reflected my interest in the study of environmental science, particularly environmental chemistry. I wanted to write a concise and affordable textbook that covered not just theoretical aspects of environmental chemistry and geochemistry, but also the essential areas of report writing and safe laboratory practice. This present textbook is the result.

The study of environmental chemistry is fundamental to an understanding of the natural and anthropogenic processes that occur on our planet. An understanding of these processes is needed if we are going to protect the fragile world in which we live and support the concept of sustainable development. Environmental chemistry is a large subject area, and identifying topics for inclusion in this book was troublesome. To complicate matters, students studying environmental science do not necessarily have academic backgrounds in chemistry. Consequently, this book includes key concepts and skills that are required for the study of environmental chemistry, as well as specific topic areas. Such topic areas include the following: the structure and composition of the Earth, key processes in the formation, extraction, purification of metals, minerals and fossil fuels, the abundance, properties and cycling of selected chemical elements, the properties and environmental chemistry of water and aqueous solutions, and selected aspects of environmental pollution. Each chapter ends with a selection of self-study exercises that are intended to help students review their progress, as well as more challenging exercises to stretch more advanced students.

This textbook is suitable for students taking a wide variety of undergraduate courses, including specialist courses in environmental science, chemistry, earth science, geology, physical geography and ecology, as well as students taking options on diverse types of joint and modular degree programmes. Some of the material may also be suitable for those students studying for pre-university qualifications.

The initial ideas and structure of this book was planned in collaboration with my friend and colleague at Middlesex University, Dr Elizabeth Mayer-Abraham, and without her input, this book would have never evolved. Sadly, Elizabeth died prematurely in 1999, and this book is dedicated to her memory. In addition, I would like to acknowledge the huge contribution of Professor Mike Revitt, also from Middlesex University. Mike has been a tremendously supportive and generous mentor, colleague and friend throughout my academic career, and several sections of this book have been influenced by his meticulous lectures and practical skills.

Many others have contributed to the production of this textbook. I would like to thank colleagues who reviewed parts or all of the manuscript and who provided many helpful suggestions throughout its development; these include Neil Ward, Hemda Garelick, Paul Fanning, Ken Hill, Mike Chapple, Dennis Elam and Malcolm Fox. Rob Newitt (of Rob Newitt Associates Ltd) and Catherine Clarke assisted in the creation of many of the original figures used in the book. I would also like to thank the staff of John Wiley & Sons, Ltd for their support, encouragement and invaluable suggestions about the structure, style and content of the text; in particular, I am indebted to Andrew Slade and David Ando.

Ian Williams
Preston, UK
November 2000

1 Concepts and Skills in Environmental Chemistry

I cannot give any scientist of any age better advice than this; the intensity of a conviction that a hypothesis is true has no bearing over whether it is true or not.

Peter Medawer, *Nobel Prize Winner*, 1979.

1.1 Introduction

Environmental chemistry is an essential component of the study of Environmental Science. Students study environmental science for a number of reasons, including general interest, career progression and development, the desire to obtain skills from a variety of subject areas and deep concern for the future of Planet Earth. Environmental science students tend to have a wider range of professional or academic backgrounds and skills than students studying other, more narrowly focused, scientific disciplines. Consequently, there is often a need at the beginning of an environmental science course to introduce students to the key concepts and skills necessary for the study of environmental chemical (and biological) processes and laboratory work. This introductory chapter aims to provide students with an understanding of these essential concepts and a step-by-step guide to fundamental practical skills such as weighing chemicals, diluting solutions and preparing calibration standards.

1.2 The Importance of Units

Numbers are meaningless without units; there is a big difference between walking one metre and walking one kilometre. Scientists have developed a systematic and internationally agreed set of units to measure quantities such as time, mass and length – the Système International (d'Unités) (SI), based on the metric system – which is used for almost all scientific work. The SI base units, or fundamental units, are the metre (m) for length, kilogram (kg) for mass, second (s) for time, ampere (A) for electric current, kelvin (K) for thermodynamic temperature,

Table 1.1 Common SI prefixes

Prefix	Name	Power
T	Tera	10^{12}
G	Giga	10^{9}
M	Mega	10^{6}
k	Kilo	10^{3}
d	Deci	10^{-1} (1/10)
c	Centi	10^{-2} (1/100)
m	Milli	10^{-3}
μ	Micro	10^{-6}
n	Nano	10^{-9}
p	Pico	10^{-12}
f	Femto	10^{-15}

mole (mol) for quantity of matter and candela (cd) for luminous intensity (luminosity). Other units, such as the watt, are recognized by the SI system if they can be expressed as products or divisions of powers of the base units.

A number of non-SI units are well established and internationally recognized. These include the minute, hour and day (time), the litre (volume) and the tonne (mass). However, in general, this book will use SI units, e.g. dm^3 rather than litre.

The Système International uses combinations of base units and prefixes. The most common SI prefixes are listed in Table 1.1. These denote multiples of powers of 10 of the SI units themselves. For example, the millimetre (mm) is $1/1000$ (10^{-3}) of a metre. The term milli (m) can be used with any units and always means $1/1000$ of the unit (e.g. mg, ms, mmol, mA, ml, etc.).

In environmental chemistry, we frequently perform experiments in order to determine the concentrations of substances in air, water and soil. The units in which the results are expressed vary with the type of sample.

Pollutants found in air can be either gaseous or particulate. The concentrations of gaseous pollutants may be expressed as follows:

- The mass of gaseous pollutant per unit volume of air, usually in $\mu g\ m^{-3}$.

- The volume of gaseous pollutant per unit volume of air, e.g. in $\mu l\ dm^{-3}$.

- The volume mixing ratio, usually in parts per million (ppm) (10^6), parts per billion (ppb) (10^9), or parts per trillion (ppt) (10^{12}). This unit expresses the concentration of a pollutant as a ratio of its volume if segregated pure, to the volume of the polluted air in which it is contained, e.g. 1 ppm is equivalent to a cubic centimetre of gas in a cubic metre of air. The volume mixing ratio is sometimes written in different ways, e.g. ppm may be also be written as ppm(v) or vpm, and you should be aware of these variations.

For particulate pollutants, only the mass of pollutant per unit volume of air is applicable, e.g. $\mu g\ m^{-3}$.

In water analyses, we are often trying to determine the mass of a substance in a known volume of water. It is convenient to report the data as a mass per litre (dm^3) of water (e.g. $mg\ dm^{-3}$, $\mu g\ dm^{-3}$, or $ng\ dm^{-3}$). Using an agreed set of units (as appropriate) makes the comparison and subsequent interpretation of experimental or environmental data much easier and more convenient. However, you will also see concentrations in water reported as ppm (grams of analyte per million grams of water). Since 1 dm^3 of water weighs 1000 g, we can write the following:

$$1\,000\,000 \text{ g of water} = 1000 \text{ dm}^3 \text{ of water}$$

and thus:

$$1 \text{ ppm} = 1 \text{ g of analyte per million grams of water}$$
$$= 1 \text{ g of analyte per } 1000 \text{ dm}^3 \text{ of water}$$
$$= 1 \text{ mg of analyte per dm}^3 \text{ of water}$$

i.e. in water, ppm units are equivalent to $mg\,dm^{-3}$. The use of ppm for concentrations in solution should be discouraged as there is potential for confusion with the volume mixing ratio for gases (this is particularly problematic when analysing solutions derived from air sampling).

In soil analyses, we usually report the mass of a substance per unit mass of dry soil (soil samples are usually air-dried and sieved prior to analysis). The data are often reported as mg of substance per kg of dry soil ($mg\,kg^{-1}$).

The time over which a particular pollutant is measured – the *averaging time* – is important with respect to atmospheric pollutant concentrations. Hourly, daily and weekly concentrations are often calculated and used when defining air quality standards.

1.2.1 Conversions between units

Although the metric system is internationally recognized, it has not been universally adopted. Many countries, including the United States and the United Kingdom, use the Imperial (or English) system of units. In addition, some metric units are used interchangeably, e.g. one litre equals 1 dm^3 (1000 cm^3). Some useful conversion factors are listed in Table 1.2. For example, 1 imperial pound (lb) is equal to 0.453 kg; consequently, 5.600 lb is equal to 5.600×0.453 kg $= 2.537$ kg.

1.3 Relative Atomic Mass

An atom is the smallest unit of matter that can take part in a chemical reaction. It is made up of protons and neutrons in a central nucleus surrounded by orbiting electrons and it cannot be broken down chemically into anything smaller. Consequently, the masses of atoms are very small – typically $10^{-24} - 10^{-22}$ g. These numbers are awkward to use mathematically and in chemical calculations. Thus, instead of using the actual masses of atoms, relative atomic masses are used.

The relative atomic mass (A_R) is defined as the mass of an atom relative to one-twelfth of the mass of an atom of carbon-12. Thus, carbon-12 is assigned a mass of exactly 12 atomic mass units (amu) and all other atoms are relative to this

Table 1.2 Useful conversion factors

Electrical	$°F = (9/5 \times °C) + 32$
1 watt = 1 joule per second	$0\,K = 0°C + 273.16$

Fluid Flow Rates

1 litre per minute = 15.85 gallons per hour
1 cubic foot per second = 448.8 gallons per minute
1 cubic foot per minute = 472 cubic centimetres per second
1 cubic centimetre per second = 3.60 litres per second

Length and Area

1 millimetre = 0.1 centimetres
1 centimetre = 0.3937 inches
1 inch = 2.54 centimetres
1 foot = 30.48 centimetres
1 metre = 39.37 inches
1 metre = 0.001 kilometres
1 metre = 100 centimetres
1 metre = 3.28 feet
1 kilometre = 0.62 miles
1 mile = 1.61 kilometres
1 square metre = 10 000 square centimetres
1 square metre = 10.76 square feet
1 acre = 4047 square metres
1 hectare = 2.471 acres

Temperature

$°C = (°F - 32) \times 5/9$

Time

1 day = 86 400 seconds
1 day = 1440 minutes
1 year = 3.1536×10^7 seconds (based on 1 year being equal to 365 days)
1 year = 5.256×10^5 minutes
1 year = 8.760×10^3 hours

Volume and Mass

1 litre = 1000 millilitres
1 gallon = 3.785 litres
1 cubic metre = 35.31 cubic feet
1 pound = 453.59 grams
1 kilogram = 1000 grams
1 metric ton = 1000 kilograms
1 cubic centimetre = 1 millilitre
1 litre = 1000 millilitres
1 cubic metre = 1×10^6 cubic centimetres
1 cubic metre = 35.314 cubic feet
1 cubic metre = 1000 litres
1 cubic foot = 7.481 gallons
1 cubic foot = 28.32 litres
1 gallon = 3.785 litres
1 milligram = 0.001 grams
1 gram = 0.001 kilograms
1 kilogram = 2.2 pounds
1 pound = 453.6 grams

standard. This can be represented mathematically by the following equation:

$$A_R = \frac{\text{mass of one atom of an element}}{1/12 \text{ mass of one atom of carbon-12}} \tag{1.1}$$

For example, the actual mass of one atom of carbon-12 is $1.992\,68 \times 10^{-23}$ g, while the mass of one atom of helium is $6.646\,58 \times 10^{-24}$ g:

$$A_R(C) = \frac{\text{mass of one atom of carbon-12}}{1/12 \text{ mass of one atom of carbon-12}} = \frac{1.992\,68 \times 10^{-23}}{1.660\,56 \times 10^{-24}} = 12 \text{ amu}$$

$$A_R(He) = \frac{\text{mass of one atom of helium}}{1/12 \text{ mass of one atom of carbon-12}} = \frac{6.646\,58 \times 10^{-24}}{1.660\,56 \times 10^{-24}} = 4 \text{ amu}$$

This shows us that one atom of carbon is three times heavier than one atom of helium.

Similarly, one atom of carbon-12(^{12}C) is (approximately):

- 12 times the mass of an atom of hydrogen ($A_R = 1.00797$);

- 1/12 the mass of an atom of neodymium ($A_R = 144.24$);

- 1/16 the mass of an atom of iridium ($A_R = 192.20$).

Consequently, we can say that 1 g of hydrogen, 4 g of helium, 12 g of carbon, 144 g of neodymium and 192 g of iridium all contain the same number of atoms. The relative atomic masses of the 109 known elements are shown in the Periodic Table on the inside cover at the back of this book. You should be aware that the relative atomic mass of an element is obtained by calculating the weighted mean of all the naturally occurring *isotopes* of the element. Isotopes are varieties of the same element, differing in the neutron content of their nuclei, and therefore in their masses. The relative atomic mass of carbon is calculated by using the relative abundances of ^{12}C (98.93%) and ^{13}C (1.07%) and is thus $(12 \times 0.9893) + (13 \times 0.01007) = 12.0107$.

1.4 The Mole

The Italian physicist, Amedeo Avogadro, calculated that the number of carbon atoms in 12 g of carbon-12 is 6.022045×10^{23}. This value is known as the Avogadro constant. It follows from the discussion in Section 1.2 above that the relative atomic mass of any element, expressed in grams, contains 6.022045×10^{23} atoms. For convenience, we call 6.022045×10^{23} units of any species a *mole* of that particular substance. A mole is the SI unit which represents the amount of a substance that contains as many elementary entities (atoms, molecules, etc.) as there are atoms in 12 g of carbon-12. The mass of one mole of any element is equal to its relative atomic mass expressed in grams.

1.5 Relative Molecular Mass

The relative molecular mass (M_R) of a compound is defined as the mass of one molecule of the compound relative to one-twelfth of the mass of an atom of carbon-12. This is represented mathematically by the following equation:

$$M_R = \frac{\text{mass of one molecule of a compound}}{1/12 \text{ mass of one atom of carbon-12}} \tag{1.2}$$

The M_R of a compound expressed in grams is a mole of the compound. For example, 28 g of dinitrogen (N_2) is a mole of N_2, and contains 6.022045×10^{23} molecules.

1.6 Molar Mass

The mass of one mole of a compound (molecular or ionic) is given the term *molar mass*. The latter may be determined if we know the relative atomic masses of the atoms that make up the molecule or the ionic substance. For example, a glucose molecule is composed of 6 carbon atoms, 12 hydrogen atoms and 6 oxygen atoms, ($C_6H_{12}O_6$). Thus, the molar mass of glucose can be calculated as follows:

Atom	Number of atoms	A_R	Total mass
C	6	12.01	$6 \times 12.01 = 72.06$
H	12	1.008	$12 \times 1.008 = 12.10$
O	6	16.00	$6 \times 16.00 = 96.00$

$$\text{Molar mass } (C_6H_{12}O_6) = 180.16 \text{ g mol}^{-1}$$

Similarly, the molar mass of aluminium oxide (Al_2O_3) is obtained as follows:

Atom	Number of atoms	A_R	Total mass
Al	2	26.98	$2 \times 26.98 = 53.96$
O	3	16.00	$3 \times 16.00 = 48.00$

$$\text{Molar mass } (Al_2O_3) = 101.96 \text{ g mol}^{-1}$$

The number of moles of a substance present in a given mass may be calculated by using the following equation:

$$\text{Number of moles of a substance (mol)} = \frac{\text{mass of substance (g)}}{\text{molar mass (g mol}^{-1})} \quad (1.3)$$

For example, the number of moles in 500 g of calcium carbonate (molar mass = 100 g mol^{-1}) is as follows:

$$\frac{\text{mass of calcium carbonate (g)}}{\text{molar mass of calcium carbonate (g mol}^{-1})} = \frac{500}{100} = 5 \text{ mol}$$

1.7 Chemical Reactions and Equations

A chemical reaction is the process of chemical change. The direction of the chemical change is symbolized by an arrow that points from the starting materials (the reactants) to the substances that are formed (the products):

$$\text{Reactants} \longrightarrow \text{Products}$$

An equation is a shorthand description of a chemical process (reaction) written by using chemical formulae (symbols) to represent the reactants and the products. Chemical formulae use the symbols for each element as outlined in the Periodic Table. An equation can describe any chemical process, whether it occurs in the laboratory or in the natural environment. Equations tell us which substances react together, which substances are formed, and the amounts involved. For example, when a small piece of potassium metal is added to water, it reacts to form potassium hydroxide and hydrogen gas. To summarize this reaction, we can write the following skeletal equation:

$$\textbf{Reactants} \longrightarrow \textbf{Products}$$

$$\text{potassium} + \text{water} \longrightarrow \text{potassium hydroxide} + \text{hydrogen gas}$$

$$K + H_2O \longrightarrow KOH + H_2 \tag{1.4}$$

A skeletal equation is a qualitative summary of a chemical reaction – it tells us which substances are present without indicating their quantities or physical states. A full, balanced, chemical equation enables us to quantitatively understand chemical processes in a shorthand (symbolic) fashion.

In order to derive a complete chemical equation for the reaction between potassium and water, we need to understand that no potassium or water molecules are created or destroyed in the process. This means that there must be the same number of potassium, oxygen and hydrogen atoms (water is composed of oxygen and hydrogen atoms) at the beginning and at the end of the reaction. The expression presented in Equation 1.4 shows that there are two H atoms on the left of the arrow, but three H atoms on the right – an *unbalanced* chemical equation. Modifying Equation 1.4 to give Equation 1.5 balances the chemical equation – there are now two K atoms, two O atoms and four H atoms on each side, as follows:

$$2K + 2H_2O \longrightarrow 2KOH + H_2 \tag{1.5}$$

The physical state of each reactant and product can be indicated using a state symbol – (s) for solid, (l) for liquid, (g) for gas, and (aq) for an aqueous solution. State symbols are usually written as subscripts after the substance whose physical state they are representing. Thus, for the reaction between potassium and water, the full, balanced, chemical equation is:

$$2K(s) + 2H_2O(l) \longrightarrow 2KOH(aq) + H_2(g) \tag{1.6}$$

A number of other examples of complete chemical equations are shown in Box 1.1.

Box 1.1 Examples of Complete Chemical Equations

nitrogen + hydrogen $\longrightarrow$ ammonia (Haber Process)
$N_2(g)$ + $3H_2(g)$ $\longrightarrow$ $2NH_3(g)$

iron + copper (II) sulfate $\longrightarrow$ iron (II) sulfate + copper
$Fe(s)$ + $CuSO_4(aq)$ $\longrightarrow$ $FeSO_4(aq)$ + $Cu(s)$

carbon dioxide + water $\longrightarrow$ glucose + oxygen (photosynthesis)
$6CO_2(g)$ + $6H_2O_{(ll)}$ $\longrightarrow$ $C_6H_{12}O_6(s)$ + $6O_2(g)$

1.7.1 Balancing chemical equations

Chemical equations obey the general rules of algebraic equations in that they must be balanced, i.e. there must be the same number of atoms of each element on both sides of the equation. The numbers in front of the formulae (the stoichiometric coefficients) may be changed to balance an equation, but the formulae themselves (subscript numbers) must not be changed, as this will change the substance involved in the chemical reaction. For example, nitric oxide is represented by the formula NO, while nitrogen dioxide is represented by the formula NO_2.

To balance a chemical equation quickly and efficiently, you can use the following stepwise procedure:

(1) Write out the skeletal equation.

(2) Balance first the element that occurs in the fewest chemical formulae.

(3) Balance last the element that occurs in the greatest number of chemical formulae.

(4) Specify the physical state of each formula.

A number of worked examples are shown in 1.1–1.4 below. The examples show that although balancing equations can seem very complex, the balancing procedure becomes much easier if you break it down into steps.

Example 1.1

When hydrogen gas reacts with chlorine gas, hydrogen chloride is formed.

1. Write out the skeletal equation.

$$H_2 + Cl_2 \longrightarrow HCl$$

2. Balance first the element that occurs in the fewest chemical formulae.
   Since both H and Cl occur in two formulae, we can balance them simultaneously by placing a two in front of the HCl, as follows:

$$H_2 + Cl_2 \longrightarrow 2HCl$$

3. Specify the physical state of each formula.

$$H_2(g) + Cl_2(g) \longrightarrow 2HCl(g)$$

Example 1.2

When zinc metal reacts with dilute hydrochloric acid, zinc chloride and hydrogen gas are formed.

1. Write out the skeletal equation.

$$Zn + HCl \longrightarrow ZnCl_2 + H_2$$

2. Balance first the element that occurs in the fewest chemical formulae.

Zn, Cl and H are each present in two formulae. However, there is only one zinc atom on each side of the equation, so we will try to leave it alone as it is already balanced. There are two chlorine and two hydrogen atoms on the right-hand side of the equation, with only one of each on the left. We can balance the equation by placing a two in front of the HCl, as follows:

$$Zn + 2HCl \longrightarrow ZnCl_2 + H_2$$

3. Specify the physical state of each formula.

$$Zn(s) + HCl(aq) \longrightarrow ZnCl_2(aq) + H_2(g)$$

Example 1.3

When iron metal reacts with pure gaseous oxygen, a blue-black solid of iron oxide is formed.

1. Write out the skeletal equation.

$$Fe + O_2 \longrightarrow Fe_3O_4$$

2. Balance first the element that occurs in the fewest chemical formulae.
   Fe and O are each present in two formulae. However, there is one iron and two oxygen atoms on the left, with three iron and four oxygen atoms on the right-hand side of the equation. Placing a three in front of the Fe on the left brings the iron atoms into balance. However, balancing the oxygen will be more complicated. The oxygen on the left will increase only in steps of two, while the oxygen on the right will increase only in steps of four. The important question to ask yourself is 'which is the least common multiple between 2 and 4?'. The answer is four, and so we will need four oxygen atoms on each side of the equation. This can be achieved by placing a two in front of the left-hand oxygen, as follows:

$$3Fe + 2O_2 \longrightarrow Fe_3O_4$$

3. Specify the physical state of each formula.

$$3Fe(s) + 2O_2(g) \longrightarrow Fe_3O_4(s)$$

Example 1.4

When methane burns in pure gaseous oxygen, carbon dioxide and water are formed.

1. Write out the skeletal equation.

$$CH_4 + O_2 \longrightarrow CO_2 + H_2O$$

2. Balance first the element that occurs in the fewest chemical formulae.
   C and H are each present in two formulae, while O is present in three. The C atoms are already balanced. Placing a two in front of the H_2O on the right balances the hydrogen atoms, as follows:

$$CH_4 + O_2 \longrightarrow CO_2 + 2H_2O$$

3. Balance last the element that occurs in the greatest number of chemical formulae. There are two oxygen atoms on the left-hand side and four oxygen atoms on the right-hand side of the equation. Placing a two in front of the left-hand oxygen balances the equation, as follows:

$$CH_4 + 2O_2 \longrightarrow CO_2 + 2H_2O$$

4. Specify the physical state of each formula.

$$CH_4(g) + O_2(g) \longrightarrow CO_2(g) + 2H_2O(g)$$

In some cases, balancing a chemical equation leads to fractional stoichiometric coefficients. For example, the equation for the combustion of ethane is as follows:

$$C_2H_6(g) + \tfrac{7}{2}O_2(g) \longrightarrow 2CO_2(g) + 3H_2O(g) \tag{1.7}$$

It is common to clear the fractional coefficient by multiplying the entire equation by a numerical factor. Multiplying Equation 1.7 by the denominator of the fractional coefficient (the number at the bottom of the fraction, in this case 2) gives the following:

$$2C_2H_6(g) + 7O_2(g) \longrightarrow 4CO_2(g) + 6H_2O(g)$$

1.8 Stoichiometry

As well as telling us which substances react together, equations also tell us the amount of each substance that reacts. For example, potassium chlorate is used in matches and fireworks as it releases oxygen when heated. The equation for its thermal decomposition is as follows:

$$2KClO_3(s) \longrightarrow 2KCl(s) + 3O_2(g) \tag{1.8}$$

Equation 1.8 shows that two moles of potassium chlorate decompose to give two moles of potassium chloride and three moles of oxygen gas. We can summarize this information by writing the following:

$$2 \text{ mol } KClO_3(s) \cong 2 \text{ mol } KCl(s) \tag{1.9}$$

$$2 \text{ mol } KClO_3(s) \cong 3 \text{ mol } O_2(g) \tag{1.10}$$

In chemistry, the '$\cong$' sign is called a stoichiometric relation and means 'is chemically equivalent to'. In calculations, this symbol is treated like an equal ($=$) sign.

We can use this stoichiometric relation to calculate the masses of material involved in the reaction. Since the molar masses are $KClO_3 = 122.55$, $KCl = 74.55$ and $O_2 = 32.00$ g mol^{-1}, from Equations 1.9 and 1.10, we can see that 245.1 g of $KClO_3$ will decompose to give 149.1 g of KCl and 96.0 g O_2.

As you can see, stoichiometry is the branch of chemistry that deals with quantitative relationships between the amounts of reactants and products in chemical reactions. In industrial and environmental chemical reactions, one reactant is usually present in excess of the stoichiometric amount required to react with the other(s). The excess material is left unused when the reaction is complete. The following worked examples illustrate the usefulness of the stoichiometric relation in industrial and environmental chemistry.

Example 1.5

More than 80% of the ammonia (NH_3) made industrially by the Haber Process is used as fertilizer. What mass of ammonia can be obtained from 20.0 kg of hydrogen during the Haber Process?

First, we write the complete chemical equation:

$$N_2(g) + 3H_2(g) \longrightarrow 2NH_3(g)$$

We then write the stoichiometric relation for this reaction:

$$3 \text{ mol } H_2 \cong 2 \text{ mol } NH_3$$

This can also be written as follows:

$$1 \text{ mole } H_2 \cong \tfrac{2}{3} \text{ mol } NH_3 \qquad (1.11)$$

The molar masses may be calculated as follows:

Molar mass of $H_2 = 2.016$ g mol^{-1}
Molar mass of $NH_3 = 17.034$ g mol^{-1}

The number of moles present in 20.0 kg of H_2 can be calculated from Equation 1.3, as follows:

$$\text{Number of moles of a substance (mol)} = \frac{\text{mass of substance (g)}}{\text{molar mass (g mol}^{-1})}$$

$$= \frac{20.0 \times 10^3 \text{ g}}{2.016} = 9920 \text{ moles}$$

By using Equation 1.11, we can see that the number of moles of NH_3 generated from 9920 moles of H_2 will be $2/3 \times 9920 = 6614$.

From Equation 1.3, the mass of NH_3 generated will be:

Number of moles of a substance (mol) $\times$ molar mass (g mol^{-1})

$$= \text{mass of substance (g)}$$

$$= 6614 \times 17.034 = 1.127 \times 10^2 \text{ kg.}$$

Example 1.6

Zinc is widely used to protect iron from corrosion (see Chapter 5), and may be obtained from its ore by reduction (also see Chapter 5). When 20.00 tonnes of zinc oxide are reduced by 15.00 tonnes of charcoal (carbon), what mass of zinc can be obtained?

The complete chemical equation is:

$$ZnO(s) + C(s) \longrightarrow Zn(s) + CO(g)$$

The stoichiometric relations show that:

$$1 \text{ mol ZnO} \cong 1 \text{ mol C} \cong 1 \text{ mol Zn}$$

The molar mass of ZnO may then be calculated as 81.37 g mol^{-1}.

The number of moles present in 20.00 tonnes of ZnO can be calculated from Equation 1.3:

$$= \frac{20.00 \times 10^6}{81.37} = 2.458 \times 10^5 \text{ moles}$$

The number of moles present in 15.00 tonnes of C is then:

$$= \frac{15.00 \times 10^6}{12.01} = 1.249 \times 10^6 \text{ moles}$$

Zinc oxide is present in the smaller amount. Consequently, the amount of zinc formed is limited by the amount of zinc oxide. In this reaction, zinc oxide is the limiting reagent – the reactant which governs the maximum yield of product in a given chemical reaction.

We can see that the number of moles of Zn generated from 20.00 tonnes of ZnO will be 2.458×10^5 moles. Thus, the mass of Zn generated will be:

Number of moles of a substance (mol) x molar mass (g mol^{-1}) = mass of substance (g)

$$= 2.458 \times 10^5 \times 65.37 = 16.07 \times 10^6 \text{ g} = 16.07 \text{ tonnes.}$$

1.9 Types of Reactions

Environmental processes utilize four types of chemical reactions: acid–base, oxidation–reduction (redox), precipitate formation and complex formation. Examples of these reactions in the environment are given in 1.9.1–1.9.4. We still study all four types in more detail in subsequent chapters.

1.9.1 Acid–base reactions

Sulfur dioxide (which is acidic), produced in the burning of coal for electricity, may be removed (neutralized) by lime (which is basic). This process is known as scrubbing:

$$SO_2(g) + Ca(OH)_2(aq) \longrightarrow CaSO_3(aq) + H_2O(l)$$

Acid–base reactions in the environment are discussed in detail in Chapter 7.

1.9.2 Oxidation–reduction reactions

Oxidation is strictly defined as the loss of electron(s) by an element, compound or ion, whereas reduction is the reverse process (i.e. the gain of electrons). A chemical change (reaction) where one reactant is oxidized and the other is simultaneously reduced is known as an *oxidation–reduction* or *redox* reaction. Such reactions are common in the environment, and many examples are given throughout the book, particularly in Chapters 4 and 5.

1.9.3 Precipitate formation

Deposition of fur (scale) in kettles and boilers from hard water is due to the formation of insoluble calcite ($CaCO_3$), as a precipitate:

$$Ca^{2+}(aq) + 2HCO_3^-(aq) \longrightarrow CaCO_3(s) + CO_2(g) + H_2O(l)$$

Tripolyphosphates are used in water softening and in detergents to bring about dissolution of insoluble metal compounds. Simple phosphates of most metals are insoluble, but the triphosphates are soluble due to complex formation.

1.9.4 Complex formation

When ammonia solution is added to a blue solution of copper (II) sulfate, a deep blue colour develops due to the formation of a complex substance, known as tetraamine copper (II) sulfate:

$$CuSO_4 + 4NH_3 \rightleftharpoons [Cu(NH_3)_4]SO_4$$

The determination of hardness in water is based on the reaction between Ca^{2+} and ethylenediaminetetraacetic acid (EDTA), in which a complex is formed (see Chapter 7).

Some reactions, for example the neutralization of acids by bases and the precipitation of chloride by silver ions, are instantaneous, while others, such as corrosion of metals (for example, on cars), may be slow. The latter, however, is speeded up by the presence of moisture and salt spray, as well as acidic gases, in the air. In industrial processes, *catalysts* are often used to speed up the rates of reactions. A catalyst is a substance that will increase the rate of a chemical reaction without being consumed in the reaction itself. Many environmental processes are catalysed by *enzymes* – biological catalysts. In some cases, slowing down a reaction is in our interest (inhibition) – negative catalysts are used for this purpose.

There are also certain reactions which are *reversible*. In these cases, a state of *equilibrium* is established in which the rate of the forward reaction equals the rate of the reverse reaction. In environmental systems, there are many examples of chemical equilibria. The dissociation of water molecules is perhaps the simplest example of chemical equilibrium:

$$H_2O \rightleftharpoons H^+ + OH^-$$

Chemical equilibria are explained in more detail in Box 1.2.

Box 1.2 Chemical Equilibria

Many chemical reactions go to *completion* – the reaction continues until one of the reactants is completely used. Many other reactions are *reversible* – they have a forward and a reverse reaction. For example, iron will react with steam upon heating to form a black solid oxide and hydrogen:

FORWARD REACTION $\quad 3Fe(s) + 4H_2O(g) \longrightarrow Fe_3O_4(s) + 4H_2(g)$

However, if we pass hydrogen gas over the heated iron oxide, iron and steam are produced:

REVERSE REACTION $\quad Fe_3O_4(s) + 4H_2(g) \longrightarrow 3Fe(s) + 4H_2O(g)$

If iron and steam are heated in a closed container (so that the products cannot escape), an *equilibrium* is established:

CONSTANT CONDITIONS $\quad 3Fe(s) + 4H_2O(g) \rightleftharpoons Fe_3O_4(s) + 4H_2(g)$

A chemical reaction where there is no further tendency for the composition to change under constant conditions (e.g. constant temperature and pressure) is at chemical equilibrium. This situation is represented in a chemical equation by using an equilibrium sign ($\rightleftharpoons$), which indicates both the forward and the reverse reactions. All chemical equilibria are dynamic; the forward and reverse reactions occur at the same rate, so the concentrations of reactants and products remain unchanged.

1.10 Solutions

A *solution* is a homogeneous mixture of two or more substances with no definite composition. A solution is *saturated* when no more of the solute will dissolve in it. *Solubility* is usually expressed as the mass of solute required to saturate 100 g of solvent at a given temperature. (Note that this is NOT a concentration term since it does not involve volume.)

Aqueous solutions play an important part in environmental processes by transporting chemicals between different environmental compartments, e.g. atmospheric pollutants such as SO_2 and NO_x are removed through dissolution in water and subsequent deposition on land via precipitation.

1.10.1 Molar solutions

The molarity of a solution is defined as the number of moles of solute in 1.00 dm^3 of solution. A number of equations can be used to calculate the molarity of a solution, for example:

$$\text{Molarity (M) (mol dm}^{-3}) = \frac{\text{number of moles of solute}}{\text{volume (dm}^3) \text{ of solution}} \qquad (1.12)$$

By using Equation 1.3:

$$M = \frac{(\text{g of solute})/(\text{molar mass of solute})}{\text{volume (dm}^3) \text{ of solution}}$$

$$= \frac{\text{g of solute}}{(\text{molar mass of solute}) (\text{dm}^3)}$$

Example 1.7

What is the molarity of an aqueous solution of 10.0 g of NaOH in 500 cm^3 of solution?

The molar mass of NaOH $= 23.0 + 16.0 + 1.0 = 40.0$ g, and therefore:

$$M = \frac{10.0}{40.0 \times 0.5} = 0.5 \text{ mol dm}^{-3}$$

1.10.2 Standard solutions

Standard solutions are solutions of known concentration. A standard solution is one against which a solution of unknown concentration can be compared to determine the concentration of the latter. They may be prepared by weighing, accurately, a solid reagent, which is then dissolved in an appropriate solvent and made up to a given volume in a volumetric flask. Although water is commonly used as a solvent for the preparation of standard solutions, not all materials are soluble in water; it is very common to prepare standard solutions of metals by dissolving the pure metal in acid, while organic materials are often dissolved in organic solvents.

Example 1.8

What weight of solid NaOH would you require to prepare 200 cm^3 of a 1.5 M aqueous solution?

$$M = \frac{\text{g of solute}}{(\text{molar mass of solute}) (\text{dm}^3)}$$

By cross-multiplying:

$$\text{g of solute} = M (\text{molar mass}) (\text{dm}^3)$$

$$= 1.5 \times 40.0 \times 0.2 = 12.0$$

1.10.3 Preparation of standard solutions

The accuracy of any analytical determination depends on the use of correct standards. It is therefore vital that standard solutions are prepared as accurately as possible. A step-by-step procedure for preparing aqueous solutions of water-soluble solids is outlined below; this procedure should obviously be adapted for the preparation of pure metal or organic standards.

(1)   Weigh the solid accurately in a suitable container on a calibrated balance (Section 1.11 outlines the procedure for weighing solids).

(2)   Transfer the solid quantitatively into an appropriately sized beaker. You may need to wash any small grains of solid that adhere to the container into the beaker by using double-distilled, deionized water.

(3)   Completely dissolve the solid in the beaker by using distilled water. You may need to use a glass rod to crush any large lumps of solid and to stir the solution.

(4)   Transfer the solution in the beaker to a clean (calibrated) volumetric flask of the desired capacity. You will need to use a funnel to avoid spilling the solution. Rinse the beaker and funnel twice with a small volume of distilled water to ensure quantitative transfer of the solution.

(5)   Make up the volume to the calibration mark of the volumetric flask with more distilled water. Take care not to exceed the calibration mark – you will have to start the procedure all over again if you do.

(6)   Close the volumetric flask with a closely fitting stopper and mix the solution thoroughly by inverting the flask a number of times.

In some cases, it is possible to purchase standard solutions of high concentration (top standards) from which laboratory standards can be made up by dilution (see Section 1.12).

1.11 Weighing Chemicals

Weighing chemicals accurately is a very important operation and must be performed with great care. A small weighing error can cause the results of any chemical analysis to become meaningless. A step-by-step procedure for the use of an electronic balance is outlined below.

(1)   Ensure that the balance to be used is in the appropriate range for weighing the sample. Weighing 10–20 g of soil requires a different balance from the one required for weighing 50–100 mg of soil (see Box 1.3). The precision of weighing also depends upon the size of the sample – a high-precision balance weighs to ± 0.01 mg, while a low-precision one may only weigh to ± 0.01 g.

(2)   Check that the balance is level, clean and free of material that could cause a false reading, e.g. loose soil or volatile liquid on the pan of the balance.

(3)   Check with a technician or tutor that the balance is calibrated. Balances should be routinely calibrated by using appropriate standard weights (e.g. 10.0000 g or 100.00 g).

(4)   Select an appropriate container to hold your sample. Sometimes, you will need to use a specialist 'weighing boat' – a plastic or glass container designed for holding samples to be weighed. Weighing boats come in many sizes and you should select an appropriate size for your sample and balance. For some samples, you will need to use a beaker or a crucible for weighing. The type

of container you select will depend on the type and nature of your sample, the quantity of material to be weighed and the type of experiment you are performing. Place your clean, empty container on to the balance.

(5) Close the door to the balance to exclude draughts.

(6) Zero the display with the container on the pan.

(7) Open the door and transfer the sample to the container by using a spatula or spoon. Ensure that you do not spill any of the sample material on to the pan.

(8) When you have transferred the desired amount of material into the container, close the door. Allow the reading to settle and then record the exact value in your logbook. If you have under- or over-estimated the amount of sample required, add or remove some sample and re-read the display until you have accurately weighed the amount you require.

Box 1.3 Types of Balance

Weighing small amounts of material (mg) with a high degree of accuracy requires an analytical balance. Analytical balances ('four-figure balances') are carefully designed to allow easy access, cleaning and reading and are completely enclosed in a draught-proof shield. Top-pan balances ('two-figure balances') are more rugged and are used for weighing larger amounts of material. All balances require calibration with a reference weight prior to use.

If you wish to weigh out tiny amounts of material – amounts in the μg range – you will need a microbalance. Such balances can weigh material down to 10^{-7} g. However, they are very expensive and require special conditions, including protection from vibration interference (using a vibration-damping mount or a special balance table) and a humidity- and temperature-controlled chamber.

1.12 Diluting Samples and Standards

In practice, we often need to dilute a solution to make it less concentrated. Dilution is a technique that is used a great deal in both wet and instrumental methods of analysis. It is carried out by adopting the following procedure:

- accurately measuring a known volume of standard solution with a calibrated pipette;

- placing the known volume into an appropriately sized, calibrated volumetric flask;

- adding the appropriate solvent (often double-distilled, deionized water) to the calibration mark.

The following equations may be used for dilution calculations:

$$M_t V_t = M_n V_n \tag{1.13}$$

or

$$C_t V_t = C_n V_n \tag{1.14}$$

where M_t, C_t and V_t represent, respectively, the molarity, concentration and volume of the top standard (the standard being used to prepare all other standards, and which has the highest concentration), and M_n, C_n and V_n represent the molarity, concentration and volume of the new standard, respectively.

Example 1.9

What volume of 6.00 M HCl is needed to prepare 250.0 cm³ of 1.50 M HCl?
    From Equation 1.13:

$$M_t V_t = M_n V_n$$

By cross-multiplying:

$$V_t = (M_n V_n)/M_t$$

$$V_t = (1.50 \times 0.25)/6.00 = 0.0625 \text{ dm}^3, \text{ or } 62.5 \text{ cm}^3$$

Example 1.10

A top standard of sodium chloride is available containing 50.0 mg dm^{-3} of Na. You need to dilute this top standard to make 100.0 cm³ standards containing 2.5, 5.0, 7.5 and 10.0 mg dm^{-3} Na.
    From Equation 1.14:

$$C_t V_t = C_n V_n$$

By cross-multiplying:

$$V_t = (C_n V_n)/C_t$$

$$V_t = (2.5 \times 100.0)/50.0 = 5.0 \text{ cm}^3$$

Thus, 5.0 cm³ of top standard is required in order to make up a 100.0 cm³ standard solution containing 2.5 mg dm^{-3} Na. Similar calculations show that 10.0, 15.0 and 20.0 cm³ of the top standard will be required for the other standards – check this for yourself.

Example 1.11

What volume of 0.1 M EDTA solution is required in order to make up 500 cm³ of 0.01 M solution?
    From Equation 1.14:

$$C_t V_t = C_n V_n$$

By cross-multiplying:

$$V_t = (C_n V_n)/C_t$$

$$V_t = (0.01 \times 500)/0.1 = 50 \text{ cm}^3$$

1.13 **Performing a Titration**

An environmental chemist will often be required to analyse an environmental sample for chemical information. In *qualitative analysis*, we try and determine which substances are present in our unknown sample. In *quantitative analysis*, we try and determine how much of a particular species (the *analyte*) is present in our unknown sample. Quantitative analysis provides numerical information, often in the form of a concentration term, about the analyte (see Section 1.2 above).

The methods used for quantitative analysis of environmental samples may be categorized as either *classical* or *instrumental* methods, although this distinction is largely historical and somewhat artificial. Classical methods of chemical analysis include traditional techniques such as gravimetry, titrimetry and volumetry; we will study some of these techniques later in this book. Instrumental methods all depend upon the use of a suitable instrument to determine some physico-chemical parameter of an analyte, such as an electrical or an optical property. Instrumental methods are generally faster than classical methods, and are perfectly suited to the performance of a large number of routine determinations. All analytical methods involve the correlation of a physical measurement with the analyte concentration.

We can use our knowledge of standard solutions to help us perform a quantitative determination, i.e. determine the unknown concentration of a species in an environmental sample or solution. One method of doing this is a *titration*, in which a standard solution is added from a burette to a sample in a conical flask (laboratory equipment is discussed in Section 2.3). A step-by-step procedure for a typical titration, for example, in determining the carbonate concentration of a lake water sample, is outlined below.

(1)   Gather together all the equipment you will need, e.g. a calibrated burette held by a boss and clamp to a retort stand, a large white ceramic tile, conical flasks, graduated and calibrated pipettes, and distilled water.

(2)   Check that all the apparatus is clean. Rinse the burette with distilled water, checking that the liquid flows smoothly down the walls of the burette and through the tap. The burette should then be rinsed thoroughly with the standard solution before being clamped in a vertical position to the retort stand. The pipette should also be rinsed with distilled water and then with the water sample to be determined.

(3)   Fill the burette with standard solution by using a funnel. Make sure you allow some standard to run into a conical flask, checking that no air bubbles become trapped.

(4)   Pipette a known volume – often 25 cm^3 – of the water sample into a conical flask. Add a few drops of an appropriate indicator. Place the conical flask under the burette so that the tip of the burette is inside the conical flask while still allowing the burette tap to turn freely.

(5)   Read and record the volume of standard in the burette. Make sure that you record the reading in your laboratory notebook. Use a white tile or a piece of lined white paper to help you to read the position of the *meniscus* – the curved surface of the liquid. The bottom of the meniscus is normally read.

(6) Slowly add the standard from the burette into the conical flask whilst swirling its contents (most analysts use their right hand for swirling the conical flask and their left for controlling the tap). You will find that the colour of the indicator changes as the titration proceeds. The desired final colour of the indicator appears at the end point of the titration. You will need to add the standard dropwise when you get very close to the end point. Overshooting the endpoint will obviously result in an erroneous reading.

(7) Read and record the final volume of the standard used in your notebook.

In order to obtain an accurate value for your titration, you will normally need to perform at least three replicate titration procedures on the water sample. Your first (or rough) titration will enable you to observe the change of indicator colour at the end point and also to estimate the volume of standard required. Thus, the rough titration value is normally an overshoot of the true value. You should then repeat the titration procedure until two concordant values have been obtained.

1.14 Calibration Procedures

The term *calibration* may be defined as the preparation of a scale that can be used to determine the values of unknown quantities. The scale (a calibration curve) is prepared by using a calibration standard (a standard solution) in order to obtain a reading from an instrument. A series of instrument readings are measured and recorded by using calibration standards of progressively higher concentrations, including a blank in order to establish a zero reading (see Section 1.15 below) (you must ensure that each instrument reading is stable before recording the value). The calibration curve is then produced by plotting the instrument response (e.g. a measured pH or absorbance value) on the *y*-axis versus the concentration of the analyte on the *x*-axis (see Section 3.8.4 later for more information about plotting graphs). An instrument reading is then obtained for the analyte and this can be compared to the response of the calibration standards. In this way, the concentration of the analyte can be determined.

The method of calibration can be illustrated by the following example.

Example 1.12

In the spectrophotometric determination of iron based on the conversion of iron (II) into a phenanthroline complex, the absorbance of the complex was measured at a wavelength of 508 nm. The results obtained with standard solutions of iron are shown in Table 1.3.

A 5.0 cm^3 aliquot of a water sample, treated in the same way and made up to 100.0 cm^3 with double-distilled, deionized water, gave an absorbance reading of 0.420.

The calibration curve for these data is plotted in Figure 1.1, together with the equation for the straight line calculated by using the method of least squares (see Chapter 3). We can see that the value of 0.420 corresponds to a concentration of

Table 1.3 Results obtained in the spectrophotometric determination of iron at a wavelength of 508 nm

Concentration of iron(II) (mg dm^{-3})	Absorbance
0.0	0.000
1.0	0.243
2.0	0.479
2.5	0.583
4.0	0.952
Unknown	0.420

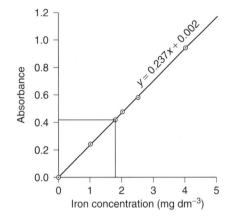

Figure 1.1 Plot of iron concentration versus absorbance obtained by using the data given in Table 1.3

1.8 mg dm^{-3}. As the 5.0 cm^3 sample was diluted to 100.0 cm^3, the concentration of iron(II) (C_1) in the original water sample can be calculated in the following way.

From Equation 1.14:

$$C_1 V_1 = C_2 V_2$$

By cross-multiplying:

$$C_1 = (C_2 V_2)/V_1$$

$$C_1 = (1.8 \times 100.0)/5.0 = 36.0 \text{ mg dm}^{-3}$$

Although, in practice, we do not usually know the concentration of the samples to be determined, it is useful to have some idea of the range of concentrations within which the samples might fall. An estimate of the expected concentration range can be made by studying the scientific literature. This will enable you to prepare calibration standards that cover the expected concentration range of the unknown samples. If the analytical method is only applicable to a limited concentration range (and many methods are), the samples will have to be diluted to bring them into the appropriate range.

1.15 Blank Determinations

Establishing an appropriate zero reading is an important feature of analytical determinations in both wet chemical and instrumental methods. It is particularly important in the analysis of environmental materials where samples with different chemical compositions and concentrations are being determined. Impurities may be introduced into the analyte through the reagents or containers being used in experiments. The error introduced via this type of contamination may be eliminated, or at least minimized, by running blank determinations.

A blank is a sample containing a negligible (or zero) amount of the analyte being determined. Blanks are used either to assess the amount of contamination in analytical determinations or to correct for constant, unavoidable contamination. There are many different types of blanks used in environmental analytical determinations, including the solvent (or reagent), trip and method blanks, as follows:

- a solvent blank is used to check the purity of solvents;

- a trip (or field) blank is used to determine any accidental contamination that may occur when samples are collected in the field;

- a method blank is used to determine any contamination that may occur during sample preparation and analysis (e.g. from laboratory bench surfaces, reagents, glassware, etc.)

In a blank determination, the analyte is omitted and replaced by double-distilled, deionized water (or another appropriate solvent). The experimental procedures are then followed exactly as with the analyte – the same amount of reagent is added, the same volume of blank is used, and the same temperature and the same period of time are employed. Any reading that is obtained from the blank can be assumed to come from contamination and a blank correction can be utilized if necessary.

1.16 Precision and Accuracy

In everyday language, the terms precision and accuracy are often used interchangeably. However, in environmental science, they have different meanings.

Precision refers to the closeness of agreement between replicate determinations, i.e. the variability of a measurement. If replicate determinations produce identical or almost identical values, we can say that the results have been obtained with high precision. However, results obtained with high precision may not have been obtained with high accuracy because there may be an error affecting all of the results in the same way – a *systematic error*. Precision is most frequently quantified in terms of the *standard deviation* of a number of replicate determinations (see Section 3.8.4 below), although other statistical indicators (relative standard deviation or coefficient of variation) are useful when comparing the precisions of analyses performed in different laboratories.

Accuracy refers to the closeness of agreement between the experimental results and the 'true result' – the ultimately correct value. However, in the analysis of

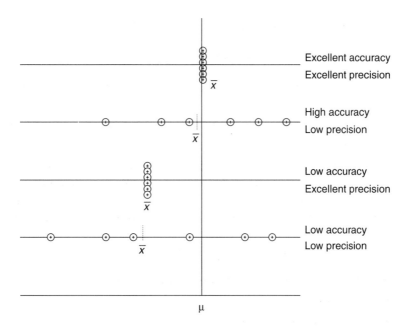

Figure 1.2 Scatter diagram illustrating the difference between precision and accuracy in environmental analysis

environmental samples, true results are rarely known; here, scientists are striving to produce the best *estimate* of the true result. This may be achieved by using a number of different analytical methods for the same determination and by checking analytical methodologies by using certified and/or standard reference materials. Accuracy can only be quantified in terms of *error*, which can similarly only be estimated (see Section 1.19 below). The difference between precision and accuracy is demonstrated as a scatter diagram in Figure 1.2.

It must be emphasized that a single analysis of a single sample, whether it is water, air or soil, will only provide a 'snapshot' of the concentration of the analyte in question. In order to obtain a better estimate of the 'true' concentration of the analyte, a series of representative samples must be analyzed in order to obtain an 'average value' (see Section 3.8.4 below). In the analysis of environmental samples, you should be aware that many factors can affect the accuracy of your analytical data, including losses or contamination during sample collection, storage, extraction and pre-concentration, as well as daily or seasonal variations. For example, the concentration of dissolved oxygen in a water sample is likely to be higher during the day (due to photosynthesis) than at night, while the organic content of topsoil is higher than that of the lower soil horizons.

1.17 Errors

Analytical determinations are subject to errors that may be reduced, but not completely eliminated. In fact, results of analytical determinations are best regarded as an *estimate* of the true value. The result therefore will only be of limited value until the magnitude of the error can be quantified by using appropriate statistical techniques.

Errors may be divided into the following:

- *Determinate errors*, which have a definite value and can, at least in principle, be measured.
- *Indeterminate errors*, which do not have a definite value and may fluctuate in a random manner.

Determinate errors include personal errors resulting from carelessness or prejudices of the analyst, instrumental errors and method errors resulting from non-ideal chemical behaviour of the reagents used. Indeterminate (or accidental) errors arise from uncertainties in a measurement that cannot be controlled by the analyst.

Errors can be minimized by the following:

- calibration of all apparatus and application of corrections;
- use of blanks;
- use of control samples;
- use of certified and standard reference materials (CRMs and SRMs);
- running of parallel determinations (i.e. duplicates and triplicates);
- use of different, independent methods of analytical determination;
- proper maintenance of equipment;
- use of skilled staff, staff training and development.

1.18 Sensitivity

The sensitivity of an analytical method or instrument is a measure of its ability to distinguish between small differences in the concentration of an analyte. Suppose we have two analyte concentrations, i.e. X and Y, which are very low and close in value. An instrument with a high sensitivity will generate a signal for X that is readily detectable and distinguishable from the signal for Y.

In effect, sensitivity is the change in signal per unit of concentration and thus may be determined by measuring the slope of the calibration curve at the concentration of interest (the calibration sensitivity). Most calibration curves are linear, and of the following form:

$$S_i = mc_i + b \tag{1.15}$$

where S_i is the signal at concentration c_i, b is the intercept (i.e. the blank) and m is the slope of the calibration curve (i.e. the sensitivity). In Example 1.12 (see Section 1.14), the calibration sensitivity equals the slope of the calibration curve, i.e. 0.237 mg dm^{-3}.

In practice, the ability to distinguish two analyte concentrations is dependent upon the precision of the determinations, as well as the sensitivity of the method.

1.19 Detection Limits

The detection limit (or limit of determination) of an analytical method or instrument may be defined as the lowest mass or concentration of analyte that can meaningfully be detected (i.e. discriminated from zero or the blank). There are several definitions of detection limit, but the one that we shall use in this present text is as follows:

$$\text{detection limit} = 3\sigma_{\text{blank}} \qquad (1.16)$$

where σ_{blank} is the standard deviation of the blank (see Chapter 3 for a fuller discussion of standard deviation). It is easy to assume that the standard deviation of the blank increases with its absolute magnitude and that low detection limits are associated with low blanks (at high sensitivity). However, this is not always the case. Suppose, for example, that an environmental scientist has used two methods for determining the concentration of cadmium in drinking water. Method 1 gives a reagent blank of 20.00 ± 0.02 (σ_{blank}) ng dm^{-3} of Cd, while Method 2 gives a reagent blank of 2.0 ± 0.4 ng dm^{-3} of Cd. The environmental scientist runs a standard solution containing 0.30 ng dm^{-3} of Cd using both methods. In Method 1, the instrument will give a response equivalent to 20.30 ng dm^{-3}. If both the reagent blank and the sample have a standard deviation of 0.02 ng, the standard can easily be distinguished from the blank. However, in Method 2, the instrument will give a response of 2.3 ± 0.4 ng dm^{-3}, which is not easily distinguished from the reagent blank. Thus Method 1, with the higher blank, has a lower detection limit than Method 2.

1.20 Significant Figures and Rounding

In calculating the results of your experiments, the numerical figures obtained in your measurements will have to be processed and recorded. It is important that the results you report have the correct number of significant figures. The accuracy of your data must not be sacrificed, but equally, it is important not to imply an unjustified degree of accuracy. Significant figures express the magnitude of the number (experimental result) to a specified degree of accuracy by including all of the digits known with accuracy plus one which is uncertain. In the case of zero, this is significant when it is part of a number, but not where it merely indicates the magnitude. This procedure may sound complicated, but after seeing some examples, you will see that it is quite straightforward. For example, 6 543 543 to three significant figures is 6 540 000, while 0.007 321 9 to two significant figures is 0.0073. In this textbook, you should treat trailing zeros as significant unless you are instructed differently.

The use of calculators and computer software enables analysts to obtain experimental results containing a string of digits; it is imperative that these are rounded to the correct number of significant figures. There are some simple rules to follow which will help you, as shown in the following:

- Non-zero digits are all significant, including zeros between numbers.

- Zeros on the left of a non-zero digit are not significant (they are only used to locate the decimal point).

- Zeros to the right of a decimal point that are preceded by a significant figure are significant.

- The final significant figure is rounded up if the following digit is greater than 5 and down if it is below 5. For numbers ending in 5, always round to the nearest even number. Rounding should only be performed at the final stage of the calculation.

Alternatively, you can express the data in scientific notation with one non-zero digit in front of the decimal point. Counting the number of digits will give you the number of significant figures.

Example 1.13

- 321.03 g of soil, which is known within limits of $\pm$ 0.01 g, has five significant figures. A mass of 321.03 g is the same as 3.2103×10^2 g, i.e. a quantity with five significant figures.

- 0.0531 g, which is known within the same absolute limits, has three significant figures.

- The number 33 has two significant figures, while 2.000 has four significant figures.

In multiplication or division, calculated data should not be expressed to more significant figures than the factor in the calculation with the least number of significant figures.

Example 1.14

Determine the result of the following calculation to the correct number of significant figures.

$$y = \frac{4.00 \times 0.064\,05 \times 293}{0.78}$$

By using a calculator, we obtain $y = 96.239\,231$. The factor 0.78 has the least number of significant figures – only two. Consequently, the calculated result should be rounded off to two significant figures, thus giving a correct answer of $y = 96$.

In addition or subtraction, the calculated result should have no more figures after the decimal point than the least number of figures after the decimal point in any of the numbers that are being added or subtracted.

Example 1.15

Determine the result of the following calculation to the correct number of significant figures.

$$y = 4.717 + 3.009\,19 = 7.726\,19 \text{ (round off to 7.726)}$$

The last two digits in 7.726 19 are not significant, because we know the value of the first number in the addition (4.717) to only three digits beyond the decimal point.

Example 1.16

Determine the result of the following calculation to the correct number of significant figures.

$$y = 7241.85 - 7240 = 1.85$$

Subtraction yields 1.85, which must be rounded off to 2 as the second number (7240) has no digits to the right of the decimal point.

There is no point in quoting more digits than the precision of any procedure allows.

1.21 Summary

After careful study of this chapter, you will have an understanding of the key concepts and skills necessary for the study of environmental chemical (and biological) processes and the performance of basic laboratory work. You can now use the following self-study exercises to test your knowledge of these basic principles. Such fundamental practical skills will be of great advantage performing when the experiments outlined in Chapters 4–7.

Self-Study Exercises

The relative atomic masses required for performing your calculations can be found on the inside cover at the back of the book.

Units

1.1 Express the following quantities in scientific notation:

(a) 0.000 000 000 096 (the approximate length of the O–H bond in water in metres);

(b) 37 000 000 000 (the number of distintegrations per second in 1 curie);[†]

(c) 0.000 548 (the atomic mass of one electron in atomic mass units);

(d) 30 000 000 000 (the velocity of light in centimetres per second);

(e) 8760 (the number of hours in one year);

(f) 602 204 500 000 000 000 000 000 (the Avogadro constant in atoms per mole);

(g) 0.000 000 000 000 000 000 000 000 001 666 (one atomic mass unit in kilograms).

1.2 Convert the following quantities into standard scientific notation:

(a) 1 200 000 g;

(b) 0.000 001 m;

[†] A curie is a unit of activity of a radioactive nuclide, being roughly equal to the activity of 1 g of radium.

(c) $15\,000 \times 10^2$ dm^3;

(d) $1\,550 \times 10^{-2}\,$°C;

(e) 0.0275×10^5 Å;[‡]

(f) 0.0005 mol;

(g) 17.5×10^{-3} kg.

Conversions between Units

1.3   Complete the following equalities:

(a) 1 µg = g;

(b) 22.3 cm^3 = dℓ;

(c) 15.2 µm = pm;

(d) 1 800 cm = nm;

(e) 0.15 dℓ = ml;

(f) 15.9 µg = ng;

(g) 0.002 ℓ = µℓ;

(h) 5 000 pm = fm.

1.4   Complete the following equalities:

(a) 1 Mm = Gm

(b) 173.51 kg = g

(c) 32.5 Ml = Gℓ

(d) 0.007 Tg = Mg

(e) 3500 kℓ = ℓ

(f) 13.752 Gg = Tg

(g) 0.087 652 m = Mm

Molar Mass

1.5   Calculate the molar mass of the following:

(a) Calcium carbonate ($CaCO_3$);

(b) Sulfuric acid (H_2SO_4);

(c) Ammonium sulfate ((NH_4)$_2SO_4$);

[‡] Å, angstrom – a unit of wavelength for electromagnetic radiation covering the visible light and X-ray regions; $1\,Å = 10^{-10}$ m.

    (d)   Ethanoic acid (CH_3COOH);

    (e)   Sucrose ($C_{12}H_{22}O_{11}$);

    (f)   Fayalite (Fe_2SiO_4);

    (g)   Chromite ($FeCr_2O_4$).

1.6   Calculate the number of moles in the following:

    (a)   250 g of calcium carbonate;

    (b)   1 710 g of sucrose;

    (c)   48.045 kg of silicic acid (H_4SiO_4);

    (d)   69.275 mg of $KClO_4$;

    (e)   25.50 µg of $(NH_4)_2S$;

    (f)   1.1 ng of sodium ethanoate (CH_3COONa);

    (g)   392 mg of albite ($Na(AlSi_3O_8)$).

1.7   Calculate the mass corresponding to the following:

    (a)   0.5 moles of sodium hydroxide;

    (b)   3.4 moles of nitric acid;

    (c)   0.012 5 moles of rhodocrosite ($MnCO_3$);

    (d)   4.5 millimoles of Fe_2O_3;

    (e)   15.78 micromoles of glucose;

    (f) 1.45×10^{-3} moles of kaolinite ($AlSi_4O_{10}(OH)_8$).

Balancing Chemical Equations

1.8   Balance the following chemical equations:

    (a) $H_2 + I_2 \longrightarrow HI$;

    (b) $Na + Cl_2 \longrightarrow NaCl$;

    (c) $K + H_2O \longrightarrow KOH$;

    (d) $S_8 + F_2 \longrightarrow SF_6$;

    (e) $Fe + O_2 \longrightarrow Fe_2O_3$;

    (f) $O_2 \longrightarrow O_3$;

    (g) $C_2H_6 + O_2 \longrightarrow CO_2 + H_2O$

1.9   Balance the following chemical equations:

    (a) $Fe_2O_3(s) + CO(g) \longrightarrow Fe(s) + CO_2(g)$ (manufacture of steel);

    (b) $N_2H_4(g) + N_2O_4(g) \longrightarrow N_2(g) + H_2O(g)$ (combustion of rocket propellant);

(c) $NH_3(g) + O_2(g) \longrightarrow NO(g) + H_2O(g)$ (first stage in Oswald Process to manufacture nitric acid);

(d) $CO_2(g) + NH_3(g) \longrightarrow H_2O(l) + CO(NH_2)_2(s)$ (preparation of urea, a commercial fertilizer);

(e) $ZnS(s) + O_2(g) \longrightarrow ZnO(s) + SO_2(g)$ (first stage in the extraction of zinc from its ore);

(f) $C_3H_5(NO_3)_3(l) \longrightarrow N_2(g) + H_2O(g) + CO_2(g) + O_2(g)$ (detonation of nitroglycerine);

(g) $CaO(s) + P_4O_{10}(s) \longrightarrow Ca_3(PO_4)_2(l)$ (use of quicklime to remove impurities from molten iron).

Stoichiometry

1.10 Explain what is meant by the stoichiometry of a chemical equation.

1.11 Explain the term 'limiting reagent'.

1.12 How many grams of ammonia $(NH_3(g))$ can be produced from 16.50 g of $H_2(g)$, assuming an excess amount of $N_2(g)$ is available? How many grams of $N_2(g)$ are required? (See Example 1.5 for the complete chemical equation.)

1.13 How many metric tons of aluminium can be obtained from 50 metric tons of bauxite, $(Al_2O_3.2H_2O)$, the principal ore of aluminium, assuming that the ore is pure and dry?

Molar Solutions

1.14 How many moles of NaOH are present in the following:

(a) 50 cm^3 of 0.75 M NaOH(aq);

(b) 150 cm^3 of 0.05 M NaOH(aq);

(c) 250 cm^3 of 0.003 75 M NaOH(aq);

(d) 3.5 dm^3 of 0.001 25 M NaOH(aq)?

1.15 How many moles of $FeSO_4$ are present in the following:

(a) 70 cm^3 of 0.35 M $FeSO_4$(aq);

(b) 0.05 dm^3 of 0.015 M $FeSO_4$(aq);

(c) 1×10^{-3} dm^3 of 2.4 M $FeSO_4$(aq);

(d) 8.5 µdm^3 of 0.125 M $FeSO_4$(aq)?

1.16 Potassium dichromate and iron (II) ammonium sulfate (FAS) are both used as reagents in the determination of the organic matter content of soil (see Chapter 4).

(a) How many moles of FAS $(FeSO_4.(NH_4)_2SO_4)$ are present in the following:

(i) 0.5 cm^3 of 0.5 M FAS(aq);

(ii) 2.0 dm^3 of 0.045 M FAS(aq);

(iii) 15×10^5 cm^3 of 0.035 M FAS(aq);

(iv) 25×10^2 µdm^3 of 1.73 M FAS(aq)?

(b)   What mass of potassium dichromate ($K_2Cr_2O_7$) is required to prepare 3.5 dm^3 of 0.167 M $K_2Cr_2O_7$(aq)?

Standard Solutions

1.17   What weight of solid KOH would you require to prepare the following:

(a)   150 cm^3 of a 0.75 M aqueous solution;

(b)   5.5 dm^3 of a 0.067 M aqueous solution;

(c)   350 cm^3 of a 2.75 M aqueous solution;

(d)   2.73 dm^3 of a 0.001 25 M aqueous solution?

1.18   What weight of sucrose ($C_{12}H_{22}O_{11}$) would you require to prepare the following:

(a)   32.68 cm^3 of a 1.5 M aqueous solution;

(b)   70 µdm^3 of a 0.067 M aqueous solution;

(c)   5.02 cm^3 of a 1.85 M aqueous solution;

(d)   0.32 dm^3 of a 0.001 25 M aqueous solution?

1.19   What volume of 5 M H_2SO_4 is required to prepare the following:

(a)   250 cm^3 of 3 M H_2SO_4;

(b)   0.5 dm^3 of 0.75 M H_2SO_4;

(c)   1.5 dm^3 of 0.025 M H_2SO_4;

(d)   37.5 cm^3 of 0.475 M H_2SO_4?

Dilutions

1.20   A top standard of lead nitrate is available containing 750 mg dm^{-3} of Pb. How much of this top standard is required to prepare 250 cm^3 standards containing 5.0, 10.0, 15.0 and 20.0 mg dm^{-3} Pb, respectively?

1.21   A top standard of zinc chloride is available containing 10 mg dm^{-3} of Zn. How much of this top standard is required to prepare 10 cm^3 standards containing 5.0, 10.0, 15.0 and 20.0 µg dm^{-3} Zn, respectively?

1.22   A top standard of cadmium nitrate is available containing 5 mg dm^{-3} of Cd. How much of this top standard is required to prepare 50 cm^3 standards containing 25, 50, 75 and 100 ng dm^{-3} Cd, respectively?

1.23   What volume of 0.5 M $CeCl_2$ solution is required in order to make up 250 cm^3 of 0.075 M solution?

1.24   What volume of 2.5 M $KMnO_4$ solution is required in order to make up 100 cm^3 of 0.375 M solution?

1.25   What volume of 0.08 M $K_2Cr_2O_7$ solution is required in order to make up 20 cm^3 of 0.001 25 M solution?

Calibration

1.26   In the spectrophotometric determination of phosphate by a colorimetric method, the absorbance of the determined solutions was measured at a wavelength of 880 nm. The results for the blank, calibration standards and three water samples (Unknowns 1–3) are shown in the table below. The water samples were diluted by pipetting 25 cm^3 of sample into a 100 cm^3 volumetric flask and making up to the mark with double-distilled, deionized water. Plot the calibration graph in an appropriate fashion and use it to determine the phosphate concentrations of the three unknowns.

Concentration of phosphate (mg dm^{-3})	Absorbance
0.000	0.000
0.125	0.101
0.250	0.193
0.500	0.394
1.000	0.806
Unknown 1	0.067
Unknown 2	0.375
Unknown 3	0.658

1.27   In the determination outlined in Exercise 1.26, a fourth unknown water sample gave an absorbance reading of 0.926. How would you determine its concentration?

Blanks

1.28   Explain how an analytical method with a high blank can have a lower detection limit than a different method which has a lower blank.

Precision and Accuracy

1.29   Explain the difference between precision and accuracy.

1.30   Explain how it is possible to have high accuracy without high precision.

Errors

1.31   Distinguish between the following terms:

   (a) Random and systematic errors;

   (b) Determinate and indeterminate errors.

Sensitivity

1.32 Explain why sensitivity is a function of precision.

1.33 Calculate the calibration sensitivity of the curve you plotted in Exercise 1.26.

Detection Limits

1.34 Distinguish between the terms 'sensitivity' and 'the limit of determination'.

Significant Figures and Rounding

1.35 State the number of significant figures in each of the following numbers:
  (a) 0.0165;
  (b) 0.016 50;
  (c) 1.6500×10^5;
  (d) the number 165;
  (e) 16 005;
  (f) 1 650 000 000;
  (g) 6.022×10^{23}.

1.36 Calculate z to the correct number of significant figures in each of the following:
  (a) $341.29 - 339 = z$;
  (b) $(42.3)^3 = z$;
  (c) $102/36.04 = z$;
  (d) $537.4689 + 11.45 + 39\,287 = z$;
  (e) $(6.022 \times 10^{23})\,(2.7 \times 10^{-2}) = z$;
  (f) $(6.626\,196 \times 10^{-34})\,(2.997\,925 \times 10^9)/(1.380\,62 \times 10^{-23}) = z$;
  (g) $(8.559\,011\,9 \times 10^{-21} + 2.527\,61 \times 10^{-17} - 2.841\,67 \times 10^{-17})\,(2.997\,925 \times 10^8) = z$.

Challenging Exercises

Molar Mass

1.37 Calculate the molar mass of the following:
  (a) Lead (II) arsenate ($Pb_3(AsO_4)_2$);
  (b) Vitamin B_{12} ($C_{63}H_{88}CoN_{14}O_{14}P$);
  (c) Phosphate rock ($Ca_{10}(OH)_2(PO_4)_6$);
  (d) Hexamminecobalt(III) chloride ($[Co(NH_3)_6]Cl_3$);
  (e) Tetraamminediaquacopper (II) chloride ($[Cu(NH_3)_4(H_2O)_2]Cl_2$);

(f)   Parathion ($C_{10}H_{14}NO_5PS$);

(g)   Vanadinite (($PbO)_9(V_2O_5)_3PbCl_2$).

Balancing Chemical Equations

1.38   Balance the following chemical equations:

(a) $Ca_{10}(OH)_2(PO_4)_6(s) + H_2SO_4(l) \longrightarrow H_3PO_4(l) + CaSO_4(s) + H_2O(l)$ (conversion of phosphate rock to phosphoric acid);

(b) $KCl(aq) + HNO_3(aq) + O_2(aq) \longrightarrow KNO_3(aq) + Cl_2(g) + H_2O(l)$ (manufacture of potassium nitrate fertilizer);

(c) $Ca_{10}F_2(PO_4)_6(s) + H_2SO_4(l) \longrightarrow HF(g) + Ca(H_2PO_4)_2(s) + CaSO_4(s)$ (reaction of the mineral fluoroapatite with sulfuric acid);

(d) $C_8H_{18}(l) + O_2(g) \longrightarrow CO_2(g) + H_2O(l)$ (combustion of octane, the main component of gasoline);

(e) $C_4H_{10}(g) + O_2(g) \longrightarrow CO_2(g) + H_2O(g)$ (combustion of butane in cigarette lighters);

(f) $As_2S_3(s) + O_2(g) \longrightarrow As_4O_6(s) + SO_2(g)$;

(g) $As_4O_6(s) + 3C(s) \longrightarrow CO_2(g) + As(l)$ (conversion of sulfides of arsenic to its oxides, followed by reduction to the element).

Stoichiometry

1.39   Propane, C_3H_8, is a common fuel which burns in oxygen according to the equation: $C_3H_8(g) + 5O_2(g) \longrightarrow 3CO_2(g) + 4H_2O(l)$. Calculate the following:

(a)   the number of grams of O_2 which are required to burn 125.0 g of C_3H_8;

(b)   the number of grams of H_2O and CO_2 that are subsequently produced.

Molar Solutions

1.40   What molarity of a $CaCl_2(aq)$ solution should you use if you want the molarity of the $Cl^-(aq)$ in the solution to be 0.200 M?

1.41   Calculate the molarity of scandium nitrate ($Sc(NO_3)_3$) and of the $Sc^{3+}(aq)$ and $NO_3^{3-}(aq)$ in a solution containing 5.72 g of scandium nitrate in 200 cm^3 of solution.

Standard Solutions

1.42   How many cm^3 of commercial phosphoric acid (14.6 M) are required to prepare 0.5 dm^3 of 0.550 M $H_3PO_4(aq)$?

1.43   How many grams of $CuSO_4.5H_2O(s)$ are required to produce 250 cm^3 of an aqueous solution that contains 300 mg dm^{-3} $Cu^{2+}(aq)$?

2 Laboratories for Environmental Chemistry

Where observation is concerned, chance favours only the prepared mind.
Louis Pasteur, *French chemist*, 1854.

2.1 Introduction

Environmental pollutants may have a detrimental effect on human health and the environment at extremely small concentrations. Advances in modern technology and in methods of environmental analysis have enabled environmental chemists to accurately determine concentrations of pollutants down to pg dm^{-3}, and even lower, thus enabling us to monitor our surroundings and protect our environment, our ecosystems and ourselves. However, in order to obtain reliable and accurate measurements, environmental chemists must follow good laboratory practice and ensure that any analytical method is suitable for its intended purpose. Developing the knowledge and skills necessary to perform environmental analyses to professional standards requires a detailed knowledge of laboratory equipment, practice and procedures.

Laboratories for environmental chemistry are hazardous places. They frequently contain chemicals and equipment that can potentially cause significant harm to laboratory workers and damage to the laboratory. In addition, environmental samples sometimes contain organisms that may constitute a microbiological hazard, e.g. river water, sewage effluent. Consequently, laboratory workers require a high level of health and safety awareness, proper training in safety procedures and appropriate safety equipment and clothing.

Laboratory health and safety is the responsibility of everybody – laboratory staff, tutors and students – and its importance cannot be emphasized enough. Imagine how you would feel if your carelessness or recklessness caused a chemical burn to a friend or started a major fire. Laboratories are generally safe places providing everybody follows the correct procedures and behaves responsibly.

Working in a laboratory also means that you have to record accurately the methodology of your experiments, the equipment and chemicals used and the results you obtain. Students are often tempted to scribble their data on scraps of paper, which they then use at home to write-up their practical reports tidily. This is a mistake for two reasons:

- Scraps of paper are easily lost. This leads to incomplete or inaccurate data sets (and thus the loss of valuable marks) and sometimes to the non-submission of coursework.

- It is unprofessional practice – the proper recording of scientific information is a professional requirement of all scientists since it is essential to demonstrate the 'traceability' of all recorded data.

This present chapter aims to:

- outline the main features of laboratories used for environmental chemistry;

- identify typical laboratory equipment, including safety equipment;

- outline the main safety rules and procedures that should be followed when performing laboratory activities;

- identify the responsibilities of laboratory technicians, tutors and students;

- explain the purpose and format of a laboratory notebook.

2.2 A Typical Laboratory for Environmental Chemistry

A typical undergraduate laboratory for environmental chemistry will have similar features anywhere in the world, although the layout of laboratory furniture and equipment can vary widely. Laboratories frequently contain several rows of work benches facing a desk from which a tutor will direct the class. Work benches in most modern laboratories are carefully designed to allow sufficient space for an individual to work safely and also allow ease of access in case of an accident or emergency. In addition, many laboratories will contain side benches along its walls and furniture such as stools, fume cupboards, incubators, drying cabinets, ovens and refrigerators. A separate room or preparation area for technical staff is often found adjacent to the laboratory.

The side and work benches are typically coated with a water-impermeable surface which will resist chemical attack from acids, bases and organic chemicals. The utilities available at each workspace may include electrical plug points, water, sinks, gas taps (for Bunsen burners), gas taps for drying glassware (compressed air) and use in preparative chemistry (e.g. nitrogen and argon), and a temporary storage area for chemicals in use (often a raised shelf). Underneath the benches, there are sometimes cupboards and drawers for storing equipment, tools and glassware.

However, laboratories for environmental chemistry will also have a number of specialist features that you may not notice immediately. These features may be divided into two categories, i.e. equipment that you will use regularly, and safety equipment. You should make a mental note of their locations when you first use the laboratory.

Equipment that you will use regularly includes the following:

- distilled and double-distilled, deionized water, available in plastic dispensing bottles which require refilling from a central still or large plastic container;
- balances for weighing samples, chemicals, etc.;
- paper towels for clearing spills;
- sinks for washing your hands and adjacent hand-dryers;
- separate bins for disposing of broken glass and paper/plastic waste (never cross-contaminate these bins – it can be dangerous for staff when they remove the waste);
- specialist locations or containers for the disposal of chemical waste such as acids, organic chemicals and wastewater samples (again, you should never cross-contaminate chemicals – if the containers are full, ask advice from a tutor or technician);
- hooks or areas for the storage of clothing (laboratory coats and jackets) and bags.

The safety equipment which is always available in laboratories used for environmental chemistry is usually signposted and includes the following:

- a shower, for use if clothing catches fire or in the event of a large spill of caustic chemicals on an individual;
- facilities for flushing chemicals out of an individual's eye with cold water or saline solution;
- fire buckets, blankets and extinguishers – it is particularly important that you know the different types of fire extinguisher and their application (see Box 2.1);
- spill kits, to prevent the spread of spilled chemicals that may be hazardous;
- a 'Health and safety' notice board that displays information such as chemical data sheets, location of fire exits, first-aid procedures and contact points for staff qualified in first aid;
- safety spectacles or goggles, full face-shields, disposable gloves and heat-resistant gauntlets, for use as appropriate (see Section 2.4);
- a dustpan, brush and mop, for clearing broken glassware and water spills respectively;
- a telephone, so that the emergency services – fire/ambulance/police – can be called if necessary.

Box 2.1 Type of Fire Extinguisher and their Application

In order to understand how fire extinguishers work, we need to know about fire. Four things must be present at the same time in order to produce fire:

- enough oxygen to sustain combustion;
- enough heat to raise the material to its ignition temperature;

- fuel or combustible material;

- the chemical, exothermic reaction that 'is' fire.

Oxygen, heat and fuel are frequently known as the fire triangle. The fourth element, the chemical reaction, makes a fire tetrahedron. Take any of these four things away and you will not have a fire or the fire will be extinguished. Essentially, fire extinguishers put out fire by taking away one or more elements from the fire tetrahedron.

There are three main types of fire extinguishers:

- *Water (air-pressurized water* (APW)) extinguishers take away the heat element of the fire triangle. They are designed for Class A (wood, paper and cloth) fires only. They should never be used to extinguish flammable liquid or electrical fires.

- *Carbon dioxide* extinguishers displace or take away the oxygen element of the fire triangle. Such extinguishers are filled with the non-flammable gas, carbon dioxide (CO_2), under extreme pressure. The CO_2 is very cold when it exits the extinguisher, so it cools the fuel as well. CO_2 extinguishers are designed for Class B and C (flammable liquid and electrical, respectively) fires only.

- *Dry chemical* extinguishers put out fire by coating the fuel with a thin layer of dust, thus separating the fuel from the oxygen in the air and interrupting the chemical reaction of fire. They come in a variety of types, which are always clearly labelled:

    - 'DC' short for dry chemical;

    - 'ABC', indicating that they are designed to extinguish Class A, B and C fires;

    - 'BC', indicating that they are designed to extinguish Class B and C fires.

Class D fires, those involving metals such as potassium, sodium, aluminium and magnesium, can be extinguished by using special extinguishing agents such as foam.

It is easy to remember how to use a fire extinguisher – simply remember the acronym 'PASS', which stands for Pull, Aim, Squeeze and Sweep:

- **P**ull the pin, allowing you to discharge the extinguisher.

- **A**im at the base of the fire. If you aim at the flames, the extinguishing agent will fly through the flames rather than hitting the fuel.

- **S**queeze the top handle or lever, which releases the extinguishing agent.

- **S**weep from side to side until the fire is completely out.

You should start using the extinguisher from a safe distance away from the fire, and then move forward. Once the fire is out, keep an eye on the area in case it re-ignites. ALWAYS inform the appropriate authorities about a fire, even if you successfully extinguish it.

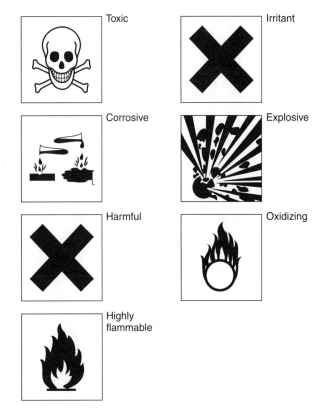

Figure 2.1 Some examples of hazard warning symbols used by chemical suppliers

Laboratory users should familiarize themselves with the hazard warning symbols used by chemical suppliers (see Figure 2.1). These symbols are normally shown as black on an orange-yellow background. In addition, you should always locate and remember the position of the fire exits in the laboratory.

2.3 Laboratory Equipment

Laboratory equipment includes general equipment such as glassware, spatulas, thermometers, balances, ice and heated water baths, heating mantles, weighing boats, basins and ultrasonic baths as well as more specialized equipment for analytical determinations. Illustrations of typical laboratory equipment, together with brief explanations of their uses, are shown in Figure 2.2.

2.4 Health and Safety in the Laboratory

2.4.1 Introduction

The potential for accidents in a laboratory is very large. The accidents can be relatively minor – it is very easy to cut yourself on broken or damaged glass – or

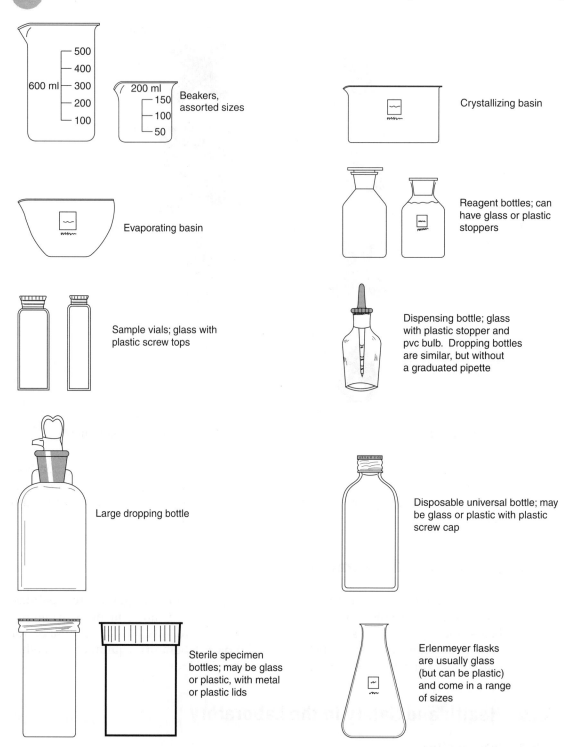

Figure 2.2 A selection of typical laboratory equipment

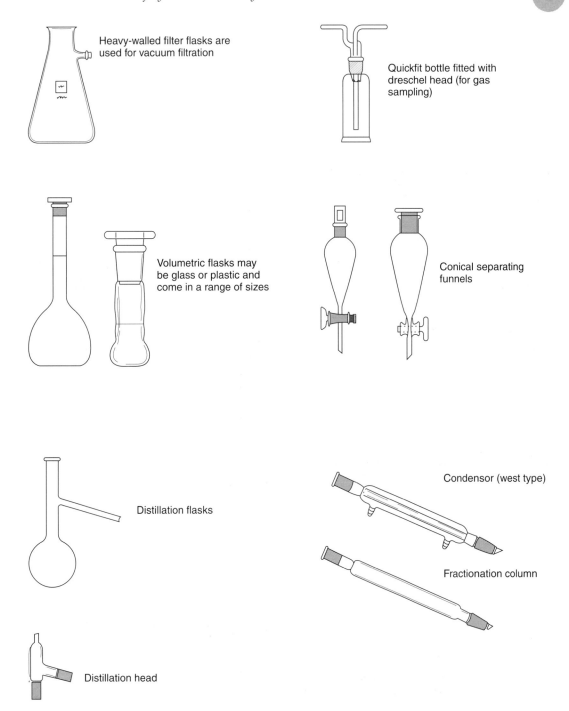

Heavy-walled filter flasks are used for vacuum filtration

Quickfit bottle fitted with dreschel head (for gas sampling)

Volumetric flasks may be glass or plastic and come in a range of sizes

Conical separating funnels

Distillation flasks

Condensor (west type)

Fractionation column

Distillation head

Figure 2.2 *Continued*

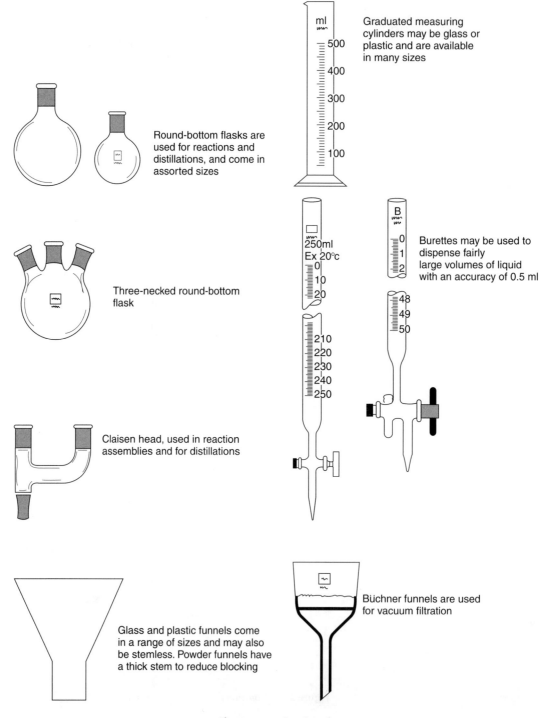

Graduated measuring cylinders may be glass or plastic and are available in many sizes

Round-bottom flasks are used for reactions and distillations, and come in assorted sizes

Three-necked round-bottom flask

Burettes may be used to dispense fairly large volumes of liquid with an accuracy of 0.5 ml

Claisen head, used in reaction assemblies and for distillations

Glass and plastic funnels come in a range of sizes and may also be stemless. Powder funnels have a thick stem to reduce blocking

Büchner funnels are used for vacuum filtration

Figure 2.2   *Continued*

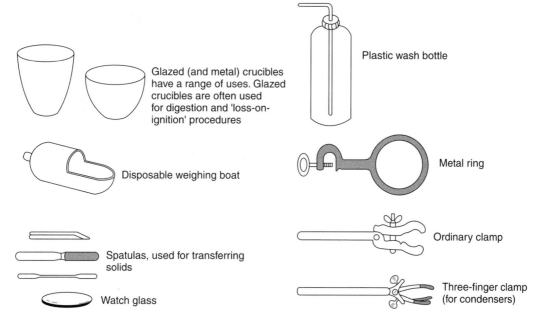

Glazed (and metal) crucibles have a range of uses. Glazed crucibles are often used for digestion and 'loss-on-ignition' procedures

Plastic wash bottle

Disposable weighing boat

Metal ring

Spatulas, used for transferring solids

Ordinary clamp

Watch glass

Three-finger clamp (for condensers)

Figure 2.2 *Continued*

extremely serious – a major explosion, fire or release of toxic fumes. Major accidents in laboratories used for environmental chemistry are very rare because of the presence of trained staff and specialized safety equipment. However, these are not reasons for complacency.

Most Universities and industrial companies have their own Health and Safety Policy that must obey the legal requirements set out by legislation, e.g. in the UK, health and safety policies are governed by The Health & Safety at Work etc. Act 1974. The Policy will outline the responsibilities of all staff and students who use laboratory facilities. The Control of Substances Hazardous to Health (COSHH) is governed by the COSHH regulations, 1988, and other similar regulations. A list of useful publications is provided in Box 2.2.

2.4.2 Laboratory safety

The overall responsibility for ensuring safety in any laboratory rests with those who have managerial responsibility in that laboratory. In general terms, this means that the technical staff are responsible for the provision of safe equipment, storage and maintenance, and teaching staff for the overall safety of the students and their activities. The students must take responsibility for their own actions and take every reasonable precaution to ensure their own safety and that of others. It is imperative that students obey the laboratory health and safety rules and any instructions given to them by tutors and technicians.

Most laboratory safety rules are nothing more than common sense. However, once you become familiar with working in a laboratory, you may be tempted to pay less attention to safety precautions that you see as restrictive or even unnecessary.

Box 2.2 Useful Health and Safety Information

There are a number of organisations which provide useful health and safety information.

The Health & Safety Executive
HSE Information Centre, Broad Lane, Sheffield, S3 7HQ, UK.
Publications include the following:

- 'Introducing COSHH' – Leaflet IND (G)65(L);

- 'Introducing Assessment' – Leaflet IND(G)64(L);

- 'Hazard and Risk Explained' – Leaflet IND(G)67(L).

The Royal Society of Chemistry
Burlington House, Piccadilly, LONDON, W1V OBN, UK.
Publications include the following:

- 'COSHH in Laboratories';

- 'Safe Practices in Laboratories'.

Her Majesty's Stationery Office (HMSO)
PO Box 276, LONDON, SW8 5DT, UK.
Publications include the following:

- 'The Control of Substances Hazardous to Health 1988' (ISBN 0-11-087657-1);

- COSHH: Guidance for Universities and Colleges of Further and Higher Education (ISBN 0- 11 -885433-X).

Institute of Occupational Medicine
Roxburgh Place, Edinburgh, EH8 9SU, UK.

For example, some people would prefer to work with open laboratory coats, untied (long) hair and without those (horribly unglamorous) safety spectacles. However, these practices are unsafe and safe practice should be part of every laboratory. Students, technicians and tutors should *always* have an awareness of safety issues when working in the laboratory and an attitude that promotes safe working practices. This may sound dogmatic and officious, but safety awareness can only become second nature if everybody works at it at all times.

The following section is not a complete list of safety rules – it is merely an indication of typical safety rules. Local safety rules may differ slightly in each laboratory depending upon circumstances, although they must always obey legal requirements and good laboratory practice. You should also remember that safety rules are only useful if they are followed and enforced.

General safety rules

- Before performing any laboratory work, students should familiarize themselves with the layout of the laboratory and make a mental note of its specialist and safety features (see Section 2.2).

- Bags, coats or any other such items must be left in an identified location within the laboratory (e.g. a clothing storage area or locker) or in a secure place outside the laboratory, ensuring that corridors/fire doors are not obstructed.

- Laboratory coats must always be worn in the laboratories and must be buttoned up. In addition, you must wear suitable clothing when working in a laboratory. Inappropriate clothing can be dangerous – loose sleeves can sweep flasks from the laboratory bench, sandals do not protect feet from spills, while shorts and short skirts do not protect legs from chemical spills.

- If you have long hair, you must tie it so that it does not present a potential safety hazard.

- Safety goggles and gloves should be worn when appropriate. If you are in any doubt, then ask the tutor or technician before starting the experiment. Wearers of spectacles or contact lens wearers should wear properly fitted goggles.

- No food or drink is allowed in the laboratory.

- Smoking is not allowed in the laboratory.

- Mobile phones are not normally allowed in the laboratory, except by prior consent from a tutor or a technician under exceptional circumstances.

- No unauthorised experiments are allowed in the laboratory. Pranks and other types of irresponsible behaviour have no place in laboratory areas.

- Chemical data sheets must always be read before the practical work is attempted. The instructions given for the handling, use and storage of all articles and substances, especially those chemicals that are flammable, toxic, explosive or radioactive, must be obeyed.

- Great care must be taken when handling and assembling glass apparatus. Broken glass must be disposed of in the special bins provided.

- When working with electrical equipment, take particular care to avoid introducing additional hazards due to liquid spillage, etc. Always seek advice if you are not sure about electrical safety.

- Apparatus must not be set up and left unattended unless permission has been obtained and clear instructions are displayed indicating the action to take in an emergency.

- Flasks, beakers, etc. must be clearly labelled (e.g. with a water-soluble-ink pen) as to their contents, in order to avoid potentially serious consequences.

- Laboratory chemicals must not be sniffed, inhaled or tasted. Mouth pipetting of chemicals is not allowed.

- You should avoid looking down the necks of flasks and test tubes or holding them above eye-level for examination. You should also be aware of the safety of others when handling chemicals in glassware – the open neck of glassware should never point towards an other individual.

- Throughout, and at the end of a practical, your working area should be cleaned and tidied. Glassware should be rinsed and left to dry on a drying rack or tray;

chemicals should be stored safely and equipment should (normally) be switched off. (Be aware that some items of equipment should never be switched off – ask if you are unsure.) Always wash your hands before leaving the laboratory.

The reporting of accidents

You should report any accident, however minor, to the tutor or technician. Accidents include spills of chemicals or water and breakage of equipment, as well as personal injury. A formal, written record is usually kept for accidents that result in property damage or personal injury, and occasionally, a written report is needed for more serious incidents. Such records are needed for COSHH, the Health & Safety Executive and insurance purposes, and also to alert staff to any unsafe practices or working conditions.

Emergency procedures

There are a number of basic actions you can take to minimize the risk of injury or property damage in the event of a laboratory emergency. In the event of a laboratory fire, explosion, release of toxic substances or similar imminent danger, you should take the following actions;

(1)   Evacuate the laboratory through a designated exit.

(2)   Report the situation immediately to an appropriate member of staff.

(3)   In some circumstances, you should sound the fire alarm so that all persons can evacuate the building.

(4)   Do not re-enter the laboratory or building (as applicable) until advised it is safe to do so by a Health and Safety Officer.

Eye protection

Whenever you are performing an experiment in a chemical laboratory, you will usually need to wear eye protection. It is relatively easy for chemicals to splash into your face, so this precaution cannot be too strongly emphasized. It has also been known for individuals to rub their eyes after they have spilled chemicals on their fingers – it is in this type of situation that safety awareness is so important.

Depending upon the nature of the planned experiment, safety glasses, goggles or full face shields may be necessary. Prescription spectacles may provide sufficient protection, but *only* if they have been designed so that they prevent splashes entering the eyes from the top, bottom and sides. Contact lenses provide no adequate protection. If you are not sure which type of eye protection you require, then take advice from a tutor or technician before starting your experiment.

Storing and handling chemicals

Although students do not assume legal responsibility for the large-scale safe storage of chemicals, you should be aware of the general principles of storing and handling chemicals.

All chemical substances used in laboratories must be stored in safe conditions. You should only bring the minimum amounts of the chemicals you need into the working area, although small quantities of commonly used substances can be stored appropriately in the laboratory. Chemicals must be suitably separated in order to minimize risks of dangerous intermixing and spillage trays should be considered for more corrosive chemicals, e.g. concentrated acids. Highly toxic substances must be locked in an appropriate store, while flammable substances must be stored in an appropriate fire-resistant storage facility.

Some solids are inherently hazardous (e.g. sodium, potassium and white phosphorous) and must be segregated from other chemicals and held in a fire-resistant store. Many of these chemicals are damped down with a suitable solvent in order to keep them safe and they must be regularly inspected to ensure they have not dried out. Large bottles used to store chemicals – Winchesters – should only be carried in proper Winchester carriers.

Use of fume cupboards

A fume cupboard is a special piece of laboratory furniture that safely removes potentially hazardous fumes that may be generated during the use of chemicals. The main safety features include an extraction system to remove (and sometimes treat) waste gases, an explosion-proof sash window that moves up and down to allow access, and external controls for gas, water, electrical and vacuum services. The extraction fan must always be switched on when the fume cupboard is in use and apparatus in the cupboard must not interfere with the efficient opening/closing of the sash window. The release of toxic or hazardous fumes should be minimized, if necessary, by the use of scrubbers or similar devices fitted to the exhausts of the apparatus used. Any reaction that results in a serious discharge of gas must be notified to the laboratory supervisor immediately.

The safety features built into fume cupboards can lead to complacency when they are being used for experiments. Normal protective clothing, including eye protection, must be used when using fume cupboards.

Project work

Projects involving practical work require careful planning and consultation with others. Project work may only be undertaken under the supervision of a tutor or qualified technician/researcher. Students undertaking individual projects are often required to make an appointment at least *one week in advance* with appropriate laboratory staff before starting project work, although each organization will have its own local arrangements. Where necessary, laboratory and instrument technicians must be made aware of the nature of your project and be given adequate warning of the required equipment/chemicals. Note that equipment may not always be available immediately.

Project work often involves becoming very proficient in the use of one or more items of equipment. You should always receive proper training in the use of new equipment and an assessment of your proficiency will be made before you are allowed to work independently. However, using expensive, complicated equipment can be daunting. If you are deemed proficient to use a piece of apparatus but you feel unsure of your new skills, ask for extra supervision. Equipment loaned to you should be carefully handled and any faults or accidents that occur during its operation must be reported. You are strongly advised to book the loan of equipment well in advance of project work – the equipment may be required by others.

2.5 Laboratory Technicians

The role of technical staff in the laboratory is to maintain, set up and clear away equipment, prepare solutions of chemicals, obtain environmental samples for testing and maintain adequate stock control. Technical staff also have responsibilities for health and safety and good laboratory practice. Some technicians can demonstrate equipment and/or procedures to students and provide technical and safety advice. Technicians may often have other roles outside the laboratory.

However, laboratory technicians are not employed to carry out students' tasks for them, nor are they expected to clear up unreasonably messy work benches. Instructions given by technicians about safety or working practices must be obeyed. Students must not expect technical staff to drop their current task immediately in order to help them, as they often work to tight deadlines. Consequently, if you require advice or assistance from a technician when you are planning project work, it is best to make an advance appointment.

2.6 Emergency First Aid

First aid is the initial assistance or treatment given to a casualty for any injury (or sudden illness) before the arrival of a medical professional. The term 'first aider' is usually applied to a person who has completed a course on first aid and passed a professionally supervised examination. All workplaces, including universities, have qualified first aiders on site, and some have fully qualified medical staff. You should always seek assistance from a medical professional or a qualified first aider in the event of a laboratory accident.

However, some laboratory accidents require immediate action, and since you may have to wait for a suitably qualified person to be found, you should be aware of certain basic first-aid procedures. In all cases:

- assess the situation quickly and calmly;

- ensure that you are not at risk;

Table 2.1 General procedures for dealing with potential laboratory accidents[a]

Potential laboratory accident	Action
Small cut	Clean the wound by rinsing under cold running water. Dry gently with clean tissue or sterile swab. Apply an adhesive dressing (plaster)
Large cut	Apply direct pressure over the wound, preferably with a sterile dressing or clean pad. Elevate injured limb – it may help to lay the casualty down. Treat for shock by keeping person warm and quiet
Minor heat burn	Flush the injured area with cold water for 10 minutes. Remove any constricting jewellery or clothing from the area. Cover with a sterile dressing
Organic chemicals on the skin	Wash the affected area thoroughly with soap and water
Caustic chemicals on the skin	Flush the affected area continuously with cold water for 15–20 minutes
Chemicals in the eye	Hold the affected eye under gently running cold water for 10–15 minutes. You may need to gently pull the eyelid open to ensure that the eye is fully irrigated. Take care that contaminated water does not splash the unaffected eye. Cover the eye with a clean non-fluffy pad
Burning clothing	Extinguish the flames by using a shower, fire blanket or wet towels. Treat minor heat burns as above; major heat burns should only be treated by qualified personnel. Treat for shock by keeping the person warm and quiet

[a]The information in this table is provided simply to *raise awareness* of the procedures that are generally followed when dealing with laboratory accidents. It is *not* meant to be used a definitive guide for the treatment of laboratory injuries. In *all* cases, you should ensure that specialist help has been summoned from suitably qualified staff.

- protect the casualty from danger;

- summon specialist help immediately;

- avoid touching wounds with your fingers – beware of blood contact;

- minimize the risk of infection (e.g. don't cough or breathe over a wound).

Some general procedures for dealing with potential laboratory accidents are outlined in Table 2.1.

2.7 The Laboratory Notebook

It seems obvious that you will need to keep accurate and detailed notes of your laboratory experiments while the work is being carried out. It may not be so obvious that you need to do this in an organized and systematic fashion. Many students scribble notes on rough bits of paper or the laboratory sheets, only to find that they can't understand their handwriting or have lost their records when they come to write up the experiment. This is clearly poor practice and can lead to inaccurately reported results or, even worse, a failure to submit the coursework. For many years, scientists have employed a laboratory notebooks to help keep records of their work.

2.7.1   The purpose of your notebook

A laboratory notebook (or logbook) is where you write an account of your laboratory work while the experiment is being performed. A notebook can serve several purposes, as follows:

- It provides a traceable record of experimental work and results.

- It acts as a personal reference book when you come to write up laboratory reports.

- It enables you or another person to repeat or revise the experiment exactly. This is especially important when you perform independent research in your university 'Final Year Project'.

- It develops the record-keeping skills all good scientists require when they work professionally in research, industry, forensic testing and as consultants, as well as in academia.

- It enables the tutor to see if any procedural mistakes or incorrect calculations have been made.

A laboratory notebook is a hard-cover bound book containing lined pages on the right-hand side and graph paper on the left. These are usually available at bookstores, stationery supply stores and students' union shops. Do not use loose-leaf or spiral notebooks, as they damage quickly and pages can be easily lost.

2.7.2   Keeping your laboratory notebook

The extent of notes taken will vary from experiment to experiment. Everything you do and observe in the laboratory could potentially be recorded in your notebook. Consequently, it is important to use an organized format to avoid cluttering and allow ease of reference. Using the conventions outlined below will enable you to keep experimental records in a clear and ordered manner.

(1)   Make sure your name is written on the front cover of the book. It is also advisable to add your address or telephone number in case it is lost.

(2)   Leave three blank pages at the beginning of the notebook for a table of contents. You can add to this table as your studies progress.

(3)   If the notebook pages are not already numbered, then number these yourself before you use the book.

(4)   Traditionally, the right-hand side of the notebook is used for writing up experiments, while the left-hand side is used for calculations, notes and graphs. At the present time, many tutors provide laboratory worksheets to aid

their students – these may be stuck in the notebook (on the right-hand side) or omitted and referred to in your text.

(5) For each experiment, the following data are usually recorded:

- The date and title of the experiment.

- The aims and objectives of the work.

- A list of the apparatus required. For some experiments, a labelled diagram of the equipment should be drawn (there is no need to draw equipment such as Bunsen burners or burettes). Below each diagram should be an explanatory title.

- A brief account of the procedure that you follow (this may just be a reference to the laboratory worksheet – which should be attached).

- Observations, e.g. 'the indicator turned from yellow to blue as the alkali was added' or 'the end point of the titration was difficult to detect', any faults or problems with instruments, times and temperatures of reactions, etc.

- Any calculations performed. Students are often afraid of performing calculations in the laboratory for fear of making mistakes. You can minimize the opportunity for mistakes by taking a three-step systematic approach, i.e. by writing down: (1) what you know, (2) what you want to know, and (3) a relationship (equation) linking (1) and (2).

  Don't forget to record the units (e.g. dm^3, $mg\ m^{-3}$ etc.) and to pay careful attention to the number of significant figures that you use.

(6) Use tables wherever possible to record your data (results). Same examples of good practice are shown in Tables 2.2 and 2.3.

Table 2.2 Example of a data table for use when performing a titration

	Titration volume (cm^3)		
Measurement	Rough titration	Titration 1	Titration 2
Final reading	21.00	41.75	22.95
Initial reading	0.00	21.00	1.20
Total titration volume	21.00	20.75	20.75

Table 2.3 Example of data table for use when weighing out a soil sample

	Weight of sample (g)		
Measurement	Sample 1	Sample 2	Sample 3
Weight of soil + beaker	5.8727	6.2439	6.0038
Weight of beaker	5.3620	5.7921	5.5537
Weight of soil sample	0.5107	0.4518	0.4501

Determination of Hardness in Water (3/3/01)

Procedure
Follow p 5-7 of Laboratory Handbook (2001)
Samples used: Tap water (Thames Water)
                     Lake water (Trent Park Lake, sample taken 2/3/01)

Observations
End point red → blue
* Clearly overshot end point during trial titration

Calculation
For details, see p7 of laboratory handbook (2001).
For tap water:
1.0 cm^3 of 0.001 M EDTA $\cong 1.0$ mg $CaCO_3$
in 100 cm^3 of tap water:
   29.9 cm^3 of EDTA $\cong 29.9$ mg $CaCO_3$
∴ Conc of $CaCO_3$ in $1 \text{ dm}^3 = 29.9 \times 10 = 299.0$ mg $CaCO_3 \text{ dm}^{-3}$

For lake water:
1.0 cm^3 of 0.001 M EDTA $\cong 1.0$ mg $CaCO_3$
in 100 cm^3 of lake water
   23.35 cm^3 of EDTA $\cong 23.35$ mg $CaCO_3$
∴ Conc of $CaCO_3$ in $1 \text{ dm}^3 = 23.35 \times 10 = 23.35$ mg $CaCO_3 \text{ dm}^{-3}$

Determination of Hardness in Water

Titration–Tap water

		Titration volume (cm^3)	
	Rough titration	T1	T2
Final reading	30.20	45.10	42.60
Initial reading	0.10	15.20	12.70
Volume of EDTA	30.10	29.90	29.90

Average of T1 and T2 $= \dfrac{29.9 + 29.9}{2} = 29.9 \text{ cm}^3$

Titration–lake water

		Titration volume (cm^3)	
	Rough titration	T1	T2
Final reading	24.10	47.60	23.80
Initial reading	0.00	24.20	0.50
Volume of EDTA	24.10	23.40	23.30

Average of T1 and T2 $= \dfrac{23.4 + 23.3}{2} = 23.35 \text{ cm}^3$

Rhiannon Davies 3/3/01

Figure 2.3 Example pages from a well-organized laboratory notebook

(7) Record your work clearly and legibly. Writing scribbled notes may seem OK when you are working quickly in the laboratory, but in reality it saves little time and may lead to inadequate reports.

(8) You must not rip out pages or use correction fluid if you make mistakes. Line out and initial errors, and then use the next available space for subsequent entries.

Having read this list, you may be thinking that keeping a laboratory notebook will be a time-consuming chore. In fact, after a couple of laboratory sessions, you will probably find that you have developed a fast, elegant method of taking notes which will serve you well in the outside (e.g. industrial) work place, as well as in academia. Some example pages taken from a well-organized laboratory notebook are shown in Figure 2.3.

2.8 Summary

This chapter has summarized the principal features of the laboratories used for environmental chemistry and identified many key items of laboratory equipment. In addition, you should now be fully aware of a reliable and accurate method of keeping a laboratory notebook, fundamental laboratory safety rules and procedures, and the responsibilities and authority of laboratory personnel. This is vitally important information; not only will it help to develop your practical skills, but it will also help to maintain a safe working environment for the laboratory users.

Self-Study Exercises

2.1 Outline the main safety features of a typical undergraduate chemistry laboratory.

2.2 List the utilities (services) you would expect to find on a typical laboratory work bench.

2.3 Under what circumstances would you use the following fire extinguishers:

(a) water;

(b) carbon dioxide;

(c) powder?

2.4 Give reasons for the following safety rules:

(a) Contact lenses should not be worn in the laboratory.

(b) Water should not be used to extinguish laboratory fires.

(c) No running is permitted in the laboratory.

(d) A laboratory experiment must not be left unattended.

(e) Broken glass should be disposed of in special bins.

(f)   A chemical spill on the skin should be washed off with water, not with an organic solvent.

(g)   The contents of flasks and beakers should be clearly labelled.

(h)   All laboratory accidents should be reported.

2.5   What action should you take in each of the following circumstances?

(a)   You see another student mouth-pipetting an acid.

(b)   You spill sulfuric acid on to your hands.

(c)   The clothing of the student next to you catches fire.

(d)   You spill a volatile solvent on a hot plate.

(e)   You splash a chemical into your eye.

(f)   The student next to you cuts his leg badly on broken glass and starts to feel faint.

(g)   You burn the palm of your hand by touching a red-hot crucible.

(h)   The contents of your beaker catch fire.

(i)   The student next to you splashes a chemical into their eye.

(j)   You spill a large quantity of distilled water on the laboratory floor.

2.6   How should prescription spectacles be designed if they are to double effectively as laboratory safety glasses?

2.7   List the conventions used when keeping a laboratory notebook.

Challenging Exercises

2.8   You are asked to give a talk on laboratory safety to Sixth Form (High School) students at a local school. Design some A4-sized overhead projection slides to demonstrate that chemistry laboratories are hazardous places to work.

2.9   What action, if any, would you take in the following situations?

(a)   When performing project work in a chemistry laboratory, you see a friend placing two broken volumetric flasks in the paper waste bin.

(b)   You take a summer job as a technician with a small environmental consultancy. You notice that the technicians eat, drink and smoke in the laboratory during tea and lunch breaks.

(c)   You accidentally spill the last 500 cm^3 of a postgraduate's expensive solvent that you were not supposed to use. You have not read the safety data sheet. Nobody sees the accident and the solvent rapidly evaporates. You know that the postgraduate will be very angry as the person concerned is close to the deadline for completing their project.

(d)   You see another student measuring concentrated acid from a Winchester into a beaker, without wearing proper protective clothing or eye protection.

(e) A tutor verbally remonstrates you after discovering you eating a bar of chocolate in the laboratory during a class.

(f) You want to start the practical work for your project in two weeks time.

(g) Your best friend works next to you in the laboratory. Your friend is an excellent scientist but very messy and untidy. You often clear and tidy his/her desk after they have left because you know the technician will complain about the mess. The whole class will have a practical examination in three weeks time.

2.10 Draw a flowchart that illustrates clearly the hierarchy of managerial responsibility for safety in an undergraduate chemistry laboratory.

3 Laboratory Report Writing

There are three reasons for becoming a writer: the first is that you need the money; the second that you have something to say that you think the world should know; the third is that you can't think what to do with the long winter evenings.

Quentin Crisp, in *The Naked Civil* Servant, 1968.

3.1 Introduction

During your scientific career, you will be required to present scientific and technical information in many different formats to many different audiences. As a student, you will be required to submit essays, laboratory and project reports to your tutors for assessment. You may also be required to write text for posters and articles for newspapers or magazines for assessment purposes. In future years, as a researcher, technician, teacher, academic, industrialist, civil servant, media correspondent, author, salesperson or politician, you may be required to write a range of scientific text targeted at a specific audience. This prospect may terrify you; many people regard writing as difficult, and something to be delayed or avoided – a chore. In fact, scientific writing is a skill, which, like tying your shoelaces or performing a titration, can be mastered with practice and perseverance. Like any other skill, scientific writing can be developed into something that will give you confidence, satisfaction and pleasure.

At undergraduate level, laboratory reports are very important components of assessed work, and consequently, it is worth trying to produce good quality reports right from the word go. Laboratory reports are intended to demonstrate some or all of the following:

- you have performed and understood an experiment;
- you have some knowledge of the theoretical basis of the experiment;
- a laboratory skill/technique has been learned or improved;
- you can process/interpret the data obtained from an experiment;
- you can relate fundamental or derived laws to the outcome of the experiment;
- you can present these ideas/results in an appropriate context and can evaluate their significance.

Having read this, you may feel that there is no scope for individuality of expression. This is not true – it would be a dull world if it were! There is no such thing as an 'absolutely correct' style of scientific writing: instructions to authors printed in scientific journals do differ from journal to journal, and formats of project reports differ from institution to institution. Nevertheless, there are some basic conventions governing the style of writing which are generally held throughout the scientific community. This chapter aims to achieve the following:

- Outline a strategy for effective scientific writing;
- Identify the basic conventions of written laboratory reports, including conventions for the format and presentation of reports and the citation of references;
- Provide a 'skeleton' upon which you can 'flesh out' your own laboratory reports;
- Outline methods for reporting and interpreting your experimental data effectively.

3.2   Effective Scientific Writing

The function of a scientific report is to communicate information clearly, concisely, unambiguously and accurately in a readable style. Writing requires more care than speech, as it is a more formal activity which produces a permanent record. Don't despair if you are worried about your writing skills – you will improve with guidance and practice. Careful use of the flowchart shown below will help you to improve your writing skills.

- Remember the purpose of your writing – communicate clearly, concisely, unambiguously and accurately.

- Consider your audience (the tutor) and the assessment criteria.

- Use the appropriate format (see Section 3.8 below).

- Plan and arrange your ideas in a logical order.

- Treat what you write first as a draft.

- Make sure your grammar, spelling and punctuation are correct.

- Ensure the first draft is legible.

- Re-read and edit your first draft as necessary (nearly everybody makes mistakes in early drafts of their reports).

- Proof-read the final draft, correcting any remaining mistakes.

3.3 Grammar and Style

All the text in your report should be grammatically correct, properly punctuated and comprise complete sentences. The overwhelming majority of scientific reports are written using the impersonal Third Person/Past Simple Tense/Passive Voice form, avoiding, if possible, the use of the personal pronoun (I, we or you). The following examples illustrate what is intended:

Preferred   The samples were stored at 0°C.

Not preferred   I stored the samples at 0°C.

Try to avoid clumsy statements such as 'The samples were stored by the writer'; for assessed reports, it will be assumed that you did all of the work, unless stated to the contrary. A statement such as 'The samples were stored by Dr Bennett' is quite acceptable.

Spelling mistakes are unacceptable and care should be used to check spelling as you write the manuscript, and during proof-reading. You must allow yourself adequate time for proof-reading the text and for attending to any consequent corrections. Most word-processing packages have spell-checking and grammar-check features to aid you with this. Poor grammar, bad spelling and an unreasonable number of uncorrected typing errors may result in assessed reports being penalized.

3.4 Presentation and Format

Laboratory reports should be good to look at; a well-presented report will please the reader, give him/her confidence in the report and will aid assessment. A careful balance of headings, paragraphs, tables and figures helps you structure your work and aids clarity and appearance. Paper size, margins, page numbering, font size and typographical style should be constant throughout the whole report, with normal lines of text being single spaced and with a line space between paragraphs. The typical font size used for reports is 11 or 12; different font sizes may be used for (a) long quotations and (b) table, figure and appendix text. Upper case or bold type is often used for sections, table, figure and appendix headings, the title page (usually the cover itself) and the names of authors in references listed in the reference section. Whatever the style used, consistency is essential.

A printed report cover will aid the presentation of your work, as well as providing important information to your assessor. The report cover should have your report title, the module number, the date, your name and your tutor's name 'typed' on it. This front cover, together with a blank back cover, may be inserted into a plastic binder or bound to the body of your work (by means of a plastic binding strip) to provide a pleasing and professional-looking report.

You may be thinking that making your report look good is a lot of extra work and may not be worth the effort. Think again. The extra time that you spend preparing structured, stylish reports is minimal when compared to the time spent performing experimental work, background research and writing the text itself. Good presentation will become second nature after a little practice and will give you an opportunity to develop your creative talents, thus individualizing your work. Poorly presented, scrappy and abbreviated work loses the reader's sympathy and may cost you marks.

3.5 Units, Symbols and Quantities

Laboratory reports should always use SI units, unless alternatives can be properly defended (see Chapter 1). The time of day should be given in the form 0630. Dates should be given in the form 14 January 2003. Numbers should be given as numerals, and not spelled out, except at the beginning of a sentence or where numerals would appear absurd, e.g. '1 test tube was placed in the centrifuge'.

3.6 Scientific Nomenclature

In the middle of the 19th century, the structures of many chemicals were unknown and compounds were given names that were illustrative of their origins or properties. In fact, some compounds were named after relatives or friends of the scientists who first discovered them. For example, the name *barbituric acid* (and hence the well-known drug classification, *barbiturates*) comes from the woman's name 'Barbara'. These names are called trivial or common names. However, as time has progressed, this system has become unworkable, and a systematic method of nomenclature has been developed. This system, known as the Geneva or IUPAC International Union of Pure and Applied Chemists (IUPAC) system, should be utilized in laboratory reports.

3.7 Pagination

Proper pagination of your reports will assist you to structure your work, as well as being good practice. It will also assist the reader/assessor to 'navigate' your report, thus making it easier to find relevant sub-sections, tables, figures, etc. Your contents page should always contain the page numbers of each section of your report; for this reason the contents page is usually written last. Pages containing preliminary information (e.g. the Title Page) are paginated in small Roman numerals (i, ii, iii, iv, etc.), whereas pages of the main body of the report are given in Arabic numerals (1, 2, 3, etc.). The pages of Appendices are also numbered in Arabic numerals, but starting again from 1.

3.8 Structure of the Laboratory Report

The structure of any piece of writing differs according to the target audience and depends upon whether it is an essay, a newspaper article, a scientific paper, or a laboratory or project report. Most academic writing is written in a logical fashion with a structure that is appropriate to the topic; consequently, it is not sensible to develop a generic skeletal framework for, e.g. essays, because they can be written by using a variety of formats and for widely differing audiences. However, it is possible to suggest generic frameworks for laboratory reports, as shown in Box 3.1, because there is wide agreement and acceptance of their basic structure.

Box 3.1 Basic Structure for Laboratory Reports

- Title page (usually the cover itself);
- Aims of the Experiment;
- General introduction;
- Materials and Methods;
- Results;
- General discussion;
- Conclusions;
- References.

Appendices are not normally included in laboratory reports.

3.8.1 Aims and objectives of the experiment

The aims of the experiment should clearly and briefly state the purpose of undertaking the experiment. They usually include the development of particular (named) skills, as well as the specific overall aims of the experiment. For example, in an experiment that measures the oxygen content of water, the principal aim may be:

- to determine the dissolved oxygen content of samples of tap water and river water using the Winkler method – see Chapter 6 (Experiment 6.3);

but *further* objectives could include:

- to further develop laboratory skills in titration and other aspects of volumetric analysis;
- to understand some of the processes by which oxygen is dissolved in, and removed from, water;
- to compare and contrast the Winkler method with other techniques for the measurement of dissolved oxygen.

You should always refer back to your aims in the Conclusions section of your report and comment upon whether they have been achieved satisfactorily.

3.8.2 General introduction

The General Introduction should establish the context of the experiment, and explain the rationale for undertaking it (i.e. why is it worth doing at all). Here, you should provide some background information on the problem under investigation, such as the sources of the pollutant under investigation and any potential health/environmental effects. This section can also involve a description of the theory relating to the experiment and the experimental technique(s) to be used. It should leave the reader with the feeling that the report has a general relevance and that to read on would be worthwhile.

3.8.3 Materials and methods

The Materials and Methods section should contain a concise but adequate description of all of your experimental materials and procedures so that your results could be verified independently. If a particular method is well known and you have merely followed a procedure already described, you may simply cite the reference, e.g. 'The soil sample was digested according to the method described by Babatunde (2000)'. There is also no need to repeat routine instructions for using apparatus or equipment where they are well-known or available in manufacturers' instructions.

Any form of sampling procedures must be very fully described – both the sampling techniques and the sampling strategy. Sampling is usually undertaken to obtain some estimate relating to a population – an average, variability, total numbers, etc., and even small changes in methodology can result in quite dramatic differences in the estimate obtained.

Materials, too (unless standardized) should be as fully described as is necessary for replication. 'Sodium chloride' may well adequately define a material, but 'leaf litter', 'software', 'bacteria', etc. are much too vague. Similarly, locations and study areas should be described well enough for a reader to duplicate, locate or visualize.

3.8.4 Reporting results

Clearly, the Results are an exceptionally important part of your report and great care should be taken in their presentation. Over the years, a number of conventions have developed in the reporting of results. It is important to open your Results section with appropriate text rather than by just presenting tables of data. A table must follow, and never precede, the first reference to it in the text. You should not leave it to the reader to interpret tables – that is **your** job. An acceptable format is of the type, 'The data presented in Table 1 show that . . .'. Indeed, the reader should be able to appreciate the significance of the results without reference to any tables of data; the data are evidence to support your statements. However, pure *repetition* of data in text and in tables should be avoided.

While **tables** are used to present the data, **figures** can be helpful in interpreting them, while **statistics** are necessary for assessing their significance. These three items are now considered separately in the following discussion.

Tables

Tables are the main vehicles for conveying data to the reader. A table can be considered as a complete entity and, in a sense, should be able to exist separately

Table 3.1 Sources of the principal pollutants in the UK in 1995

Source	Total emissions (%)				
	Sulfur dioxide	Black smoke	Nitrogen oxides	Carbon monoxide	Volatile organic compounds[a]
Road transport	2	50	46	75	30
Power stations	67	5	22	4	–
Other industry	24	6	12	1	58
Domestic	3	19	3	4	1
Other	4	20	17	4	11
Total (Kilotonnes)	2365	356	2295	5478	2337

[a]Volatile organic compounds does not include methane. (Adapted from Department of the Environment, 1997).

in the text. A well-constructed table does not need a lengthy explanation on how it is to be interpreted but should be self-explanatory and be characterized by its simplicity and unity. The legend (title) of a table is clearly important if the table is to stand as a separate entity: it is a common mistake to make table captions too brief. Each table should also have its own number that should accompany the legend at the top of the table.

Tables should be kept within the normal page margin and tables which are longer than half a page are often centred on a page of their own. A very wide table may be typed lengthways (i.e. on its side) so that the top of the table is towards the spine of the report. Very large tables may be photo-reduced until they fit on a page but the text must remain sufficiently large to be legible. It is desirable to insert the table as near as possible to the first mention for ease of reference. Table 3.1 is a well laid out and clear example. It is not necessary to isolate every cell of a table with horizontal and vertical lines. Lines are required to group data in particular ways. Row and column headings are usually separated from the data, as also are totals.

Figures

Well-drawn figures can greatly enhance the effectiveness of display and interpretation of the results presented in a report. However, figures should not be used to duplicate data that are already presented in tabulated form. Be sure to leave the same margin all round the figure as on a page of text, and allow sufficient space underneath for the caption. Figures that have been drawn to fill an A4 page can be conveniently reduced by using a photocopier; this orientates the diagram vertically, giving an appropriate margin and plenty of room for a legend. Clearly, the availability of computer software has aided the production of graphs, figures, etc. but students should beware of plagiarizing other people's work (see Section 3.10 below). The lettering on any figure should be typed in, but be sure that the lettering is appropriate in relation to the size of the figure – see, for example, Figure 3.1.

Care should be taken to avoid cluttering up figures with unnecessary lettering or text. Explanation should go in the caption or, if lengthy, in the text. As with tables, figures should follow the first textual reference to them, and should be ordered

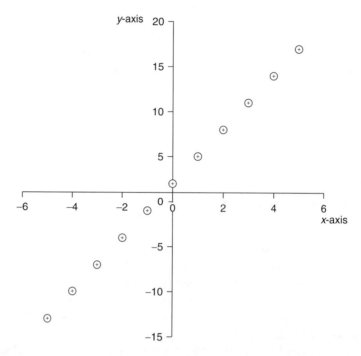

Figure 3.1   Straight-line graph of $y = 3x + 2$, illustrating a direct relationship

numerically throughout the report. Figure legends should always be placed below the figure. You also need to take care with the placement of figures on the page, e.g. if the figure is narrower than the page margins it should be placed centrally between these margins.

Graphs

Laboratory exercises will often involve the production of graphs from the data collected. A graph can provide much more information than a set of data. It gives a visual representation of trends and relationships, and permits the prediction of what happens between the known points.

When plotting graphs, the general convention is that the independent variables (i.e. the variable whose values you choose) are plotted on the x-axis and the dependent variables (i.e. the variable whose values you measure) are plotted on the y-axis. The value of the dependent variable depends upon the value of the independent variable. It is also important to choose a sensible scale – a large square should be 1, 2, 5 or (a multiple of) 10 whole units, as other units can be difficult to plot. The graph should be as large as you can conveniently make it so as to minimize errors. The points of the graph should be marked with crosses (+); do not use 'kisses' (crosses) ($\times$), points ($\cdot$), rings (O) or circles with dots ($\odot$). The formal reason for this approach is that the horizontal and vertical lines represent the x- and y-components of the coordinates, but it is also because the shape of a cross usually does not interfere with any lines you draw. It is also important to label the axes clearly and provide a number and title for the graph.

It is now common practice for students to display their graphs by using computer software, rather than drawing them. This is entirely acceptable providing you know what you are doing and you don't let the default settings of the software take all the decisions for you – this can lead to errors and inappropriate data presentation.

Many experiments will produce data which, when plotted on graph paper, will produce a graph with a straight line (although this will often require some judgement about where the 'best-fit' line falls between plotted points). The *formula for a straight-line graph*, shown in Equation 3.1, contains information about the gradient (m) and the intercept (c):

$$y = mx + c \qquad (3.1)$$

The gradient represents the amount that y increases or decreases as x changes, while the intercept is the point where the graph line cuts the y-axis, i.e. the intercept is the value of y when $x = 0$. Equation 3.1 is also known as the *simple linear regression equation*, and is used to predict the relationship between the independent variable, x, and the dependent variable, y.

The graphs illustrated in Figures 3.1 and 3.2 show the lines obtained from the equations $y = 3x + 2$ (where $m = +3$) and $y = -3x + 2$ (where $m = -3$), respectively. You will see that an upward slope, represented by a positive value for m, implies a *direct relationship* between x and y, whereas a downward slope (a negative value for m), reveals an *inverse relationship* . In both cases, the value of the intercept is $+2$, i.e. $y = 2$ when $x = 0$. Graphs are often used in place of tables to represent

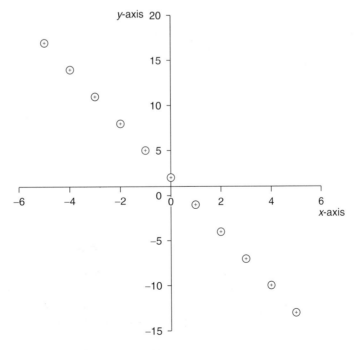

Figure 3.2   Straight-line graph of $y = -3x + 2$, illustrating an inverse relationship

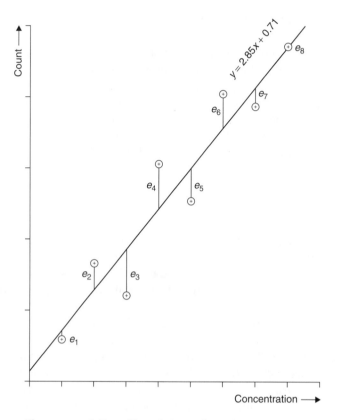

Figure 3.3   A line of best fit is used to minimize errors

data, especially when they illustrate straight-line relationships, e.g. when they are used to display calibration curves.

However, real experimental data rarely fall exactly upon a perfect straight line because of the presence of random errors. Some random errors are positive, while others are negative, as shown in Figure 3.3. Overall, the expected error should average out to zero. Thus, in order to fit the best straight line to a set of experimental data, we should draw the line so that some of the errors are positive (above the line) and some are negative (below the line), with an average error of zero, i.e. in Figure 3.3, $e_1 + e_2 + e_3 + e_4 + e_5 + e_6 + e_7 + e_8 = 0$. The *Method of Least Squares* is frequently used to estimate the line of best fit for a set of experimental data that show a linear relationship (see Box 3.2 and Example 3.1).

Box 3.2 The Method of Least Squares

The *Method of Least Squares* is used to estimate the best straight line for a set of experimental data. If we have a data point (x_i, y_i), the observed value of y is y_i. However, we can predict the value of the dependent variable, y, by using a linear regression equation of the following form:

$$\hat{y}_i = m_1 x_i + c_0$$

where m_1 and c_0 are estimates of m (the slope) and c (the intercept), respectively.

The error of the prediction (the residual), e_i, is the difference in the actual y_i and the predicted y_i:

$$e_i = y_i - \hat{y}_i$$

The method of least squares chooses estimates of m_i and c_0 that mimimizes the total error of the prediction. A single value for the total error is obtained by squaring the residuals and summing them for all cases. This single value is known as the sum of the squares for error (SSE):

$$SSE = \sum(y_i - \hat{y}_i)^2 = \sum(y_i - c_0 - m_1 x_i)^2$$

The formulas for the least squares estimates of m_1 and c_0 are as follows:

$$m_1 = SS_{xy}/SS_{xx} \tag{3.2}$$

$$c_0 = \bar{y} - m_1 \bar{x} \tag{3.3}$$

where:

$$SS_{xy} = \sum x_i y_i - \left(\sum x_i\right)\left(\sum y_i\right)/n$$

$$SS_{xx} = \sum x_i^2 - \left(\sum x_i\right)^2/n$$

Although the formulae look very complicated, they are straightforward to use, as shown in Example 3.1.

Example 3.1

The data used to generate Figure 3.3 are listed in Table 3.2.

You can see from Figure 3.3 that the data show an approximately linear relationship between x and y. To use this information in order to determine the least squares line, we must calculate the values of Σx_i, Σy_i, Σx_i^2 and of $\Sigma x_i y_i$ and then write the results in a table, as shown in Table 3.3.

By substituting the values from Table 3.3 into Equations 3.2 and 3.3 (from Box 3.2):

From Equation 3.2, $m_1 = SS_{xy}/SS_{xx}$

$$SS_{xy} = 607.12 - (36.00 \times 108.32)/8 = 119.68$$

$$SS_{xx} = 204.00 - 36^2/8 = 42$$

Thus, $m_1 = SS_{xy}/SS_{xx} = 119.68/42 = 2.85$

From Equation 3.3, $c_0 = y - m_1 x$

$$x = 36.00/8 = 4.50$$

$$y = 108.32/8 = 13.54$$

Table 3.2 Data used to plot Figure 3.3

X	1.00	2.00	3.00	4.00	5.00	6.00	7.00	8.00
Y	2.95	8.30	6.02	15.27	12.68	20.22	19.33	23.55

Table 3.3 Data table used to determine the line of least squares

N	x_i	y_i	x_i^2	x_iy_i
1	1.00	2.95	1.00	2.95
2	2.00	8.30	4.00	16.60
3	3.00	6.02	9.00	18.06
4	4.00	15.27	16.00	61.08
5	5.00	12.68	25.00	63.40
6	6.00	20.22	36.00	121.32
7	7.00	19.33	49.00	135.31
8	8.00	23.55	64.00	188.40
	$\Sigma x_i = 36.00$	$\Sigma y_i = 108.32$	$\Sigma x_i^2 = 204.00$	$\Sigma x_iy_i = 607.12$

Thus, $c_0 = 13.54 - (2.85 \times 4.50) = 0.71$

Hence, the line of least squares is $y = 2.85x + 0.71$.

If the results of an experiment produce a graph curve with either of the characteristic shapes shown in Figures 3.4 or 3.5, then you are almost certainly dealing with exponential data. Applying logarithms to the base e (ln x on your calculator) to your data will produce a graph of the form $y = mx + c$, i.e. the formula for a straight-line graph. This has been carried out on the data shown in Figure 3.5, thus resulting in Figure 3.6. As you can see, this treatment allows easier interpretation of the data.

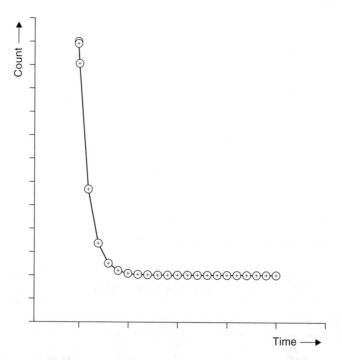

Figure 3.4 Graph showing an exponential decrease

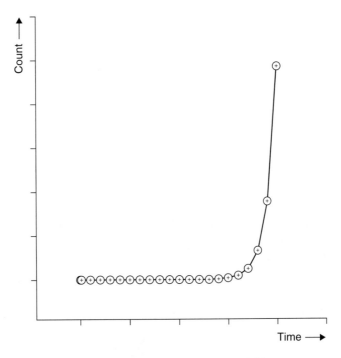

Figure 3.5 Graph showing an exponential increase

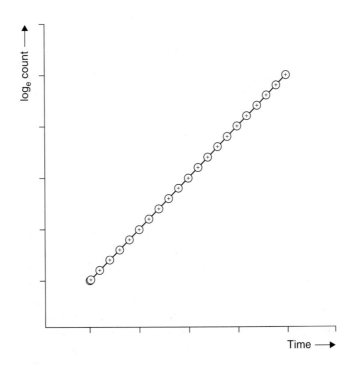

Figure 3.6 Log$_e$ treatment of the data shown in Figure 3.5

Use of statistics

The statistical analysis of data can be an important component of experimental results and, used intelligently, can enhance the value and authority of the results. Some examples of statistical information that are often used in scientific reports are listed below.

- Number of measurements (n).
- *Mean*, or average, of the data (μ or $\bar{x}$). The mean is defined by the following equation:

$$\bar{x} = \sum_{i}^{n} x_i/n \qquad (3.4)$$

  where x_i represents the individual results. The mean is often used to represent the estimated experimental result of a set of data.
- The *standard deviation* (σ or s) is a measure of the dispersion of a data set. The standard deviation is the positive square root of the variance. The sample variance, s^2, is an average squared distance of the sample values from the sample mean. The variance is calculated by using the following relationship:

$$s^2 = \sum_{i}^{n} [(y_i - y)^2/(n-1)] \qquad (3.5)$$

  and therefore the sample standard deviation is calculated from $s = \sqrt{s^2}$. Most experimental data will form a 'bell-shape' when plotted as a histogram, as shown in Figure 3.7. In these circumstances, approximately 68% of the measurements will fall within one standard deviation of the mean, while approximately 95% of measurements will fall within two standard deviations of the mean. Measurements that are far removed from the rest of the data, i.e. more than two standard deviations from the mean, are known as *outliers*.

The mean and standard deviation of a data set may be obtained by using a scientific calculator rather than by hand, i.e. by using Equations 3.4 and 3.5.

- The *range* (or spread) of the data is the numerical difference between the highest and lowest values in a data set (maximum value – minimum value).
- As we have already seen, two variables may be related to each other in such a way that they fall in a straight line (a linear correlation). The strength of the association is commonly measured by using the *Pearson correlation coefficient*, r. Box 3.3 shows how r may be determined. The value of r will always be between -1 and $+1$; the closer it is to -1 or $+1$, then the stronger the linear relationship between x and y. When there is a direct relationship between x and y, as shown in Figure 3.1, r is positive and we say that there is a positive linear relationship between x and y. When there is an inverse relationship between x and y, as shown in Figure 3.2, r is negative and we now say that there is a negative linear relationship between x and y. The correlation coefficient is particularly useful for demonstrating the degree of association between a measured value (e.g. the absorbance of a solution) and the concentration of a solution when plotting a calibration curve. In general, $|r| > 0.99$ indicates a strong linear relationship.

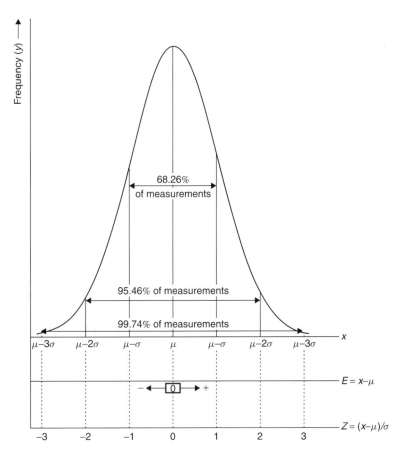

Figure 3.7 Typical 'bell-shaped' curve which is followed by empirical measurements (a Gaussian, or normal, distribution curve)

Box 3.3 The Pearson Correlation Coefficient

When two variables are related to each other in such a way that they fall in a straight line, the strength of the association may be measured by using the *Pearson correlation coefficient*, r. The latter may be determined by using the following expression:

$$r = \frac{SS_{xy}}{\sqrt{(SS_{xx}SS_{yy})}} \qquad (3.6)$$

where:

$$SS_{xy} = \sum x_i y_i - \left(\sum x_i\right)\left(\sum y_i\right)/n$$

$$SS_{xx} = \sum x_i^2 - \left(\sum x_i\right)^2/n$$

$$SS_{yy} = \sum y_i^2 - \left(\sum y_i\right)^2/n$$

We can illustrate the use of statistics by using the following examples.

Example 3.2

Two environmental analysts, 'P' and 'Q', were asked to check their technique for the determination of phosphate in river water by a colorimetric method. They were both given 5 ℓ of the same river water and asked to perform 20 determinations. The results that they obtained are shown in Table 3.4.

The real value of phosphate in the river water was estimated to be 0.49 mg dm^{-3}. The statistics suggest that Analyst P has a better technique than Analyst Q – P's mean is closer to the 'real value' and P's range (spread) of data and standard deviation from the mean are small. We would need a more complex statistical test to confirm that Analyst P has a better technique than Analyst Q, although the basic statistics have given us a guide.

Example 3.3

Analyst P was asked to determine the strength of the linear relationship between the concentration of standard phosphate solutions and the values measured by using a colorimetric method. Analyst P obtained the results shown in the first two columns of Table 3.5, and then calculated the Pearson correlation coefficient by using Equation 3.6.

Substituting the values from Table 3.5 into the appropriate equations from Box 3.3 we obtain:

$$SS_{xy} = 1.917 - (4.500)(2.697)/9 = 0.568$$

$$SS_{xx} = 3.188 - (4.500)^2/9 = 0.938$$

$$SS_{yy} = 1.152 - (2.697)^2/9 = 0.344$$

From Equation 3.6, $r = 0.568/\sqrt{(0.938 \times 0.344)} = 1.000$

Table 3.4   Phosphate in river water (mg dm^{-3}) as determined by Analysts P and Q

Analyst 'P'	Analyst 'Q'
0.42	0.38
0.45	0.47
0.46, 0.46	0.50
0.48	0.52
0.49	0.56, 0.56
0.50, 0.50, 0.50, 0.50, 0.50, 0.50, 0.50, 0.50, 0.50, 0.50	0.58, 0.58, 0.58
0.51	0.60, 0.60, 0.60, 0.60, 0.60
0.52	0.62
0.55	0.63
0.56	0.64
	0.66
	0.72
	0.81
Mean determination = 0.50	Mean determination = 0.59
Standard deviation = 0.03	Standard deviation = 0.09
Range = 0.56–0.42 = 0.14	Range = 0.81–0.38 = 0.43

Table 3.5 Data table used to determine the Pearson correlation coefficient

Phosphate (mg dm^{-3}) x_i	Absorbance y_i	x_i^2	y_i^2	$x_i y_i$
0.000	0.000	0.000	0.000	0.000
0.125	0.070	0.016	0.005	0.009
0.250	0.145	0.062	0.021	0.036
0.375	0.223	0.141	0.050	0.084
0.500	0.302	0.250	0.091	0.151
0.625	0.378	0.391	0.143	0.236
0.750	0.451	0.562	0.203	0.338
0.875	0.522	0.766	0.272	0.457
1.000	0.606	1.000	0.367	0.606
$\Sigma x_i = 4.500$	$\Sigma y_i = 2.697$	$\Sigma x_i^2 = 3.188$	$\Sigma y_i^2 = 1.152$	$\Sigma x_i y_i = 1.917$

The value of r clearly suggests that there is an extremely strong linear relationship between phosphate concentration and absorbance when using this colorimetric method.

The use of statistics is full of pitfalls and if there is doubt about the application of a particular technique, then do seek advice. Most of the statistical techniques used by undergraduate students are well known and thus it is not necessary to present full formulae or step-by-step workings of each test. Where presentation of the detailed workings of a statistical method can be justified, they can be included in tabular form in the Results section or committed to an appendix.

The ready availability of scientific calculators and computer software for executing statistical tests can present temptations to create a set of data in order to obtain an answer which might appear to give the results 'respectability'. Because a program *can* be used to process a set of data, it is by no means always certain that it *should* . You should always be prepared to defend the use of any statistical treatment you have used, just as you would an experimental technique.

3.8.5 General discussion

The General Discussion should draw all the threads of the report together and is, arguably, the most important part of the report. The discussion offers the widest scope for individual freedom of expression, and may include items such as the following:

- a discussion of the possible limitations of the methods;

- comments upon the precision (reproducibility or repeatability) of the results, as well as on their accuracy (closeness to 'true' values), if known;

- a discussion of effectiveness and limitations of the experiment and any statistical treatment of the data;

- a comparison of the results with those obtained or published elsewhere;

- a discussion of the significance of the data in an appropriate context;

- comments on the value of the results in a wider scientific, environmental or even commercial context.

No *new* data and results should be included in the discussion. In addition, attention should be drawn to any faults/problems with the chemicals or equipment used and to any deficiency in the assumptions upon which the experiment is based. Modifications and improvements should be included if appropriate.

3.8.6 Conclusions

The Conclusions section should summarize the main findings of the research. It is *not* a summary of your work programme or a description of the research carried out. It is often helpful to use 'bullet points', each no more than two or three lines, to summarize your results. This enables you (and your tutor) to see, at a glance, whether you have addressed all of the important areas and helps you to check that you have covered everything that you wanted to.

3.8.7 References

Some students are confused by the difference between a bibliography and a list of references. A bibliography is usually a list of all of the books that you have consulted in the production of your report but have not quoted from in the text. References are statements that you have found in the literature and used in the text of your report. Bibliographies are widely used in many subjects; if you study environmental law as part of an environmental science course, a full bibliography will probably be expected. However, a bibliography is not normally required as part of a laboratory report, *unless* you are recommending that the reader should refer to the texts listed. The use of up-to-date and appropriate references shows the tutor that you have consulted the literature properly and adds gravitas to your report.

Citing references

References may be cited in the text in a number of ways, depending upon your style of writing or the context of your reference. However, there are conventions that should be followed, as shown below – note the use of brackets.

- Natural levels of carbon monoxide are low, typically in the range 10–200 ppb (Grimes and Clement, 1993).

- Kinnear (1998) describes a system for sampling PM_{10} on an hourly basis, while Hegarty, Scanlon and Chan (2001) describe a system for the continuous sampling of PM_{10}.

If a reference has more than two authors, all of the authors are mentioned only at the first citation; at subsequent citations, only the first is mentioned; Thus, the 2001 reference above is subsequently cited as Hegarty *et al.* (2001) (*et al.* translates as 'and others'). If more than one paper by a particular author in the same year is cited, then distinguish them by appending letters to the date in the order that you cite them, e.g. (de Souza, 1996a), (de Souza, 1996b), etc. If the papers span more than one year, they should be cited in chronological order.

You may want to cite an official or company report, or government paper, where there is no specified author or the authorship belongs to a committee. In such cases, you normally cite the body responsible for publishing the paper or report. Thus, in the text, the body responsible for publishing the paper is cited with the year of publication, e.g. (Department of Environment, Transport and the Regions, 2001), (EnviroTech Ltd, 2000), etc. If you are writing to an informed audience, then it may be acceptable to abbreviate names, e.g. (DETR, 2001), although you should always use a 'key' for your abbreviation the first time that you use it.

Some journals use a system of numbering to identify references. In such journals, each reference is given an identifying number according to its first citing in the text, e.g. the first reference cited would be number '1', the second would be number '2', etc. The numbers may appear in brackets or as a superscript and are subsequently placed in ascending order in the reference section, followed by full details of each reference (see *The Reference section*, below), as follows:

- Natural levels of carbon monoxide are low, typically in the range 10–200 ppb (9).

- Natural levels of carbon monoxide are low, typically in the range 10–200 ppb[9].

It is very unwise to cite papers that you have not read (or at the very least for which you have not seen an abstract). First, you may get caught out in a *viva voce* (oral examination) and, secondly, there is no need. It is perfectly acceptable to write these as follows, e.g. (Clarke 1999, quoted by Yu 2000), or to cross-refer to an author's references. The best advice is, of course, to read all of the relevant literature and to cite it properly.

Sometimes it is necessary to cite information that has not yet been published but which has been obtained by word of mouth. This is cited in the following form: (M. Boyd, *pers. comm.*), but it is *not* listed in the references.

The Reference Section

The Reference section must include details of all references that have been cited in the text. It does not include peripheral reading. The protocols outlined below should also be used in bibliographies.

The details of each reference include the following: name(s) of the author(s) (surname first, with a comma), the year of publication, and the title of the publication. In the case of books and reports, the name of the publisher and place of publication is also given. There is more than one way of presenting this information; the following examples illustrate the use of upper and lower case letters, italics, punctuation marks and general layout.

Books By convention, book titles are usually written in italics (or underlined) when listed in a reference section or bibliography.

(i) Single edition.
  Brimblecombe, P. (1986). *Air Composition and Chemistry*. Cambridge University Press, Cambridge, UK.

(ii) Multiple editions.

When books have been published in subsequent editions it is important to specify the edition number because there are often considerable differences between editions. The edition number is shown in brackets after the date, as follows:

Roberts, M.B.V. (1984). Biology: *A Functional Approach* (3rd Edn). Nelson Publishers, London.

Collected works

(i) Citation of complete works.

Publications that consist of writings by a number of authors are identified under the name of the editor(s). The editor(s) are designated by the abbreviation (Ed.) or (Eds) after the name, as follows:

Leggett, J. (Ed.), (1990). *Global Warming – the Greenpeace Report*. Oxford University Press, Oxford.

(ii) Citation of specific chapter or paper.

References to specific chapters or papers in collected works are identified under the names of the particular authors, and then reference is made to the whole publication. It is usual to indicate the page numbers or identification number of the paper, as follows:

Williams, I.D., Hamilton, R.S. and Revitt, D.M. (1992). 'Public attitudes to road traffic pollution and nuisance'. Proceedings of the Ninth World Clean Air Congress, Volume **6**, *Risk Assessment, Strategies and Pollution Prevention*, Paper No. IU-14B.07.

Papers in journals or periodicals The main difference in citing references to papers in journals or periodicals is that the *journal* title is written in italics (or underlined) – not the paper title – and no publisher or editor is listed. The titles of most journals have standard abbreviations, but if in doubt give the whole title. The numbers of the issue and pages are also given, as follows:

Williams, I.D., Revitt, D.M. and Hamilton, R.S. (1996). 'A comparison of carbonyl compound concentrations at urban roadside and indoor sites'. *Sci. Total Environ.*, **189/90**, 475–483.

Official reports/Government papers When you list the references for official or company reports or government papers, the name of the body responsible for the paper or report takes the place of the author, as follows:

Department of the Environment (1994). *Sustainable Development. The UK Strategy*. HMSO, London, UK.

3.9 Plagiarism

Plagiarism is the representation of another person's published or unpublished work or ideas as your own by using an extensive unacknowledged quotation. In academia, plagiarism carries heavy penalties; your mark for any assessed work may be significantly reduced and you may be open to accusations of academic misconduct. However, this does not mean that all of your work must be completely original; expressing views that are influenced by other authors is a consequence of shared knowledge and a reflection of wide reading. In order to avoid accusations of plagiarism, you should clearly reference sources by using the conventions outlined above in Section 3.8.7.

3.10 Deadlines

All academic institutions will set a deadline for coursework. There is usually a penalty for failing to meet this deadline, such as a specified reduction in the grade awarded. It is obviously important therefore to plan your work carefully so that you meet the set deadline. It is also advisable to retain an extra copy of each report you write as a safeguard against accidental loss or theft and for your own use during revision.

3.11 Summary

After carefully studying this chapter, you will have a clear understanding of how to write professional and readable laboratory and scientific reports by using recognized conventions, and you will also be able to present and report your experimental data proficiently. As with any skill, you will need practice and perseverance in order to master the art of writing and reporting scientific information. However, given the importance of effective communication in the modern world, these skills are worth mastering.

Self-Study Exercises

3.1　List the key features of a good laboratory report.

3.2　Describe in your own words the main functions of a scientific report.

3.3　A friend in your laboratory class always loses marks for poor presentation of his laboratory reports and asks you for advice. Design a checklist for him that lists the key features of a well-presented report.

3.4　Are the following statements about the presentation of scientific reports correct or incorrect? Give reasons for your answers.

(a)　Different margin and font sizes in a report are acceptable because they make scientific reports more interesting to read.

(b)　Page numbers are unnecessary in a laboratory report.

(c)　The General Introduction should explain the rationale for the experiment and establish the context for the work.

(d)　It is not necessary to note small changes in experimental methodology.

(e)　Experimental materials should be described as fully as is necessary for repetition.

(f)　Every table should have a title and a number written above the table.

(g)   Every figure should have a title and a number above the figure.

(h)   When plotting graphs, the convention is that the dependent variables are plotted on the y-axis while the independent variables are plotted on the x-axis.

(i)   The points of a graph may be marked with crosses (+), kisses (×), points (·), rings (○) or circles with dots (⊙).

(j)   The Conclusions section of a laboratory report should include a description of the research performed.

3.5   What material is suitable for inclusion in the General Discussion section of a laboratory report? What material should not be included?

3.6   Explain the difference between a bibliography and a reference list.

3.7   A research scientist in an environmental consultancy firm regularly determines the potassium and sodium content of liquid effluent from a fertilizer factory. Over the years, he has discovered that his calibration curves always follow the following same straight line graphs:

Potassium $y = 5.2x + 3.4$

Sodium $y = 2.7x + 4.6$

The instrument reading is plotted on the y-axis and the metal concentration (in ppm) on the x-axis. Use these equations to calculate the instrument readings for both potassium and sodium when $x = 0, 1, 2, 3, \ldots, 10$ ppm and plot your calibration curves in an appropriate fashion.

Challenging Exercises

3.8   You are asked to give a talk on writing laboratory reports to Sixth Form (High School) students at a local school. Design some A4-sized overhead projection slides to highlight the key principles of good laboratory report writing.

3.9   A postgraduate student was asked to determine the sodium concentration of a soil solution using a flame photometer, prior to an undergraduate laboratory class. First, the student prepared standard solutions containing 0.0, 2.5, 5.0, 7.5 and 10.0 ppm of sodium solution and then took five readings of each solution. Next, he averaged the measurements to obtain a mean reading for each standard solution, which he then used to plot an appropriate calibration graph before determining the sodium concentration of the unknown soil solution. Use his measurements (summarized below) to plot an appropriately presented graph and accurately determine the concentration of the unknown.

Solution (ppm)	Photometer reading[a]				
	Reading 1	Reading 2	Reading 3	Reading 4	Reading 5
0.0	0.00	0.00	0.00	0.02	0.02
2.5	4.91	4.93	4.93	4.95	4.96
5.0	9.98	9.99	9.99	10.01	10.00
7.5	15.03	15.07	15.09	15.02	15.04
10.0	19.78	19.94	19.88	19.83	19.92
Unknown	12.33	12.45	12.30	ND	ND

[a]ND, not determined.

3.10 Four students were given a sample of lake water and asked to determine its hardness in mg dm^{-3} (by using the method described late in Chapter 7). Each student was given an hour to perform as many accurate determinations as they could. Their titration results (in cm^3) were as follows:

Student A	22.8, 22.9, 22.9, 23.0, 23.0, 23.0, 23.1, 23.1, 23.2
Student B	22.6, 22.8, 22.8, 22.9, 23.0, 23.0, 23.1, 23.2, 23.2, 23.4
Student C	22.4, 22.4, 22.5, 22.5, 22.6, 22.6
Student D	22.0, 22.4, 22.5, 22.6, 22.7, 22.7, 22.7, 23.0, 23.0, 23.1, 23.2

(a) Using graph paper, display the data for each student on a scatter diagram (see Chapter 1).

(b) Complete the missing sections of the following table.

Student	Number of titrations (n)	Mean titration (cm^3)	Standard deviation	Range
A				
B				
C				
D				

(c) The true value of the titration is 23.0 cm^3. Use your answers to (a) and (b) (and your previous reading – see Chapter 1) to classify the work of each student according to the criteria listed below:

- accurate and precise;

- accurate but not precise;

- precise but not accurate;

- inaccurate and imprecise.

3.11 A student was asked to determine the association between two methods for the determination of chloride in tap water. The student obtained the results shown in the following table (in mg dm^{-3}) for a number of water samples by using the standard method (Method A) and a new method (Method B). Calculate the Pearson correlation coefficient for the association between Methods A and B.

N	1	2	3	4	5	6	7
Method A	9.0	18.2	17.5	14.2	11.0	10.1	12.2
Method B	7.5	15.5	14.3	12.2	9.0	8.5	9.8

References

Department of the Environment (1997). *Digest of Environmental Statistics, Number 19*, 1997. HMSO, London, UK.

4 The Structure and Composition of the Earth

How inappropriate to call this planet Earth when quite clearly it is ocean.

Arthur C. Clarke, *Nature*, 1990.

4.1 Introduction

Where did chemical elements originate? Everybody has a different view on this depending on his or her religion or philosophy. As scientists, we can point to the ever increasing body of evidence which supports the contention that chemical elements were, and still are, formed by stellar nuclear synthesis, and that the Earth's formation can be explained by chemical and physical processes. We also know that the Universe is expanding; this suggests that there was a starting point – the so-called Big Bang.

The Big Bang is the hypothetical explosive event that marked the origin of our Universe some 10–20 billion years ago. The Big Bang theory states that at some definite time the Universe came into being and was squeezed into a hot super-dense state. The extremely condensed matter (elementary particles) exploded as a result of high temperatures and pressures, dispersing matter through space and producing the expanding Universe. Some material then regrouped under gravitational influences into stars such as the Sun and other bodies such as the Earth.

The Sun is at the centre of our solar system and is about 4.7 billion years old. It is about 1 392 000 km in diameter and is composed of approximately 70% hydrogen and 30% helium, with other elements making up less than 1%. The hydrogen is converted into helium at the Sun's centre by nuclear fusion reactions that generate temperatures of 15 000 000 K. The Sun's radiant energy is colossal; sufficient energy is produced to support life on Earth from a distance of almost 150 000 000 km.

The Earth is the third planet from the Sun and is about 4.6 billion years old. It is almost spherical in shape – it is ovoid and flattened slightly at the poles – and has an equatorial diameter of 12 750 km. In this chapter, we will study several fundamental processes that influence the structure and composition of the Earth, including those listed below.

- The Earth's physical environment and the complex relationships between its physical, chemical and biological components.

- The nature of chemical bonds and their influence on the structure and properties of materials found on Earth, particularly in the crust.

- Important physical and chemical processes that occur in Earth's environment, their impacts on environmental processes and the interaction between the different environmental compartments.

- Physical and chemical processes that occur in the crust and soil.

We will also further develop your laboratory skills by studying methods for the quantitative determination of widely measured environmental parameters.

4.2 Structure of the Earth

We can consider the Earth as three separate areas – the solid material, the hydrosphere and the atmosphere, and we will discuss each area in turn. A fourth category, the biosphere, spans all regions of the Earth that support life. The structure of the Earth is shown in Figure 4.1 and its composition is given in Table 4.1.

4.2.1 The solid material (The internal structure of the earth)

Using seismology, geologists have determined that the Earth is made up of layers. At the centre of the Earth is a very dense and very hot *core*; this is covered by a region known as the *mantle*, which in turn is surrounded by a thin *crust*.

The core

The core is the innermost central part of the Earth below a depth of 2900 km, and may be divided into the inner and outer cores. Both the solid inner core, which has a radius of 1700 km, and the molten (semi-solid) outer core, which has a radius of 1820 km, are thought to consist of an alloy of iron and nickel.

The mantle

The mantle is the intermediate zone of the Earth between the crust and the core, accounting for 82% of the Earth's volume. The principal minerals in the mantle are olivine (Mg_2SiO_4–Fe_2SiO_4) and pyroxene ($MgSiO_3$–$FeSiO_3$). Geologists have identified that the mantle is divided into two regions containing zones with characteristic chemical compositions.

The higher region, the *upper mantle*, may be divided into six zones (A–F), as shown in Figure 4.1. The lithosphere (zone A) is the topmost region of the Earth and comprises the crust and the top zone of the upper mantle. The strong and solid lithosphere extends to a depth of about 100 km and rides on top of the

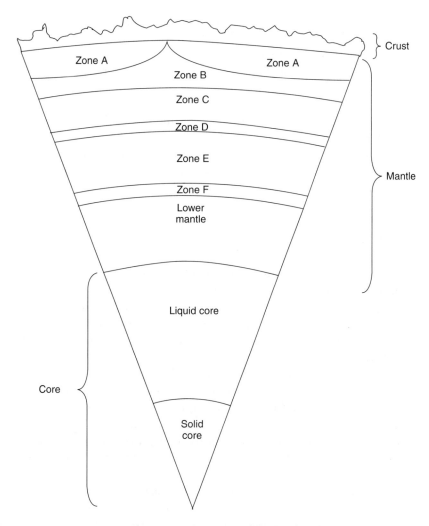

Figure 4.1 Structure of the Earth

Table 4.1 Structure of the Earth

Zone	Major chemical constituents	State of matter
Atmosphere	N_2, O_2, H_2O, CO_2, noble gases, particles	Gas, liquid, solid
Hydrosphere	H_2O (liquid), ice, snow, dissolved minerals, Na^+, Cl^-, Mg^{2+}, Br^-, etc., particles	Liquid (solution), suspended solids
Biosphere	Organic material, mineral skeleton, H_2O, trace elements	Solid, liquid, colloidal, gas
Crust	Silicate rocks, oxide and sulfide minerals	Solid (heterogeneous) (intermixed with water and air)
Mantle	Silicate minerals, particularly olivine and pyroxene (Fe and Mg silicates)	Solid
Core	Fe–Ni alloy	Liquid (top) Solid (bottom)

relatively weak and partially molten asthenosphere (the main source of volcanic magma, shown as zone B). Zone C begins at a depth of about 200 km. It consists of solid igneous rock (primarily peridotite) and extends to a depth of 400 km, where zone D begins. Zone D is called a transition zone, because in this region increases in pressure and temperature cause peridotite to undergo a physical change of phase, resulting in a more dense crystalline structure. The consequent structure (called a spinel structure) dominates zone E, which extends to a depth of 670 km. A second transition zone (F) marks the boundary between the upper and lower mantle and the change from the spinel to the perovskite, iron and magnesium oxide structures. The *lower mantle* extends from 700–2900 km and changes little in composition and phase with depth.

The crust

The crust is important with regard to reactions between itself, the hydrosphere and the atmosphere, and is composed mainly of silicates (58.7%, corresponding to an empirical formula of SiO_2) and aluminates (15%, corresponding to Al_2O_3). The crust constitutes <0.4% of the total mass of the Earth and <0.0001% of the Earth's volume and may be divided into two distinct parts, namely the continental and oceanic crusts.

The *continental crust* varies in thickness from 40–70 km, being deeper below mountain ranges. It has a complex structure and composition, consisting of a surface layer containing sedimentary and igneous rocks and a lower layer of metamorphic and granitic (silicon-rich) rocks (we will discuss these terms in more detail in Chapter 5). *Si*licon and *al*uminium oxides dominate the chemical composition of the continental crust, with continental crustal material being given the acronym *sial*.

The *oceanic crust* is an average of about 10 km thick and consists of a layer of surface sediment covering basaltic rocks. *Si*licon and *ma*gnesium oxides dominate the chemical composition of the oceanic crust, with oceanic crustal material being given the acronym *sima*.

The crust is separated from the underlying mantle by an abrupt change in chemical composition called the Mohorovičić discontinuity boundary, or Moho, which follows variations in the thickness of the continental and oceanic crust. The Moho is present everywhere, except beneath spreading ridges, and its depth varies considerably.

The composition of the Earth's crust shows that only a few elements are abundant and most are very rare (see Table 4.2). Over 100 chemical elements have been identified on Earth. There are 90 naturally occurring elements, 81 of which have at least one or more stable isotopes, while the other nine elements exist as unstable radioactive isotopes. The remaining elements do not occur naturally on Earth today, but have been synthesized by man. It is probable that some of these elements did exist in the past, but owing to short half-lives such elements have since decayed.

The abundance of each naturally occurring element varies widely. For example, oxygen is found as a major structural component in no fewer than 1300 classes of

Table 4.2 Relative abundance (by wt%) of elements in the whole Earth and in the Earth's crust

Relative abundance in the whole Earth (%)	Relative abundance in the Earth's crust (%)
Iron 35	Oxygen 46
Oxygen 30	Silicon 28
Silicon 15	Aluminium 8
Magnesium 13	Iron 6
Nickel 2.4	Magnesium 4
Sulfur 1.9	Calcium 2.4
Calcium 1.1	Potassium 2.3
Aluminium 1.1	Sodium 2.1
All other elements < 1	All other elements < 1

minerals, whereas rhenium (Re), rhodium (Rh) and osmium (Os) each form only one naturally occurring mineral, although trace amounts may substitute into other minerals to replace more abundant elements. The abundance of an element within minerals also depends on its properties, since mineral stability is dependent on temperature, pressure, pH, redox potential, etc.

We can imagine the manner in which the Earth was created from our knowledge of its physical structure and chemical composition. When the disc of material that gave rise to the Earth started to condense, the temperature was probably around 2 000 K, and hence all of the material was present in the gaseous state. Consequently, the early Earth was probably a homogeneous mixture of gases with no solid or liquid phases.

On cooling, the least volatile stable combinations of the elements in the disc would condense out, thus forming solid material. As the temperature continued to drop, more gaseous material condensed on top of the existing solid material. This period of growth (known as accretion) would have been followed by gravitational compression of the Earth into a smaller volume. This compression would have led to a rise in the temperature of the inner parts of the planet as the accumulating outer material squeezed the internal material. This heat would have accumulated in the interior of the planet faster than it could escape since rock is a poor heat conductor. Eventually, the solid material at the centre of the planet would start to melt, forming a molten core. When the iron in a layer began to melt, large drops would probably have formed. These drops would have fallen towards the centre of the planet, as iron is heavier than the other common elements of the Earth. Siderophiles ('iron-loving' metals, such as nickel) would also have been pulled towards the centre of the Earth with molten iron (as a consequence, siderophiles tend to be of low abundance in the Earth's crust). Lighter material would have been displaced upwards, thus creating a molten iron core at the centre of the Earth.

The Earth is known as a *differentiated planet*, i.e. one that is chemically zoned because during its creation heavy materials sank to the centre, light materials accumulated at the surface and the remaining materials accumulated in between, so producing the major components of the core, crust and mantle, respectively. The differentiation of the Earth probably led to the escape of gases from the interior and hence eventually to the formation of the atmosphere and the oceans.

4.2.2 The hydrosphere

The hydrosphere is the water component of the Earth and includes the oceans, seas, rivers, streams, lakes, swamps, groundwater and atmospheric water vapour. It would have been formed as the Earth cooled and water condensed – while hot, the water would have readily dissolved soluble ions to produce the mineral content of the oceans. The composition of the hydrosphere is shown later in Table 7.1. Approximately 71% of the Earth's surface is covered by oceans, which thus play a major factor in shaping the physical and chemical nature of the Earth's surface.

- Oceans modify climate by absorbing solar radiation and transporting it around the world, as well as through the evaporation/precipitation cycle that starts at the air–sea interface. Well known examples of energy transfer via the oceans include El Niño and the Gulf Stream. El Niño is a warm surface current that typically appears around Christmas in the Pacific Ocean off South America and disappears by the end of March. However, every 3–7 years, El Niño persists for up to 18 months as part of a large-scale climatic fluctuation of the tropical Pacific Ocean. The Gulf Stream is a warm ocean current of the Atlantic Ocean off North America, originating in the Gulf of Mexico and moving towards the north-east. At its beginning, the temperature of Gulf Stream water is about 27°C, although it decreases as it moves north.

- Oceans play a regulating role in controlling the abundances of O_2 and CO_2, which are necessary for life processes.

- Oceans are important participants in the hydrological cycle, in which water evaporates from oceans into the atmosphere, falls as rain/snow on land and returns via rivers and lakes to oceans again.

We will study the environmental chemistry of water in Chapter 7.

4.2.3 The atmosphere

The atmosphere is the envelope of gases around the Earth. The elemental composition of the atmosphere has fascinated natural philosophers since the earliest times – the Greeks regarded air as one of the four elements (i.e. fire, earth, water and air). However, it was not until the late 17th century, when Robert Boyle described air as a 'confused aggregate of effluviums', that air was first regarded as a mixture. Even then, while oxygen and nitrogen were known to be the principal components of air, the question remained as to whether or not it was a mixture. Sir Humphrey Davy (1778–1829) thought that air was a compound, partly because if it were not a compound, then the heavier gas (oxygen) should sink below the lighter one (nitrogen) and, therefore, oxygen should be found in slightly higher concentrations at the very bottom of the atmosphere. The strength of the mixing process, caused by atmospheric turbulence, was not appreciated.

The development of understanding that air is a mixture as opposed to a compound was based on the following reasons:

Table 4.3 The composition of dry, unpolluted air at sea level

Gas	Volume (%)	Concentration (ppm or $\mu l\ dm^{-3}$)
Nitrogen	78.09	780 900.0
Oxygen	20.94	209,400.0
Argon	0.93	9300.0
Carbon dioxide	0.03	325.0
Neon	trace	18.0
Helium	trace	5.2
Methane	trace	1.2
Krypton	trace	0.5
Hydrogen	trace	0.5
Xenon	trace	0.08
Nitrogen dioxide	trace	0.02
Ozone	trace	0.01

- the ratio of oxygen to nitrogen does vary, just slightly, from place to place;
- if air were a compound, the formula would be $N_{15}O_4$, which is very unlikely;
- the physical properties of air are identical to those of the appropriate mixture of nitrogen and oxygen (which make up about 99% of air);
- it is possible to separate the nitrogen and oxygen, which would not be possible if air was a compound;
- there is no volume change or heat release on mixing oxygen and nitrogen, which suggests that a compound is not formed.

Of course, we now know that air is a mixture of many gases, and not just oxygen and nitrogen. The typical composition of dry, unpolluted air at sea level is shown in Table 4.3.

We have already suggested that the Earth formed as hot gases condensed into solid material. The formation of Earth's atmosphere is believed to be a combination of two processes, as follows:

- As the Earth developed, the solid material degassed via volcanoes and geysers, etc. to produce a primitive atmosphere containing H_2, He, N_2, CO_2, CO, H_2O, NH_3 and CH_4. The Earth's temperature would have been quite high, thus allowing many of these gases to escape from the Earth's surface because their vertical velocities would allow them to overcome the Earth's gravitational pull; hydrogen and helium can still escape.
- As the Earth cooled, $N_2(g)$ in the primitive atmosphere would have had insufficient energy to escape into space, and because of its low reactivity, it would have become trapped. This early primordial atmosphere would have had a different composition to the later oxygen-rich atmosphere; reactive O_2 would not have been initially present. Dioxygen gas was probably produced later as a result of chemical reactions, such as the photochemical decomposition of water and biochemical photosynthesis.

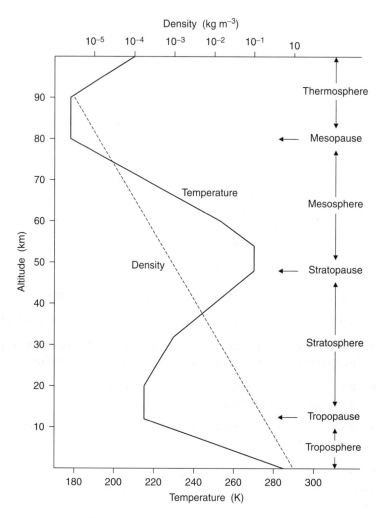

Figure 4.2 Structure of the atmosphere

The atmosphere is not so much one layer of gas, but rather a mixture of gases whose concentrations vary with altitude (height above ground). The Earth's current atmosphere consists of four different regions, namely the troposphere, the stratosphere, the mesosphere and the thermosphere, which may be distinguished by their physico-chemical characteristics. These regions are shown in Figure 4.2. Scientists are generally most interested in the two regions closest to the earth, i.e. troposphere and the stratosphere.

The troposphere is the inner layer of the atmosphere and extends for about 17 km above sea level at the equator and 8 km over the poles. It contains about 75% of the mass of the earth's air. The stratosphere is the atmosphere's second layer and extends from about 17–48 km above the earth's surface. Although the stratosphere contains less matter than the troposphere, its composition is similar, with two key exceptions:

- its volume of water vapour is about 1000 times less;
- its volume of ozone is about 1000 times greater.

Ozone in the stratosphere absorbs ultraviolet radiation from the sun, preventing UV-C radiation – which is lethal to man – reaching the earth's surface and reduces the amount of harmful UV-B radiation passing through the stratosphere. This process also warms the lower reaches of the stratosphere, thus creating the temperature inversion that can be seen in Figure 4.2. In this way, the stratosphere may be regarded as our global sunscreen.

Figure 4.2 also shows the temperature and density profiles of the atmosphere with increasing altitude. These profiles are important as they influence the movement of gases through the atmosphere. The following three characteristics illustrate this.

- In the troposphere, the temperature *decreases* with increasing altitude (called a positive lapse rate), which leads to relatively rapid mixing of atmospheric components.

- In the stratosphere, the temperature *increases* with increasing altitude (negative lapse rate), which inhibits vertical mixing – when mixing occurs it is due to diffusion and turbulence. Ozone in the lower stratosphere is very efficient at absorbing solar heat directly, thus making the air there warmer than in the upper troposphere.

- The tropopause is the boundary separating the troposphere and the stratosphere. The tropopause varies in depth from 10–17 km according to latitude and time of year. The rate of vertical mixing falls sharply here, and thus the transport of gases between these regions is very slow.

The general circulation (physical movement) of gases in the atmosphere is driven by the input of solar energy and modified by the rotation of the earth. Winds are obviously part of this general circulation, and it has long been known that certain winds prevail for much of the year. In the days of sailing vessels, these winds were given fanciful names such as the 'trade winds' or the 'doldrums'. Wind clearly plays a part in the movement of atmospheric substances, for example, the well known movement of desert dust from the Arabian Peninsular across the Atlantic. The other main determinant of the movement of substances in the atmosphere is the chemical reactions that cause the formation and removal of air pollutants.

4.2.4 Air pollutants

Air pollutants could, in principle, be termed as any atmospheric constituents that exceed the concentrations listed in Table 4.3. However, strictly speaking, a pollutant is defined as a substance that is potentially harmful to the health or well-being of human, animal or plant life, or to ecological systems.

Air pollutants may take gaseous or particulate forms. We can further divide air pollutants into regulated and unregulated, based on their treatment by environmental agencies in the United States, the European Union and major Asian countries. The main examples of unregulated air pollutants are carbon dioxide (CO_2) and nitrous oxide (N_2O). The major regulated pollutants are as follows:

- sulfur dioxide (SO_2);
- nitrogen oxides (NO_x), particularly nitric oxide (NO) and nitrogen dioxide (NO_2);

- carbon monoxide (CO);

- trace metals such as lead (Pb), cadmium (Cd) and platinum (Pt);

- organic compounds such as benzene and polyaromatic hydrocarbons (PAHs);

- photochemical oxidants such as ozone (O_3) and peroxyacetyl nitrates (PANs);

- particulate matter.

Particulate matter includes a wide range of sizes (typically $0.01\,\mu m->100\,\mu m$) and may be composed of organic or inorganic materials (or mixtures). The types of particulate matter monitored include 'total suspended particulates', smoke and particles of a specific size, such as PM_{10} and $PM_{2.5}$; these are particles with aerodynamic diameters of less than $10\,\mu m$ and $2.5\,\mu m$, respectively. The atmospheric concentrations, chemistry and environmental effects of key air pollutants are discussed later in Chapter 6.

Air pollutants may also be classified according to the way in which they are formed in the atmosphere. *Primary pollutants* are those which are emitted directly into the atmosphere; for example, CO comes directly from the incomplete combustion of fossil fuels in motor vehicles, while SO_2 is emitted from power stations and industrial plants. *Secondary pollutants* are formed in the air as a result of chemical reactions with other pollutants and atmospheric gases; for example, ozone is generated by photochemical reactions within the atmosphere. Note that pollutants do not necessarily fall into one or other of these categories; some can be both primary and secondary pollutants. For example, NO_2 is emitted directly into the atmosphere from power stations and vehicle exhaust, and some is also formed from the oxidation of NO in the air.

The distinction between primary and secondary pollutants is important for understanding air pollution and devising appropriate control strategies. For primary pollutants, there is likely to be a proportional relationship between emissions and ambient concentrations. However, with a secondary pollutant, reducing the emissions of its precursors may not necessarily lead to a proportional reduction in its ambient concentrations; indeed, in some circumstances, it may actually lead to an increase in concentrations.

Air pollutants usually have identifiable sources and these emission sources may be classified as either natural or man-made. *Natural* sources include volcanic eruptions, sand storms, lightning and forest fires, while *man-made* (or *anthropogenic*) sources are dominated by the combustion of fossil fuel for energy, particularly in power stations and motor vehicles. There are, though, many non-combustion-related anthropogenic sources, including industrial processes, coal mining, domestic and industrial solvent use, natural gas leakage in the national distribution network, and landfill. Non-combustion sources are particularly important for volatile organic compounds and methane.

Many national and international authorities issue legislative standards that aim to control the emissions of pollutants from source. These are known as *emission standards*. Important international organizations that issue emission standards include the European Commission (EC), the United Nations' Economic Commission for Europe (UNECE) and the United States Environmental Protection Agency (US

Table 4.4   Air quality limits and guide values for NO_2 (ppb)

Organization	Hourly mean		24-hour average	1-hour average
	50th percentile	98th percentile		
EC (Directive 85/203/EEC)				
Limit value	–	105	–	–
Guide value	26	71	–	–
World Health Organization	–	–	78	209

EPA). National governments participate in the process of setting an emission standard and then sign a document agreeing to implement measures which will limit emissions in their country to the designated standard. Such agreements are often legally binding. The emission standards usually apply to named processes.

Air quality standards are used to define a pollutant concentration that is regarded as acceptable, that is, the concentration of the pollutant in the air is unlikely to cause any significant detrimental effects on human health or the environment. National governments and international authorities, such as the World Health Organization (WHO), the EC and the US EPA, issue air quality standards. Two types of air quality standard are commonly issued:

- *limit values*, which are mandatory and should not be exceeded anywhere, being designed to improve the protection of human health;

- *guide values*, which are intended to serve as long-term precautions for health and the environment.

As an example, the main internationally recognized air quality standards for NO_2 are shown in Table 4.4.

4.2.5   The biosphere

The biosphere is essentially the narrow zone that supports life on Earth. It is limited to a fraction of the Earth's crust, the hydrosphere and the lower regions of the atmosphere.

4.3   Chemical Bonds

Before we look in detail at the physical structure and chemical composition of the Earth, we need to remind ourselves about the nature of atoms and chemical bonds and their influence on the structure and properties of substances.

An atom consists of a central nucleus surrounded by a three-dimensional cloud of electrons. An electron is said to occupy a three-dimensional space around the nucleus – called an *atomic orbital* – in which there is a 95% chance of finding the electron. Electrons can occupy four types of orbital, known as the s, p, d and f orbitals, which can be represented by three-dimensional shapes. The shapes of the s and p atomic orbitals are shown below in Figure 4.4. Electrons are arranged

Table 4.5 Ground-state electronic configurations of the first twenty elements in the periodic table

Atomic number	Element	Electronic configuration (by shell)	Electronic configuration (by sub-shell)
1	Hydrogen	1	$1s^1$
2	Helium	2	$1s^2$
3	Lithium	2.1	$1s^2 2s^1$
4	Beryllium	2.2	$1s^2 2s^2$
5	Boron	2.3	$1s^2 2s^2 2p^1$
6	Carbon	2.4	$1s^2 2s^2 2p^2$
7	Nitrogen	2.5	$1s^2 2s^2 2p^3$
8	Oxygen	2.6	$1s^2 2s^2 2p^4$
9	Fluorine	2.7	$1s^2 2s^2 2p^5$
10	Neon	2.8	$1s^2 2s^2 2p^6$
11	Sodium	2.8.1	$1s^2 2s^2 2p^1 3s^1$
12	Magnesium	2.8.2	$1s^2 2s^2 2p^1 3s^2$
13	Aluminium	2.8.3	$1s^2 2s^2 2p^1 3s^2 3p^1$
14	Silicon	2.8.4	$1s^2 2s^2 2p^1 3s^2 3p^2$
15	Phosphorus	2.8.5	$1s^2 2s^2 2p^1 3s^2 3p^3$
16	Sulfur	2.8.6	$1s^2 2s^2 2p^1 3s^2 3p^4$
17	Chlorine	2.8.7	$1s^2 2s^2 2p^1 3s^2 3p^5$
18	Argon	2.8.8	$1s^2 2s^2 2p^1 3s^2 3p^6$
19	Potassium	2.8.8.1	$1s^2 2s^2 2p^1 3s^2 3p^6 4s^1$
20	Calcium	2.8.8.2	$1s^2 2s^2 2p^1 3s^2 3p^6 4s^2$

around the nucleus in groups of orbitals known as *shells*. These shells are assigned whole-number values known as *principal quantum numbers* (n) that represent their increasing energy levels; the higher the value of n, then the higher the energy level of the shell. Shells may be further split into sub-shells with their own associated energy level; an s sub-shell consists of an s orbital, whereas a p sub-shell consists of up to three p orbitals (p_x, p_y, p_z). The *electronic configuration* of an element summarizes how the electrons of its atoms are arranged in their shells, sub-shells and orbitals. The ground-state electronic configurations of the first twenty elements in the periodic table are given in Table 4.5.

A *chemical bond* is a force that holds together two or more atoms, ions, molecules or any combination of these. This force is the electrostatic force of attraction between positively charged nuclei and negatively charged orbiting electrons. The strength of the attractive force depends largely upon the outer-shell electronic configurations of the atoms. In their ground states, noble gases have stable (full) outer-electron shells and do not readily form chemical bonds, whereas hydrogen (with one electron in its outer shell) readily forms bonds. In general, atoms form bonds in order to achieve the most stable (lowest-energy) electronic configuration.

Chemical bonding has a strong influence on the structure and physical properties of both elements and compounds. In elements, bonding can be either metallic or covalent. For compounds, it is either ionic or covalent. In a pure covalent bond, for example, in $O_2(g)$ or $H_2(g)$, electrons are shared equally between two atoms. At the other extreme, a pure ionic bond would involve complete transfer of an electron or electrons from one atom to another, with no sharing. In practice, there are probably no ion-pair structures that have a completely ionic bond; ionic molecules are probably composed of interacting ions bonding via the unequal sharing of electrons.

4.3.1 Ionic bonding

Some atoms achieve their most stable electronic configuration by gaining or losing electrons to form *ions*. Atoms that gain electrons are called *anions*. In gaining electrons (ne^-), they (usually) attain a stable electronic configuration and achieve an overall negative charge (A^{ne-}). Atoms that lose electrons are called *cations*. In losing electrons (ne^-), they (usually) attain a stable electronic configuration and achieve an overall positive charge (M^{ne+}). An *ionic bond* is the electrostatic force of attraction between opposite charges of cations and anions. The most common ionic compounds consist of metallic cations from Groups I and II of the Periodic Table and non-metal anions from Groups VI and VII.

Example 4.1

The transfer of an electron from a lithium atom to a fluorine atom forms the ionic bond in lithium fluoride. This transfer results in a stable (full) electronic configuration for both ions and an electrostatic force of attraction between the oppositely charged ions, as follows:

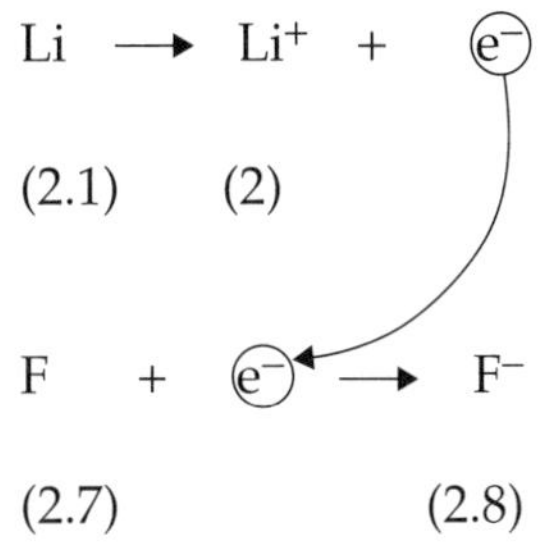

$$Li \longrightarrow Li^+ + \; \boxed{e^-}$$

$$(2.1) \qquad (2)$$

$$F \; + \; \boxed{e^-} \longrightarrow F^-$$

$$(2.7) \qquad\qquad (2.8)$$

(The electronic configurations are written in brackets under the atoms and ions.)

Example 4.2

The transfer of two electrons from a calcium atom to an oxygen atom forms the ionic bond in calcium oxide, as follows:

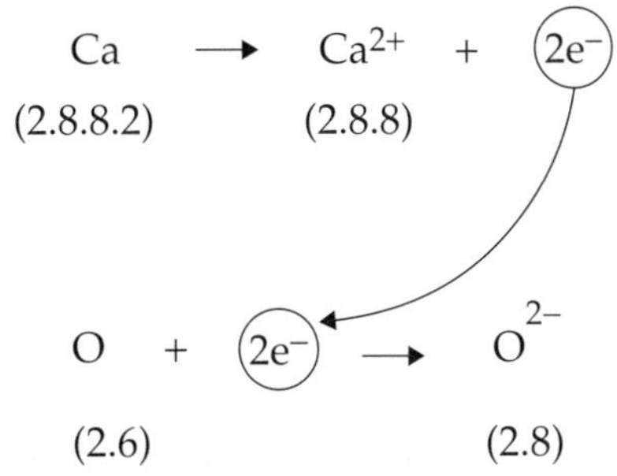

$$Ca \longrightarrow Ca^{2+} + \; \boxed{2e^-}$$
$$(2.8.8.2) \qquad\quad (2.8.8)$$

$$O \; + \; \boxed{2e^-} \longrightarrow O^{2-}$$
$$(2.6) \qquad\qquad (2.8)$$

The existence of ions has been demonstrated experimentally by using X-ray analysis to generate electron-density maps. Such maps of ionic compounds show regions (circles) of charge isolated from other regions of charge – exactly what you would expect from substances that contain oppositely charged ions. Thus, truly ionic bonds are *non-directional* because ions radiate a spherically symmetrical positive or negative field of charge. The technique also enables scientists to count the number of electrons associated with certain ions, thus allowing confirmation of the movement of one or more electrons from one atom to another.

A feature of ionic compounds is that they form *crystals* – substances with orderly three-dimensional arrangements of atoms or molecules and smooth external surfaces. Sodium chloride (table salt, NaCl) is the best known common crystal structure. When a solution of sodium chloride is evaporated to the point of crystallization, the sodium ions (Na^+) and the chloride ions (Cl^-) are much closer together than when they are in a dilute solution. At this point, the negatively charged chloride ions attract positively charged sodium ions, and vice versa, thus building up a regular three-dimensional structure of ions, as shown in Figure 4.3. Consequently, no pair of Na^+ and Cl^- ions exists independently as a molecule; the formula Na^+Cl^- represents the ratio in which the ions are present in the crystal structure (a *formula unit*).

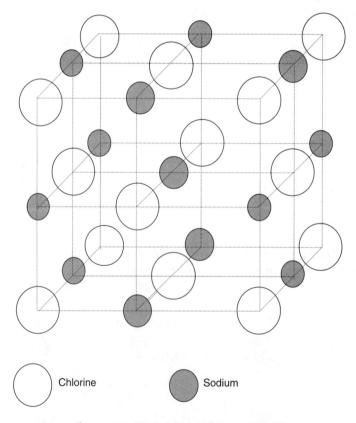

Chlorine   Sodium

Figure 4.3   The crystal structure of NaCl

4.3.2 Covalent bonding

Some atoms achieve their most stable electronic configuration by sharing electrons. This is called *covalent bonding*. A *covalent bond* consists of two electrons shared between two adjacent atoms, with each atom contributing one electron. However, in some molecules and complex ions, both the shared electrons come from one atom, so forming a *co-ordinate bond*. The electrons in covalent compounds exist in molecular orbitals formed by the overlapping of two atomic orbitals, as shown in Figure 4.4.

The American chemist, G.N. Lewis, devised a simple way of illustrating covalent bonding by using dots to represent electrons. Using *Lewis formulae*, a single covalent bond is represented by a pair of dots between two atoms. Covalent bonding can also be represented by *dot- and -cross diagrams*, where each dot and cross represents an electron in the valence shell of an atom.

Example 4.3

The covalent bond in oxygen gas ($O_2(g)$ or O–O) consists of a shared pair of electrons from two oxygen atoms.

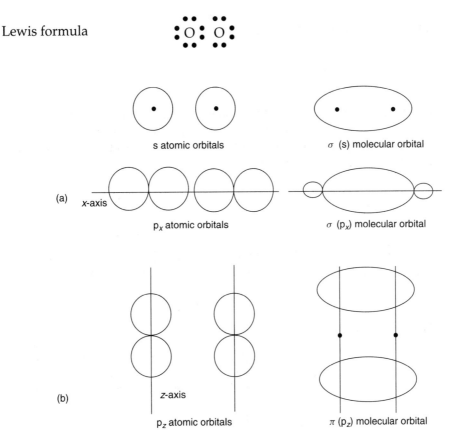

Figure 4.4 Representations of bonding orbitals: (a) sigma (σ) orbitals; (b) pi (π) orbitals

Dot-and-cross diagram

$$\overset{\bullet\bullet}{\underset{\bullet\bullet}{:}} O \overset{x}{\underset{\bullet}{x}} \quad \overset{xx}{\underset{xx}{O}} \overset{x}{\underset{x}{}}$$

Molecules consisting of two identical atoms, such as O_2, N_2, H_2 and Cl_2, are always purely covalent. However, when two different atoms are joined by a covalent bond, the bonding electrons are not shared equally. This is because the atoms have different abilities to attract electrons (see Section 4.4). The resulting bond has a small positive and a small negative pole – a *polar covalent bond*. The charged poles, known as *dipoles*, are indicated by the symbols $\delta+$ and $\delta-$ (the Greek letter δ indicates that the charges are partial).

Example 4.4

The covalent bond in hydrogen chloride gas ($HCl(g)$) consists of a shared pair of electrons, one from the hydrogen atom and one from the chlorine atom. However, the chlorine atom has a greater affinity for electrons than the hydrogen atom, and hence the shared pair is pulled towards the chlorine atom. This creates charged poles, i.e. $H^{\delta+}$ and $Cl^{\delta-}$.

Dot-and-cross diagram

$$H \overset{x}{\underset{\bullet}{}} \overset{xx}{\underset{xx}{Cl}} \overset{x}{\underset{x}{}}$$

Polar covalent bond

$$H^{\delta+} - Cl^{\delta-}$$

In a co-ordinate bond, one atom donates and shares a pair of electrons with another atom. The atom that donates the electron pair is called the electron-pair donor and the atom accepting the electron-pair is called the electron-pair acceptor. An arrow usually represents the co-ordinate bond.

Example 4.5

When boron trifluoride (BF_3) reacts with ammonia (NH_3), a white molecular solid is formed (NH_3BF_3). The co-ordinate bond in NH_3BF_3 consists of a shared pair of electrons, both of which are donated by the ammonia atom, as follows:

$$NH_3(g) + BF_3(g) \longrightarrow NH_3BF_3(s) \tag{4.1}$$

Dot-and-cross diagram

(4.2)

Shared electrons

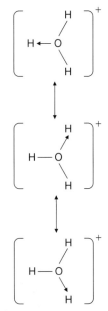

Figure 4.5 Possible positions of the co-ordinate bond in an oxonium ion, H_3O^+

Co-ordinate bond (shown by the symbol $\longrightarrow$)

$$
\begin{array}{ccc}
\text{H} & & \text{F} \\
| & & | \\
\text{H}\!-\!\!-\!\text{N} & \longrightarrow & \text{B}\!-\!\!-\!\text{F} \\
| & & | \\
\text{H} & & \text{F}
\end{array}
\qquad (4.3)
$$

Some polyatomic ions contain both co-ordinate and covalent bonds. Examples of polyatomic ions include the ammonium ion, NH_4^+, and the oxonium ion, H_3O^+. We can see from the structures presented in Figure 4.5 that the co-ordinate bond in the oxonium ion can reside in one of three positions.

Similarly, the co-ordinate bond in the ammonium ion can reside in one of four positions. In fact, it is impossible to distinguish the covalent from the co-ordinate bonds in both of these ions. This is because the electrons are *delocalized*, or spread over the entire ion. Consequently, the charge on the ion is also delocalized, or distributed over the whole ion. The same situation occurs in polyatomic ions containing multiple bonds that can be written in several equivalent positions. Consider the carbonate ion, CO_3^{2-}. The three structures shown in Figure 4.6 differ only in the position of the double bond. If one of these structures were correct, we would expect one short double bond and two long single bonds. However, experimental evidence shows that the bond lengths in the carbonate ion are all exactly the same. Again, this is because the electrons are delocalized, thus forming a resonance structure, as shown in Figure 4.7. A *resonance structure* is a hybrid (blend) of the various possible distributions of electrons. The different distributions of electrons are known as *limiting* or *canonical forms*. Many common environmental

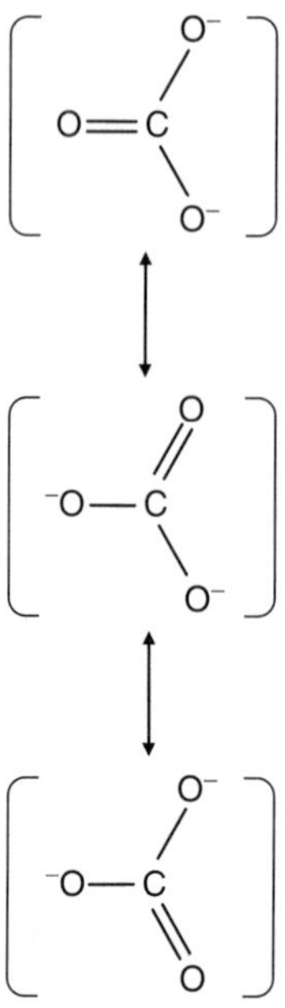

Figure 4.6 Limiting forms of the carbonate ion, CO_3^{2-}

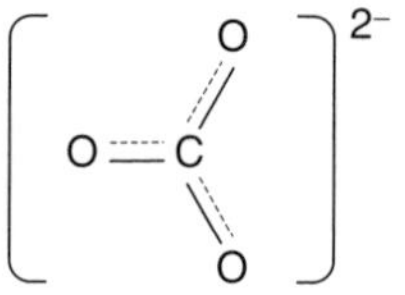

Figure 4.7 Resonance structure of the carbonate ion, CO_3^{2-}

polyatomic ions exist as resonance structures with limiting forms (e.g. CO_3^{2-}, NO_3^{-}, SO_4^{2-}, etc.).

These polyatomic ions are also known as *covalent ions*, and are part of a larger group of ions known as *complex ions*.

4.3.3 Complex ions

A *complex ion* is formed when one or more molecules or negatively charged ions become attached to a central atom. The molecule or negatively charged ion, known as a *ligand*, is normally attached to the central atom by a co-ordinate bond. Consequently, these types of compounds are called *co-ordination compounds* and the number of ligands attached to the central atom is known as the *co-ordination number* (CN). Co-ordination compounds include electrically neutral complexes such as [Ni(CO)$_4$] and ionic complexes such as K$_4$[Fe(CN)$_6$]. In Ni(CO)$_4$, the ligand is CO and the co-ordination number of the complex is four; in K$_4$[Fe(CN)$_6$], the ligand is CN$^-$, the co-ordination number is six and the complex ion is [Fe(CN)$_6$]$^{4-}$. Complexes may be neutral, positive or negative, and their formulae are usually enclosed in square brackets, as follows:

Neutral complex	tetracarbonylnickel(0)	$[Ni(CO)_4]$
Anionic complex	tetrachlorocuprate (II) ion	$[CuCl_4]^{2-}$
Cationic complex	hexaaquachromium (II) ion	$[Cr(H_2O)_6]^{3+}$

Simple ligands such as H_2O and NH_3 can only form one co-ordinate bond and are known as *monodentate ligands*. A *polydentate* ligand, such as the ethanedioate ion ($^-O_2C-CO_2^-$), can simultaneously form more than one co-ordinate bond with the central atom. Some ligands can form ring structures with central atoms, producing compounds known as chelate compounds. A common example is the ethylenediaminetetraacetate ion, $EDTA^{4-}$. Chelating ligands are very common in the environment; for example, mosses and lichens use chelating ligands to capture metal ions that are essential for their survival from the rocks that they live on.

4.3.4 Metallic bonding

Metallic elements have characteristic properties that cannot be explained by either pure ionic bonding or pure covalent bonding, and so a third type of bonding must be considered – *metallic bonding*. While elements to the right of the Periodic Table favour electron-pair sharing and are covalently bound in their elemental form, those to the left, which readily lose their valence (bonding) electrons, are metallic elements. Any theory of the metallic bond must account for the special properties of metals – that they are malleable, ductile and good conductors of heat and electricity.

In a metal, the outer-shell (valence) electrons are relatively easily removed, forming metal cations. When two metal atoms approach each other closely, their outer shells overlap to form *molecular orbitals* . The atomic orbital of a third metal atom can overlap with those of the other two metal atoms to form a new molecular orbital. For a large number of atoms, a large number of molecular orbitals are formed, thus creating a giant crystal lattice that extends over three dimensions. Consequently, the outer electrons from each atom come under the influence of a large number of atoms rather than being located in the outer shell of any single atom and are free to move through the structure – delocalized electrons. Cations are not forced apart by repulsion because they are attracted to the cloud of delocalized electrons between them. For these reasons, a metal may be described as 'an array of cations in a sea of electrons'.

This theory of the metal bond fits with the observed special properties of metals. Metals are malleable and ductile because the delocalized electrons can move away from an applied force, attracting the cations, and hence allowing the metal to change shape without fracturing. The high thermal and electrical conductivity of metals is also due to the presence of mobile delocalized electrons. If we heat an iron bar at one end, delocalized electrons at the high-temperature end will gain in kinetic energy and rapidly transfer their energy to other electrons, so causing the heat to spread to the cooler parts of the metal. Similarly, if we apply a potential difference between the ends of an iron bar, the electrons will flow towards the positive potential. The shiny appearance of metals is caused by the presence of a large number of molecular orbitals with a large number of different energy levels. This means that light at a wide range of different wavelengths (energies) can be absorbed. Thus, when light falls on a metal, a large number of electron transitions between energy levels are possible. When the electrons return to the lower energy levels, light is emitted, thus causing the metal to shine.

4.4 Electronegativity

Covalent bonding is a good model for describing bonds between non-metals, while ionic bonding is a good model for describing bonds between a metal and a non-metal. However, in practice, pure ionic and pure covalent bonding are extreme descriptions, and actual bonds tend to have both ionic and covalent characteristics. The percentage of ionic character of any bond can be calculated from the difference in the electronegativities of the constituent atoms. The *electronegativity* of an atom represents its ability to attract electrons. There have been several attempts to define an electronegativity scale. The most commonly used is that devised by the American chemist, Linus Pauling, as shown in Table 4.6. Pauling derived his scale by arbitrarily assigning an electronegativity value of 4.0 to the most electronegative of all elements–fluorine – and using bond enthalpy values to estimate electronegativities for all other elements. (Note that electronegativity values have no units.)

From Table 4.6, you can see that electronegativity increases from left to right across a period in the Periodic Table, and, with the exception of part of Group 3, it also decreases down every group. The higher the difference in values, then the greater the percentage ionic character. As a general rule, an electronegativity difference of greater than about 2.0 means that there is so much ionic character present in a bond that it is best regarded as a truly ionic compound. When the electronegativity difference is less than about 1.5, the bond is best regarded as covalent. Bonds with in-between values are best regarded as polar covalent.

There are limitations to the use of Pauling's electronegativity values. They can be applied when considering a bond between two atoms either in a diatomic molecule or between two atoms considered in isolation from other atoms and bonds. However, they should not be applied to bonds in crystal structures with regular formula units (such as $NaCl(s)$ or $CaCl_2(s)$) because the electron densities of each ion in a crystal lattice are affected by several surrounding ions. Electronegativities are most useful for non-metals, but are hardly ever used for d-block metals.

Table 4.6 Pauling electronegativity values $(c)^a$

		H 2.10				
Li 0.98	Be 1.57	B 2.04	C 2.55	N 3.04	O 3.44	F 4.00
Na 0.93	Mg 1.31	Al 1.61	Si 1.90	P 2.19	S 2.58	Cl 3.16
K 0.82	Ca 1.00	Ga 2.01	Ge 2.01	As 2.18	Se 2.55	Br 2.96
Rb 0.82	Sr 0.95	In 1.78	Sn 1.96	Sb 2.05		I 2.66

[a] NB – These values have no units.

4.5 The Shapes of Molecules and Ions

The Lewis structures we studied in Section 4.3 show the approximate locations of bonding electrons and lone pairs in a molecule. However, these two-dimensional diagrams do not show the shapes of molecules and ions since they do not show the arrangement of atoms in space. Clearly, we need a means of representing molecules in three dimensions.

The molecular orbital diagrams shown in Figure 4.4 form regular geometrical shapes. In fact, it is convenient to describe the shapes of molecules and ions by using recognized geometrical shapes, as shown in Table 4.7. In order to do this precisely, we visualize the bonds as straight lines that join the centres of the bonding atoms and report the bond angles. The *bond angle* in a molecule or ion is the angle between the two straight lines joining the centres of the atoms under consideration. In order to decide which geometrical shape is appropriate to accurately describe a given molecule or ion, we use *the valence-shell electron-pair repulsion* (VSEPR) *theory*, first devised by Nevil Sidgewick and Herbert Powell and later developed by Ronald Gillespie and Sir Ronald Nyholm.

Table 4.7 Geometrical shapes

Shape		Common environmental examples
Linear		$O=C=O$ $H-C\equiv N$
Angular		H_2O (104.5°) H_2S (92°) NO_2^- (115°)
Trigonal planar		NO_3^- CO_3^{2-} SO_3
Trigonal Pyramidal		NH_3 SO_3^{2-} ClO_3^-
T-shape		ClF_3

(continued overleaf)

Table 4.7   *(Continued)*

Shape		Common environmental examples
Tetrahedral		CH_4 NH_4^+ ClO_4^- PO_4^{3-} SO_4^{2-}
Square planar		ICl_4^-
Trigonal bipyramidal		PCl_5
Octahedral		SF_6
Pentagonal bipyramidal		IF_7

The VSEPR theory essentially assumes that regions of high electron concentration repel each other, forcing them as far apart as possible in order to minimize their repulsion. Regions of high electron concentration include bonding electron pairs in molecular orbitals and *lone pairs* of electrons (pairs of electrons that do not participate in bonding) belonging to the central atom. Experimental evidence shows that a lone pair of electrons will repel other pairs more strongly than a bonding pair of electrons. The strength of repulsion between electron pairs decreases in the following order: lone pair–lone pair > lone pair–bonding pair > bonding pair–bonding pair. Consequently, it is the *number and type of electron pairs* that primarily determines the shapes and bond angles of molecules and ions. We can visualize VSEPR theory in practice by using the methane molecule as an example.

Example 4.6

Methane consists of a central carbon atom surrounded by four hydrogen atoms, as shown in the Lewis formula below:

$$
\begin{array}{c}
\text{H} \\
\bullet\,\text{x} \\
\text{H} \;\overset{\text{x}}{\underset{\bullet}{}}\; \text{C} \;\overset{\text{x}}{\underset{\bullet}{}}\; \text{H} \\
\bullet\,\text{x} \\
\text{H}
\end{array}
$$

Imagine that all of the bonding pairs of electrons in methane lie on the surface of an invisible sphere that surrounds the central atom. The pairs of electrons move as far apart as possible on the surface of the sphere until they are at their most distant locations. At this point, we can visualize where the electrons lie and draw lines from the central atom to the pairs of electrons to represent the bonds, as follows:

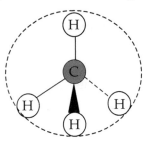

We can then compare this shape with the geometrical shapes shown in Table 4.7. For methane, the repulsion between the pairs of bonding electrons (i.e. regions of high electron concentration) will be at a minimum if they take a tetrahedral shape.

This visualization technique obviously has limitations, but it is very useful for molecules without lone pairs on the central atom. For more complicated molecules, the VSEPR theory states that:

- electron pairs in multiple (double or triple) bonds are treated as a single unit equivalent to one electron pair;

- lone pairs on the central atom of a molecule contribute to the shape of the molecule, but are ignored when we name the shape;

- lone pairs distort the shape of a molecules to minimize electron repulsions, remembering that the strengths of electron repulsions are in the order, lone pair–lone pair > lone pair–bonding pair > bonding pair–bonding pair.

Thus, in order to predict the shapes of molecules and ions accurately by using the VSEPR model, we use the following stepwise procedure:

(1) Write the Lewis structure.

(2) Identify the geometrical arrangement of electron pairs and bonds in which the lone pairs are farthest from each other and from bonds.

(3) Name the molecular shape using only the locations of the atoms (i.e. ignoring the locations of the lone pairs).

Examples 4.7 and 4.8 highlight the utility of this procedure for predicting the shapes of simple molecules and ions that are common in the environment.

Example 4.7

Suggest a shape for the carbonate ion, CO_3^{2-}.

(1) Write the Lewis structure.

(2) Treat the electron pairs in the double bond as a single unit equivalent to one electron pair. This effectively means that all three O atoms have three lone pairs each.

(3) Repulsion of the lone pairs – regions of high electron concentration – will be at a minimum if they take a trigonal planar shape (see Table 4.7 and Figure 4.7).

Example 4.8

Suggest a shape for the water molecule, H_2O.

(1) Write the Lewis structure.

(2) The lone pairs will repel each other more strongly than the bonding pairs. This will give the water molecule a slightly angular shape.

(3) Repulsion of regions of high electron concentration will be at a minimum if they take an angular shape (see Table 4.7) with a *bond angle* of 104.5° between the two O–H bonds.

4.6 The Structures of Metallic Crystals

Metal atoms pack closely together in a regular structure. There is no way of packing spheres to fill a space completely without leaving gaps between them. Arrangements in which the gaps are kept to a minimum are called *close-packed structures*. In such structures, the atoms occupy 74% of the space, whereas in non-close-packed structures, the atoms occupy 68% of the space.

Since all of the atoms in a crystal of a metal element will have identical electronic structures, we can use identical spheres to represent the atoms. By arranging the spheres in layers and then stacking the layers on top of each other, we can visualize the structure of metal crystals. For close packing, a layer of spheres can be arranged in only one way, as shown in Figure 4.8(a), where each atom is in contact with six neighbours. A second layer can be arranged on top of the first in only one way if close packing is to be preserved, and this is shown in

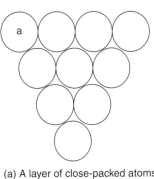

(a) A layer of close-packed atoms (shown here as spheres)

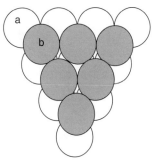

(b) Addition of a second layer on top of the first layer

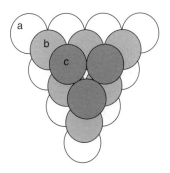

(c) Addition of a third layer to layer 'b' (abcabc arrangement)

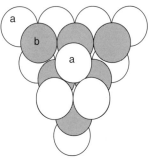

(d) Addition of a third layer to layer 'b' (abababab arrangement)

Figure 4.8   Close packing of spheres

Figure 4.8(b). From this diagram, we can see that within close-packed structures, there are two types of gap (hole) between the metal spheres, i.e. a tetrahedral gap and an octahedral gap.

In order to visualize a tetrahedral gap, take a triangle of close-packed spheres, lie it flat and place a single sphere in the hollow created by the three other spheres. The gap in the centre of the four spheres will be tetrahedrally co-ordinated. To visualize an octahedral gap, arrange four spheres into a square of close-packed spheres. Place a single sphere above the hollow created by the four other spheres and one below. The gap in the centre of the six spheres will be octahedrally co-ordinated.

There are two possible ways of adding a third layer to the arrangement shown in Figure 4.8(b) while still preserving close packing, as shown in Figure 4.8(c) and 4.8(d). This gives rise to two types of close-packed structures – cubic close-packed (otherwise known as a face-centred cubic) and hexagonal close-packed, respectively.

A *hexagonal close-packed* structure is shown in Figure 4.9. The arrangement of the layers may be labelled as ab and the repeated pattern is then abababab, etc. Thus, the fourth layer is a repeat of the second layer, while the third and fifth layers repeat the first, and so on. In this structure, every metal atom is

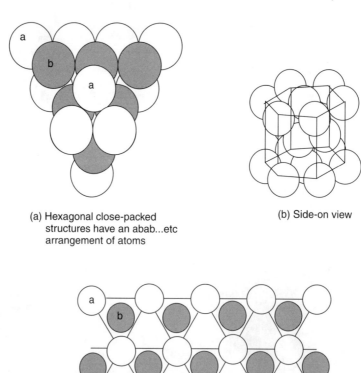

(a) Hexagonal close-packed structures have an abab...etc arrangement of atoms

(b) Side-on view

(c) View from above

Figure 4.9   Schematic representations and alternative views of a hexagonal close-packed structure

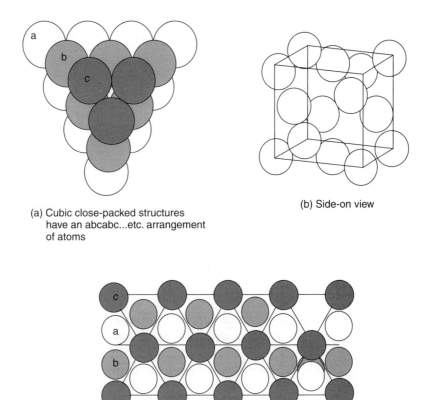

(a) Cubic close-packed structures have an abcabc...etc. arrangement of atoms

(b) Side-on view

(c) View from above

Figure 4.10 Schematic representations and alternative views of a cubic close-packed structure

in contact with 12 other metal atoms (six in the same layer, three in the layer above and three in the layer below) and hence the *co-ordination number* is 12. The co-ordination number for metals is the number of nearest neighbours of each atom. Metals that form this type of structure include beryllium, magnesium and zirconium.

A *cubic close-packed* structure is shown in Figure 4.10(a). The arrangement of the first three layers may be labelled as abc and the repeated pattern is then abcabcabcabc, etc. In this structure, the close-packed layer is not parallel to the base of the cube, but lies along the body diagonal of the cubic array, as illustrated in the alternative views shown in Figures 4.10(b) and 4.10(c). You can see that these close-packed layers have to be turned through 45 degrees to give the unit cube. As in the hexagonal close-packed structure, every metal atom is in contact with 12 others and hence the co-ordination number is 12. Metals that form this type of structure include copper, silver, gold and iron.

The third type of common metallic structure – a body-centred cube – is not close-packed. An example of this structure is shown in Figure 4.11. You can see here that the atoms are not close-packed and that the central atom is eight co-ordinated. Metals that form this type of structure include sodium, potassium and molybdenum.

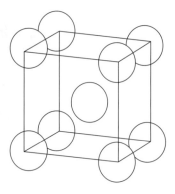

Figure 4.11 Schematic representations of a body-centred cubic structure

4.7 The Structures of Ionic Crystals

We have already seen that the ions in ionic crystals are held together by electrostatic forces. The crystals have no overall electric charge (they are electrically neutral) and are made up of stacked unit cells. The *unit cell* is the basic building block that is repeated to create a crystal lattice. Billions of unit cells are required to create a single crystal of an ionic solid. There are seven primitive unit cells and seven multi-primitive unit cells, thus giving a total of fourteen types of crystal lattice. Ionic crystals can therefore have a huge range of extremely complicated structures.

In an ionic solid, the co-ordination number means the number of ions of one electrical charge that immediately surround those of opposite charge. The co-ordination and crystal structure of ionic crystals is largely determined by two factors, as follows:

- the ratio of cations to anions;

- the radius ratio of the ions.

The *ratio of cations to anions* is simply the number of cations in a crystal lattice relative to the number of anions. For example, this ratio in sodium chloride (NaCl) is 1:1, while in calcium fluoride (CaF_2) it is 1:2. The ratio of the ionic radius of the cation (r^+) to the ionic radius of the anion (r^-) is called the *radius ratio*. The radius ratio of an ionic crystal can be calculated from the following formula:

$$\text{Radius ratio} = \frac{\text{ionic radius of cation}}{\text{ionic radius of anion}} = \frac{r^+}{r^-} \tag{4.1}$$

Table 4.8 Typical structural arrangements of compounds with differing co-ordination numbers

Co-ordination number	Structural arrangement
3	Trigonal or triangular
4	Tetrahedral
6	Octahedral
8	Cubic
12	Hexagonal or cubic close packed

The value of the radius ratio controls the co-ordination number (CN), while the co-ordination number of an ionic compound controls the geometric arrangement of ions around the central ion, as shown in Table 4.8. The radius ratio is useful for predicting the shapes of oxides found in the Earth's crust.

4.8 Bonding in Silicates, Aluminates and Aluminosilicates

Now that you are familiar with the nature of chemical bonds and their influence on the structure and properties of substances, we can look in detail at the physical structure and chemical composition of the Earth's crust. We will start by looking at the principal constituents of rocks in the Earth's crust – the *silicate* and *aluminosilicate* minerals.

Silicates are the main rock-forming minerals – most rocks (with the exception of limestone) are composed wholly, or in part, of silicates. These can form substances with an extraordinary range of physical properties-from fibrous asbestos to sheet-like mica, soft talc to hard quartz, deep blue (ultramarine) to deep red (garnet), and sand and clay to granite. This is because silicates contain silicon and oxygen in tetrahedral formula units of SiO_4, bound together in various ways to form specific structural types.

The Si–O bond has approximately 46% ionic character and 54% covalent character, thus making it polar covalent. However, you will remember from Section 4.4 that electronegativity values should not be applied to bonds in crystal structures with regular formula units. In general, it is best to treat the elements in silicate structures as being in their ionic states, as follows:

$$Si^{4+} + 4O^{2-} \longrightarrow (SiO_4)^{4-}$$

silicate anion

According to VSEPR theory (see Section 4.5), the silicate anion will have a tetrahedral shape. This is supported by using the radius ratio rule to predict the co-ordination numbers for oxides, as shown in Table 4.9. (Table 4.8 tells us that an ion with a co-ordination number of four typically has a tetrahedral structure.)

Table 4.9 Predicted and observed co-ordination numbers for oxides

Radius ratio	Cation	Radius ratio, R^+/R^-	Predicted co-ordination	Observed co-ordination
Up to 0.225	C^{4+}	0.16	3	3
	B^{3+}	0.16	3	3,4
0.225–0.414	Si^{4+}	0.30	4	4
	Al^{3+}	0.36	4	4,6
0.415–0.732	Mg^{2+}	0.47	6	6
	Na^+	0.69	6	6,8
	Ca^{2+}	0.71	6	6,8
0.733–1.000	Sr^{2+}	0.80	8	8
	K^+	0.95	8	8,12
	Cs^+	1.19	12	12

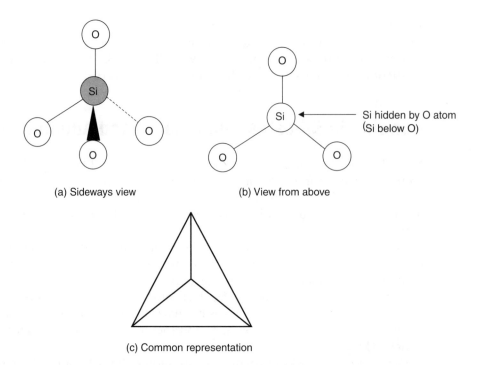

(a) Sideways view   (b) View from above

(c) Common representation

Figure 4.12   Different representations of the silicate anion, $(SiO_4)^{4-}$

Many different diagrammatic methods are used to represent the silicate anion, as shown in Figure 4.12.

The Si–O bond is the strongest bond in silicates, and therefore the silicate anions with their tetrahedral structure make up the skeleton of silicate structures. The $(SiO_4)^{4-}$ tetrahedra may exist as discrete units or may be joined via the O atoms, sharing their vertices with other tetrahedra, as shown in Figure 4.13. The number of shared vertices on each tetrahedron can vary between one and four.

Aluminium is the other element abundant in the Earth's crust that can form a complex anion:

$$Al^{3+} + 4O^{2-} \longrightarrow (AlO_4)^{5-}$$

aluminate anion

The radius ratio rule predicts that Al^{3+} can exhibit both in four and six co-ordination (see Table 4.8), giving either a tetrahedral or an octahedral structural arrangement (see Table 4.7). This is because Al^{3+} has a radius ratio with O^{2-} that is close to the borderline. The four co-ordinated tetrahedral arrangement is found in aluminate ions, $(AlO_4)^{5-}$, which are analogous to $SiO_4{}^{4-}$, whereas the six co-ordinated octahedral arrangement occurs for the Al^{3+} cation when it replaces other cations such as Fe^{2+} and Mg^{2+}.

Aluminium has the property of being able to exist as both Al^{3+} and $(AlO_4)^{5-}$ in naturally occurring minerals and in some cases both the cation (Al^{3+}) and anion ($(AlO_4)^{5-}$) can be found in the same structure e.g. muscovite mica. Because

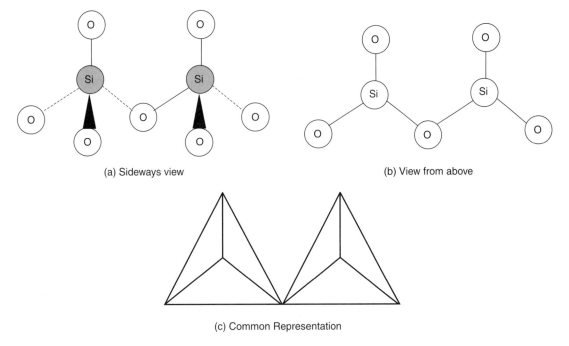

Figure 4.13 Sharing of silicate tetrahedra in $(Si_2O_7)^{6-}$

the silicate and aluminate ions have similar structures, the aluminium cation is able to replace some of the Si^{4+} ions in silicate structures, giving rise to a group of minerals called the *aluminosilicates*. This exchange of Al^{3+} for Si^{4+} has important implications for the properties of minerals, as we shall see later in this chapter.

The ability of silicates and aluminosilicates to form polyanions of varying sizes distinguishes them from all other complex anions (e.g. PO_4^{3-}, SO_4^{2-}, and ClO_4^{-}). These polyanions may be held together within a crystal lattice structure by metal anions, forming giant three-dimensional structures. In order to understand the geometry of silicate crystal structures, it is best to consider the elements present as ions whose size and charge determine their spatial relationships. In other words, the way in which the ions pack together to form a crystal depends upon their size and charge.

4.9 Isomorphism and Polymorphism

Isomorphism is the replacement of one atom or ion in a crystal structure by another atom or ion of similar size in such a way that the geometrical arrangement of the crystal lattice remains intact. Ions involved in isomorphous exchange do not have to have identical electronic charges. Isomorphism results in crystals with different chemical formulae having similar crystal structures. It is an important process during the weathering of rocks and is widespread in the silicate lattices present in the Earth's crust.

Example 4.9

Calcite ($CaCO_3$) and rhodocrosite ($MnCO_3$) are isomorphous because they have a similar crystal structure and similar sized cations, thus allowing the two cations (Ca^{2+} and Mn^{2+}) to exchange freely in these compounds. This process may be indicated by writing $(Ca, Mn)CO_3$.

Example 4.10

Ions with similar ionic sizes (similar ionic radii) are freely interchangeable as they can replace each other directly without altering the radius ratio and hence the crystal structure (co-ordination number). Exchange of $Fe^{2+}(r^+ = 0.074\,nm)$ and $Mg^{2+}(r^+ = 0.066\,nm)$ occurs in many silicates, e.g. olivines, pyroxenes and amphiboles.

Example 4.11

As we have already seen, Al^{3+} may replace Si^{4+} ions in silicate structures. However, in order to maintain an electrically neutral crystal lattice, some balancing of the charges will be required. This can be achieved by further isomorphous exchange, e.g. if Na^+ is also present in the crystal structure, it will need to be replaced by a doubly charged positive ion such as Ca^{2+}. As a general rule, one ion may replace another if the size difference does not exceed 15% of the smaller one.

    Polymorphism refers to a mineral of fixed chemical composition existing in different crystal forms, e.g. diamond and graphite (polymorphs of C), and calcite and aragonite (polymorphs of $CaCO_3$).

4.10 Classification of Silicate Structures

Silicon is the second most abundant element in the Earth's crust (after oxygen). Silicon does not occur as a free element, but occurs widely in the form of silica and silicates (over 1 000 silicate minerals are known). Silicates make up about 75% of the Earth's crust, and as we have already seen, the basic building block of all silicates is the silicate tetrahedron. Common silicates may consist of these single tetrahedra or they can be linked together to form polymeric rings, chains, sheets and giant three-dimensional structures, as shown in Table 4.10. In this section, we will study the different structures and properties of silicates.

4.10.1 Orthosilicates

Orthosilicates contain single $(SiO_4)^{4-}$ tetrahedra linked by divalent atoms in sixfold co-ordination, e.g the olivines. In the (Mg,Fe)-olivines, there is a continuous series between the two extreme structures, forsterite, Mg_2SiO_4, and fayalite, Fe_2SiO_4. A whole range of structures is possible between these extremes due to replacement of Mg^{2+} by Fe^{2+} or vice versa. The (Mg,Fe)-olivines are common and important rock-forming minerals, often found in basic igneous rock. Olivines are named after

Table 4.10 Structures of silicate minerals

Silicate type	Empirical formula	Anionic or molecular formula	Structure	Examples
Orthosilicates (no corners shared)	$[SiO_4]^{4-}$	$[SiO_4]^{4-}$		Forsterite, $Mg_2(SiO_4)$ Olivine, $FeMg(SiO_4)$ Zircon, $Zr(SiO_4)$
Disilicates (one corner shared)	$[Si_2O_7]^{6-}$	$[Si_2O_7]^{6-}$		Thortveitite, $Sc_2(Si_2O_7)$
Ring structures	$[SiO_3]_n^{2n-}$	$[Si_3O_9]^{6-}$		Benitoite, $BaTiSi_3O_9$
		$[Si_4O_{12}]^{8-}$		Axinite, Si_4O_{12}
		$[Si_6O_{18}]^{12-}$		Beryl, $Be_3Al_2Si_6O_{18}$
Single chains (Two corners shared)	$([SiO_3]^{2-})_n$	$[SiO_3]^{2-}$		Diopside, $CaMg(SiO_3)_2$
Double chains (Two corners shared)	$([Si_4O_{11}]^{6-})_n$	$(Si_4O_{11})^{6-}$		Tremolite, $Ca_2Mg_5(Si_8O_{22})(OH)_2$

Table 4.10 *(Continued)*

Silicate type	Empirical formula	Anionic or molecular formula	Structure	Examples
Sheet silicates (Three corners shared)	$[Si_4O_{10}]_n^{4n-}$	$(Si_4O_{10})^{4-}$	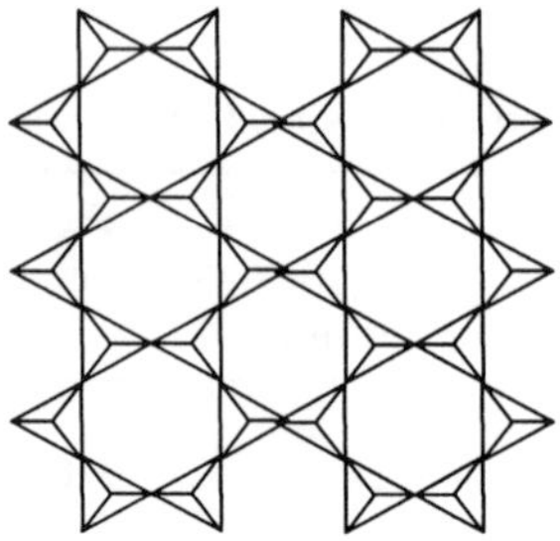	Talc, $Mg_3Si_4O_{10}(OH)_5$ Phlogopite, $KMg_3(AlSi_3O_{10})(OH)_2$ Kaolinite, $AlSi_4O_{10}(OH)_8$ Chrysotile, $MgSi_4O_{10}(OH)_8$
Framework silicates (Four corners shared)	$(SiO_2)_n$	SiO_2		Quartz, $(SiO_2)_n$ Orthoclase, $K(AlSi_3O_8)$ Albite, $Na(AlSi_3O_8)$

their colour (olive green), which is due to the presence of the Fe^{2+} ion. The structure of forsterite (as determined by X-ray diffraction) is shown in Figure 4.14 (note that the octahedral arrangement for Mg^{2+} ions is predicted by the radius ratio rule in Table 4.9).

Other examples of orthosilicates include the following:

- the humite group of minerals, with the general formula $nM_2[SiO_4].M(OH,F)_2$ (where M is a metal ion and $n = 1–4$);

- the garnet group of minerals, with the general formula $X_3Y_2[SiO_4]_3$ (where X and Y are metal ions);

- zircon, $Zr[SiO_4]$;

- sphene, $CaTi[SiO_4](O,OH,F)$;

- sillimanite, $Al_2O[SiO_4]$;

- topaz, $Al_2[SiO_4](OH,F)_2$.

4.10.2 Disilicates

Disilicates contain the $(Si_2O_7)^{6-}$ anion, as illustrated in Table 4.10. They are a small class of relatively rare minerals that include the epidote group of minerals (e.g. zoisite, $Ca_2Al_2O.AlOH[Si_2O_7][SiO_4]$), the melilite group of minerals (e.g. akermanite, $Ca_2[MgSi_2O_7]$) and lawsonite, $CaAl_2[Si_2O_7](OH)_2.H_2O$.

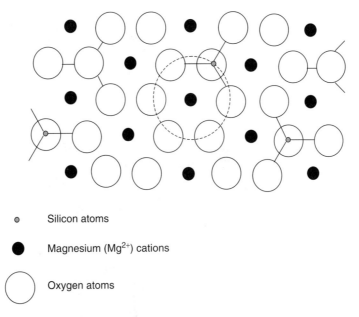

⊙   Silicon atoms

●   Magnesium (Mg^{2+}) cations

◯   Oxygen atoms

Figure 4.14 Structure of forsterite, $MgSiO_4$. Two rows of SiO_4 tetrahedra are shown with their oxygen atoms in planes parallel to the page. Each tetrahedron is associated with three metallic cations (Mg^{2+}): These are alternately behind tetrahedra pointing up and in front of tetrahedra pointing down. The dashed circle is drawn through an octahedron of oxygen atoms around Mg^{2+}

4.10.3   Ring structures (*Meta*-silicates or cyclosilicates)

Cyclosilicates may be represented by the formula $(SiO_3)_n^{2n-}$. These minerals are fairly rare, and may contain three, four or six tetrahedra, as shown in Table 4.10. Beryl (normally regarded as $Be_3Al_2Si_6O_{18}$) is a ring silicate that has rings containing six SiO_4 tetrahedra. The $(SiO_3)_6^{12-}$ rings are at two different heights, and are represented in Figure 4.15 as rings of shaded or unshaded tetrahedra. The rings at different heights are staggered, with Be^{2+} and Al^{3+} lying in a plane midway between them. The oxygen atoms inside the ring are all shared and neutral, so there are no cations inside the rings.

Beryl predominantly forms crystals in granitic rocks and is the main ore of beryllium. It also forms gemstones, including aquamarine (light-blue crystals) and emerald (dark-green crystals).

4.10.4   Single chains of indefinite extent (pyroxenes or chain silicates)

Pyroxenes are common in basic igneous and metamorphic rocks, and have the general formula, $XYSi_2O_6$. Pyroxenes have an internal structure based on single chains of silicate tetrahedra; two vertices of each tetrahedra are shared and the repeating unit consists of two tetrahedra (see Table 4.10). Pyroxenes are present in many rocks of the crust and upper mantle. A simple example would be diopside, $CaMgSi_2O_6$ (Ca as Ca^{2+}, Mg as Mg^{2+}). Although pyroxenes are fairly dense

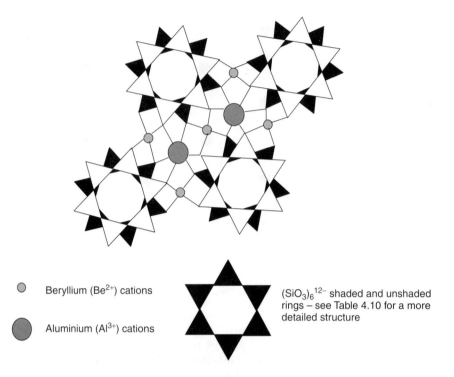

○ Beryllium (Be^{2+}) cations

● Aluminium (Al^{3+}) cations

$(SiO_3)_6^{12-}$ shaded and unshaded rings – see Table 4.10 for a more detailed structure

Figure 4.15   Structure of beryl

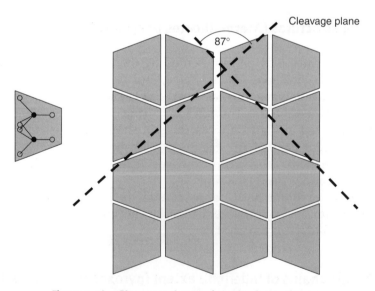

Cleavage plane

87°

Figure 4.16   Cleavage planes of single-chain silicates

and hard, they possess a characteristic plane of weakness or *cleavage*, parallel to the chains – in planes at 93° and 87°to each other – along which crystals prefer to break or split (see Figure 4.16). This is because the ionic bonds holding the different chains together are not as strong as the Si–O bonds within the chains.

4.10.5 Double chains of indefinite extent (amphiboles or band silicates)

Amphiboles occur in a wide range of igneous and metamorphic rocks and have the general formula $X_2Y_5(Si_8O_{22})(OH)_2$. Amphiboles have an internal structure based on double chains of linked pyroxene type chains, as shown in Table 4.10. The cations present in the structure (e.g. Ca, Mg, etc.) neutralize the unshared oxygens on the tetrahedra and the hydroxide ions. It is possible for some $(SiO_4)^{4-}$ tetrahedra to be replaced by aluminate, $(AlO_4)^{5-}$, ions, in which case the cations will also need to neutralize the extra negative charge.

Amphiboles show a more pronounced cleavage than pyroxenes due to the fact that the cations hold the layers together less strongly. The cleavage planes are at angles of 56° and 124° to each other, as shown in Figure 4.17. An example of an amphibole is tremolite, $Ca_2Mg_5(Si_8O_{22})(OH)_2$, which can occur in a very fibrous

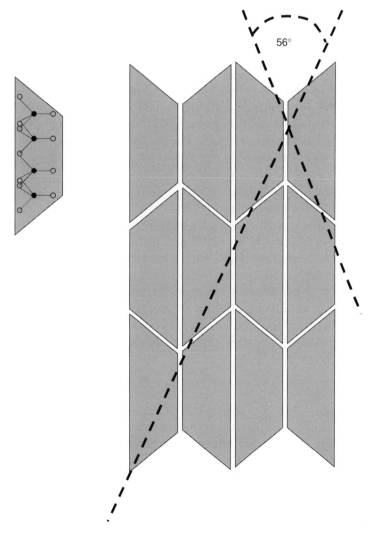

Figure 4.17   Cleavage planes of double-chain silicates

form, thus reflecting the layer structure. The fibres can be twisted into string or woven into cloth, which, because tremolite does not melt readily, can be used for safety curtains and protective clothing. This amphibole is also a thermal insulator because of the air pockets between the fibres. However, tremolite has only rarely been used for commercial purposes as there have been concerns about its potential detrimental human health effects.

4.10.6   Layer or sheet silicate structures (Phyllosilicates)

Phyllosilicates form giant two-dimensional sheets that are arranged in a honeycomb pattern (see Table 4.9). The general formula for phyllosilicates is $(Si_4O_{10})_n^{4n-}$. Examples of this structure include talc, the micas, the serpentine minerals and all clay minerals.

Talc (or soapstone)

Talc is one of the softest minerals known and is a hydrous magnesium silicate, $Mg_3Si_4O_{10}(OH)_2$. It occurs as tabular crystals, but the impure form, known as soapstone, is more common. Talc is formed by the alteration of magnesium compounds and is usually found in metamorphic rocks. Its overall structure, illustrated in Figure 4.18, consists of double layers like a sandwich.

The 'slices of bread' in the sandwich are silicate sheets, with OH^- ions in each of the hexagonal holes formed by the unshared oxygens at the apex of each tetrahedron. The 'filling' in the sandwich is the Mg^{2+} ions that hold the top and bottom layers firmly together. Although the two layers of the sandwich are mirror images of one another, they are slightly displaced with the result that the Mg^{2+} ions occupy octahedral holes relative to unshared 'Os' and 'OHs' in the two layers. All the negative charges (from the unshared Os and OHs) are inside the sandwich and are neutralized by the Mg^{2+} ions within the sandwich – they hold the sandwich together. These forces, known as van der Waals forces, are weak, and therefore the layers can readily slide across each other. This structure explains the properties of talc – its greasy feel and lubricant properties. It is widely used in powdered form in cosmetics and lubricants, and as an additive in paper manufacture.

The micas

The mica group of minerals is found in many igneous and metamorphic rocks. They are glossy, have a pearly lustre and possess good thermal and electrical insulation properties. Like talc, the micas possess a sandwich- or layer-type structure, as shown in Figure 4.19. However, the mica structure has important differences to talc, so resulting in different properties. For example, a thin sheet of talc can be easily folded and unfolded, but thin mica sheets are more elastic, and spring back to their original shape unless broken. Mica can be cleaved to give strong transparent sheets of <10 mm thickness, e.g. phlogopite, $KMg_3(AlSi_3O_{10})(OH)_2$.

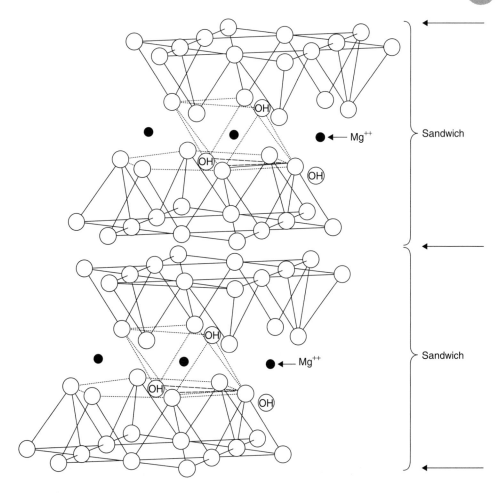

Figure 4.18 Structure of talc (a hydrous magnesium silicate), $Mg_3Si_4O_{10}(OH)_2$

The important structural difference between talc and mica is that aluminate as well as silicate tetrahedra exist in micas – one in four of the mica tetrahedra is aluminate, $(AlO_4)^{5-}$, as opposed to silicate, $(SiO_4)^{4-}$. This creates extra negative charge and the outer surfaces of the sandwiches are no longer neutral but are negatively charged. The extra positive charge required to balance this is provided by K^+ ions between the sandwiches – the K^+ ions are 12 co-ordinated, which is just possible with ions of this size. Hence, in mica, the layers will not slide because of K^+ ions holding them together. Therefore, rather than stay folded, they will spring back or break if bent far enough. However, the ionic bonding between the different layers is weak, as K^+ is *univalent* and 12 co-ordinated. As a result, cleavage along these layers is relatively easy.

Clay minerals

Clay minerals are a group of hydrous silicate minerals that constitute the bulk of the silicate minerals present in soils and sedimentary rocks, e.g. shales and siltstone.

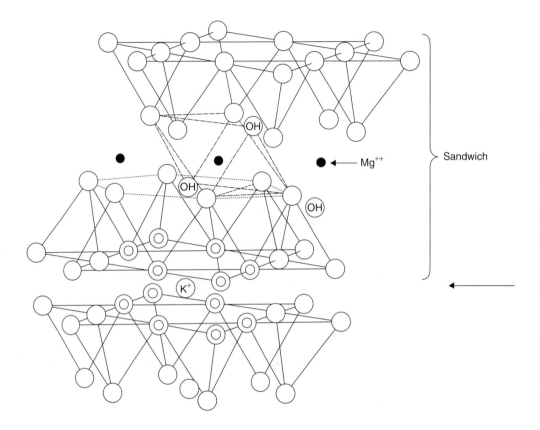

○ Oxygen coordinating K^+

Figure 4.19   Structure of mica (phlogopite), $KMg_3 (AlSi_3O_{10}) (OH)_2$

Clay minerals may form as either *primary* or *secondary* minerals. Primary minerals are essential constituents of the rocks in which they occur; their presence is typically indicated by the rock name. Secondary minerals result from the decomposition of earlier minerals. Individual crystals of clay minerals are minute and can only be seen by using an electron microscope. As we have already seen, such minerals belong to the sheet silicate classification and have essentially two types of structure, as follows:

- 1:1 clay minerals, e.g. kaolinite;

- 2:1 clay minerals, e.g. montmorillonite.

The number ratios refer to the combination of (SiO_4) tetrahedra with hexagonal and octahedral arrangements of OH^- and metal ions, respectively.

Kaolinite, $AlSi_4O_{10}(OH)_8$, is a white/cream coloured clay used in its crude form as a raw material for pottery and china. Kaolinite is made up of alternate tetrahedral and octahedral layers (see Figure 4.20); each pair, $Si_2O_3(OH)_2$ and $Al_2(OH)_6$, with loss of water becomes $Al_2Si_2O_5(OH)_4$. It is typical of all sheet silicates in that all tetrahedra point in the same direction. The OH^- ions occupy hexagonal holes, while

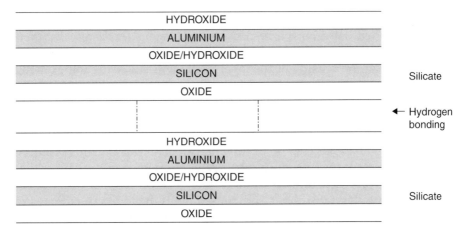

Figure 4.20 Sheet-silicate structure of kaolinite (a 1:1 clay mineral)

octahedral Al^{3+} ions neutralize the negative charges. Kaolinite forms the greater part of kaolin (china clay) deposits and is the main constituent of fire clays.

A second example of a clay mineral is montmorillonite, which is chemically very complex. Montmorillonite is built up of three-layer units comprising two tetrahedral layers separated by an octahedral layer (see Figure 4.21), and has an ideal formula of $Al_4Si_8O_{20}(OH)_4$. A typical calcium–montmorillonite would be represented by the formula $Ca_{0.5}(MgAl_3)Si_8O_{20}(OH)_4.xH_2O$; the calcium is replaced by sodium in sodium–montmorillonite.

In montmorillonite, the layers are negatively charged due to substitution of $(SiO_4)^{4-}$ by $(AlO_4)^{5-}$ and of Al^{3+} by Mg^{2+}. This overall negative charge is balanced

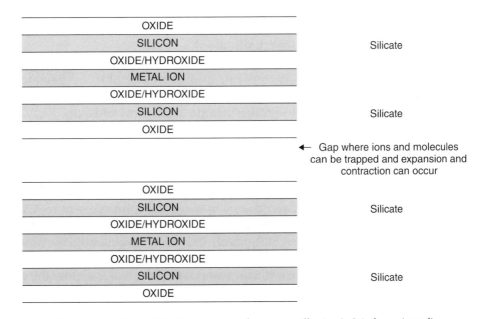

Figure 4.21 Sheet-silicate structure of montmorillonite (a 2:1 clay mineral)

by interlayer cations such as Al^{3+}, Ca^{2+}, Mg^{2+}, K^+, Na^+ and H^+, which are positioned between the respective sandwiches (*exchangeable cations*). In addition, water (H_2O) molecules can be held by hydrogen bonds (see Chapter 7) to the O atoms on the outside of the layers. (We will study the ion-exchange properties of 2:1 clay minerals in Section 4.11.)

As you can see from Figure 4.21, no hydrogen bonding exists between the sandwiches in 2:1 structures because of the structural arrangement of the silicate tetrahedra. The adjacent sandwiches are *not* therefore held closely together. As a result, water and other polar solvents can penetrate into this position. This structural arrangement thus produces absorptive properties in the clay and a swelling of the structure, which returns to its normal size on drying. Montmorillonite clays are therefore very absorbent and consequently have many commercial uses. Fuller's earth, for example, is used in the purification of oils and for removing oil and grease from textiles, while bentonite is used in papermaking and drilling muds for oil wells.

Thus, the structural features of clay minerals give them a number of useful properties – loss of water on heating, swelling and shrinking in different conditions, plasticity when wet, and cation exchange with other media – and they are therefore economically important.

Serpentine minerals

The serpentine group of minerals, i.e. hydrous magnesium silicate, $Mg_3Si_2O_5(OH)_4$, occur in soft metamorphic rocks and are usually dark green in colour. A good example of a serpentine mineral is chrysotile, $Mg_6Si_4O_{10}(OH)_8$, a layer silicate which forms bundles of parallel fibres in the shape of 'Swiss rolls'. Chrysotile is an important constituent of most commercial asbestos.

4.10.7   Framework silicates (tektosilicates)

In framework silicates, each tetrahedron shares all of its oxygen atoms with other SiO_4 or AlO_4 tetrahedra to form large, three-dimensional frameworks. Typical examples of framework silicates include quartz and feldspar.

Quartz

Quartz is a hard, crystalline form of silica and one of the most abundant minerals in the Earth's crust. Quartz occurs in many different types of rock and is resistant to mechanical or chemical breakdown. In quartz, each tetrahedron shares all of its oxygen atoms with other SiO_4 tetrahedra, and consequently, the number of oxygen atoms per silicon atom is halved (i.e. SiO_2 rather than SiO_4). In addition, an oxygen atom shared between two silicon atoms is neutral, and therefore the overall formula for quartz is $(SiO_2)_n$.

The quartz structure is much more open than the olivine structure, in which the SiO_4 tetrahedra are in a close-packed structure. Therefore, olivine is denser and is more abundant in the high-pressure mantle, while quartz is much more abundant in the crust.

Pure quartz, often known as rock crystal, is coarse, colourless and transparent. However, the semi-precious stones, e.g. agate, amethyst, onyx and jasper, are forms of finely crystalline quartz coloured with various impurities (mainly Fe). Amethyst and rose quartz are coloured purple and pink respectively, as they contain trace amounts of Fe, Mn and Ti.

Feldspars

Feldspars are the most abundant constituent of rocks and form about 60% by volume of the Earth's crust, and only slightly less by weight. They have an analogous structure to quartz, except that some of the $(SiO_4)^{4-}$ tetrahedra are replaced by $(AlO_4)^{5-}$ tetrahedra. In alkali feldspars, one in every four tetrahedra is an $(AlO_4)^{5-}$ structure. The extra negative charge is neutralized by the presence of one Na^+ or K^+ per four tetrahedra (e.g. orthoclase, $K(AlSi_3O_8)$ and albite, $Na(AlSi_3O_8)$), or in some cases by the presence of Ca^{2+} or Ba^{2+}. This strongly bonded framework gives feldspars their hardness and high melting temperatures. Feldspars typically form white, grey or pink crystals.

4.11 Ion-Exchange Processes in Clay Minerals and Soil

We have already discussed the ion-exchange properties of certain minerals due to isomorphous exchange (see Section 4.9) and the presence of interlayer cations in montmorillonite (see Section 4.10.6). (The principles of ion exchange are discussed in Box 4.1.) Many naturally occurring clay minerals possess an ion-exchange capacity and can hold ions temporarily on their surfaces – where they are resistant to leaching – but from which they may be replaced by other ions, e.g. when they are in contact with a solution containing different ions. Examples of ion-exchange processes in 2:1 clay minerals include the cycling of plant nutrients and the applications of fertilizers to soils.

Box 4.1 Ion Exchange

The term *ion exchange* is generally understood to mean the exchange of ions of like sign between a solution and a solid, highly insoluble body in contact with it. The solid must contain ions of its own and commercial ion exchangers must have an open, permeable molecular structure so that ions and solvent molecules can move freely in and out. An ion exchanger is of a complex nature and is polymeric, although some are composed of natural crystalline material. An ion exchanger carries an electric charge that is exactly neutralized by the charges on the counter ions. These active ions are cations in a cation exchanger and anions in an anion exchanger. Thus, a cation exchanger consists of a polymeric anion and active cations, while an anion exchanger comprises a polymeric cation with active anions.

- Cation-exchange resins contain free cations which can be exchanged for cations in solution:

$$(\text{Res. A}^-)\text{B}^+ + \text{C}^+(\text{soln}) \rightleftharpoons (\text{Res.A}^-)\text{C}^+ + \text{B}^+(\text{soln})$$

- Anion exchange resins contain free anions that can be exchanged for anions in solution:

$$(\text{Res. Z}^+)\text{Y}^- + \text{X}^-(\text{soln}) \rightleftharpoons (\text{Res.Z}^+)\text{X}^- + \text{Y}^-(\text{soln})$$

If the solution contains several ions, the exchanger may show different affinities for them, thus making separations possible.

In water softening, Ca^{2+} and Mg^{2+} ions are removed by passing water through a *zeolite* (aluminosilicate – natural or synthetic) column in the Na form. The reaction is reversible (regeneration); by passing a solution containing sodium ions through the product, the calcium ions may be removed from the resin and the original sodium form regenerated.

$$\text{Na-zeolite} + \text{Ca}^{2+}(\text{aq}) \rightleftharpoons \text{Ca-zeolite} + 2\text{Na}^+(\text{aq}) \quad \text{(softening)}$$

Clay minerals possess a permanent surface negative charge due to ionic substitutions. The edges of the mineral crystal lattice structures possess residual charges due to bond breakage, i.e. surface negative charges due to unco-ordinated oxygen atoms, or surface positive charges due to unco-ordinated metal atoms and/or partly unco-ordinated Si atoms. Surface hydroxyl groups exposed at the edges of a crystal may dissociate and produce an additional surface negative charge, as follows:

$$\text{Mineral}-\text{O}-\text{H}(s) + \text{M}^{2+} \rightleftharpoons \text{Mineral}-\text{O}-\text{M}(s) + 2\text{H}^+(\text{aq})$$

This reaction is clearly dependent on pH and favoured by high pH conditions.

However, not all of the cation-exchange sites in soils are associated with clay minerals. Cation exchange may occur due to salt formation by organic acids present in soil organic matter. Examples of such organic acids include humic and fulvic acids, which contain carboxyl ($-\text{COOH}$) and phenolic ($-\text{OH}$) groups. Carboxylic groups dissociate readily under soil pH conditions as the dissociation constant, K_d, is normally in the range 10^{-3}–10^{-5} (see Section 7.7 for a discussion of K_d):

$$-\text{COOH} \rightleftharpoons \text{H}^+ + -\text{COO}^-$$

If we assume that $K_d = 10^{-5}$, then for $[\text{H}^+] < 10^{-5}$ (i.e. for a soil pH > 5), $[-\text{COO}^-]/[-\text{COOH}] => 1$, indicating that dissociation readily occurs. Phenolic groups, which have K_d values in the range 10^{-7}–10^{-10}, are only weakly acidic and may not be appreciably dissociated in many soil systems.

Ions adsorbed on clay and organic surfaces can be replaced by other cations from soil solution. In particular, cations are attracted to the surface negative charge on clay particles and these will be in equilibrium with the dissolved cations in the soil water. If ions such as K^+ are extracted from soil water by plant roots, more K^+ will be released into the solution by the clay in order to restore the equilibrium. On the other hand, if K^+ ions are added to soil water (e.g. by weathering of potassium feldspar or by the addition of a potassium fertilizer), then the K^+ may displace other cations from the clay mineral surface until equilibrium is reached. Hence, the adsorbed cations constitute a nutrient reservoir within the soil system, and their cation-exchange properties are of fundamental importance to plant growth in terms of controlling the release of cations. Important participants in these processes, in addition to K^+, include Ca^{2+} and Mg^{2+}.

Anion exchange in soils is less important, but can occur by displacement of OH^- groups exposed at the edges of clay flakes. The main soil cations that would be expected to participate in these processes are nitrate, NO_3^-, and phosphate, PO_4^{3-}. Of these, the NO_3^- ion is the least strongly held and tends to be washed through the system very easily. PO_4^{3-} is strongly held by soil particles, partially as an adsorbed anion, and principally as insoluble iron and aluminium phosphates at low pH and calcium phosphates at high pH. Iron and aluminium oxides and hydroxides produced by weathering fix phosphate very strongly, as do alkaline soils containing high concentrations of Ca^{2+}.

4.12 Weathering Processes

The solid rocks that compose the bulk of the Earth's crust do not remain compact, coherent masses when exposed to the atmosphere. In many areas they are covered with soil, and even where the solid rocks appear uncovered by soil or other fragmented material, close examination reveals that the surface layers have a slightly different appearance and composition from those of the inner regions. In addition, over a period of time, the outer layers are being removed. This disintegration of solid rocks is part of the *weathering* process in which the rocks and minerals are altered until they are chemically and physically in equilibrium with their surroundings. Weathering converts chemically inactive, relatively hard non-porous rocks (which are typically formed at high temperatures and pressures, often in the absence of air and water) into soils which are soft, porous and chemically active.

There are essentially two weathering mechanisms that operate together and reinforce each other's actions – *physical and chemical weathering*.

4.12.1 Physical (or mechanical) weathering

Physical weathering is essentially erosion (fragmentation) of rocks that exposes new material to weathering. No chemical changes occur during erosion, although the process increases the surface area that may be exposed to chemical action. A large number of factors contribute to the rate and extent of physical weathering, including the following:

- the type of rock (mineral composition, structure, hardness, etc.);
- climate (temperature, humidity, etc.);
- the presence or absence of vegetation;

- geographical conditions (slope, exposure, etc.);
- time;
- heating (expansion);
- cooling (contraction);
- freezing (frost);
- thawing;
- mechanical hammering;
- human factors (farming, construction, landfill, mining, etc.);
- the action of animals;
- the action of rain, etc.

Biotic factors, such as the decay of trees and plants and the actions of burrowing animals (rabbits, moles, earthworms, etc.), may also make a contribution to mechanical weathering. The latter may be the dominant mechanism of weathering in arid areas.

4.12.2   Chemical weathering

During chemical weathering, the chemical composition of surface rocks and fragments is changed by chemical reactions, thus producing more stable compounds or species. The products of weathering may be *soluble* or *insoluble* in water. One of the most important features of weathering is that it *mobilizes* many chemical substances. In the Earth's crust, substances tend to be mobilized by dissolution (or sometimes suspension) in moving water. Soluble products, together with colloidal and fine-grained materials, are removed and transported to oceans, whereas insoluble products tend to remain in soil systems as a prior stage in a much slower journey to oceans. The presence of water and the solubility of the weathered products are therefore important factors when considering the rate and extent of weathering. For example, stone monuments in dry (arid) climates tend to be better preserved than similar monuments in more temperate regions. In addition, the lower the solubility of a compound, then the smaller will be the interaction between that compound and a live organism. Consequently, chemical weathering processes that affect the solubility of elements are crucial in governing the viability of organisms.

4.12.3   Chemical weathering mechanisms

There are four main chemical weathering mechanisms, as follows:

- dissolution;
- oxidation;
- hydrolysis;
- acid hydrolysis.

We shall look at each of these mechanisms in turn.

Dissolution

Ionic crystals often dissolve in solvents that can form an electrostatic association with the ions. For instance, water is composed of polar molecules that are able to associate with the ions from the solid and consequently can break up minerals by solvation. The solvation (hydration) of ionic compounds in water is discussed in detail in Chapter 7.

Many minerals present in the Earth's crust are soluble, and will dissolve completely in water moving through or over soils and rocks. For example, halite ($NaCl$) and gypsum ($CaSO_4.2H_2O$) are readily dissolved and leached away (except in arid regions). With increasing leaching, less-soluble minerals may also be dissolved, leaving behind a deposit of non-soluble material. Some minerals, such as feldspar, dissolve partially only after reacting with air and water, leaving behind a solid residue of clay, e.g. the weathering of sodium feldspar to kaolinite, according to the following:

$$2NaAlSi_3O_8(s) + 9H_2O(l) + 2H_2CO_3(aq) \rightleftharpoons Al_2Si_2O_5(OH)_4(s) + 2Na^+(aq)$$
$$+ 2HCO_3^-(aq) + 4H_4SiO_4(aq)$$

Oxidation

Oxidation is strictly defined as the loss of an electron or electrons by an element, compound or ion:

$$B \rightleftharpoons B^{n+} + ne^-$$

Oxidation often occurs in rocks containing reduced substances (e.g. Fe^{2+}). It is generally a slow process, but can be speeded up by the presence of water, which removes small amounts of material by dissolution, e.g. the oxidation of fayalite, Fe_2SiO_4, according to the following:

$$Fe_2SiO_4(s) + \tfrac{1}{2}O_2(g) + 5H_2O(l) \rightleftharpoons 2Fe(OH)_3(s) + H_4SiO_4(aq)$$

Fe^{2+} Fe^{3+} silicic acid

In environmental situations, the end product of such a reaction is often a mixture of iron oxides. The latter are common end products because of their stability and insolubility, e.g.

$$Fe_2O_3(s)(haematite) \rightleftharpoons FeO(OH)(s)(hydrated\ oxide$$
$$-\ several\ forms,\ e.g.\ goethite)$$

Hydrolysis

Hydrolysis is a process in which water both reacts with and dissolves mineral constituents, usually forming smaller and less complex products. In both Examples 4.12 and 4.13 below, the acid products are very weak and therefore effectively undissociated. Significant dissociation only occurs for the hydroxides which are moderately strong alkalis and thus the formation of OH^- ions results in a pH > 7. This is a general pattern for minerals, i.e. hydrolysis produces a reasonably strong alkali but a very weak acid. There are a few exceptions, however, as shown in Example 4.14.

Example 4.12

$$Mg_2SiO_4(s) + 4H_2O(l) \rightleftharpoons 2Mg(OH)_2(aq) + H_4SiO_4(aq)$$

forsterite   silicic acid

$$2Mg^{2+}(aq) + 4OH^-(aq)$$

Example 4.13

$$CaCO_3(s) + 2H_2O(l) \rightleftharpoons Ca(OH)_2(aq) + H_2CO_3(aq)$$

calcite   carbonic acid

$$Ca^{2+}(aq) + 2OH^-(aq)$$

Example 4.14

In old mine workings, iron sulfate, $Fe_2(SO_4)_3$, produced from the oxidation of iron pyrites (FeS_2), will undergo hydrolysis to produce very acidic waters (pH 1–2):

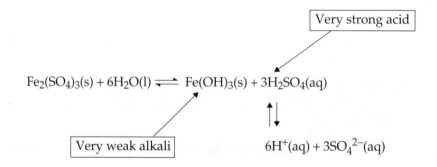

Very strong acid

$$Fe_2(SO_4)_3(s) + 6H_2O(l) \rightleftharpoons Fe(OH)_3(s) + 3H_2SO_4(aq)$$

Very weak alkali $6H^+(aq) + 3SO_4^{2-}(aq)$

Acid hydrolysis

Acid hydrolysis differs from conventional hydrolysis in that the active agent is a dilute acid (rather than water), which consequently produces more efficient weathering (acids and bases are discussed in detail in Chapter 7). Dilute acids are readily formed in the environment, e.g. carbonic and sulfuric acids:

$$CO_2(g) + H_2O(l) \rightleftharpoons H_2CO_3(aq)$$

$$SO_2(g) + H_2O(l) \rightleftharpoons H_2SO_3(aq) \xrightarrow{\text{oxidation}} H_2SO_4(aq)$$

Let us now consider the acid hydrolysis of forsterite and calcite by carbonic acid.

Example 4.15

$$Mg_2SiO_4(s) + 4H_2CO_3(aq) \rightleftharpoons 2Mg^+(aq) + 4HCO_3^-(aq) + H_4SiO_4(aq)$$

forsterite   silicic acid

Example 4.16

$$CaCO_3(s) + 2H_2CO_3(aq) \rightleftharpoons Ca^{2+}(aq) + 2HCO_3^-(aq) + H_2O(l) + CO_2(g)$$

calcite $\downarrow$ hydrolysis $\downarrow$

Relatively strong alkali $\longrightarrow$ $Ca(OH)_2(aq)$ $H_2CO_3(aq)$

$\uparrow\downarrow$ carbonic acid   Weak acid

$$Ca^{2+}(aq) + 2OH^-(aq)$$

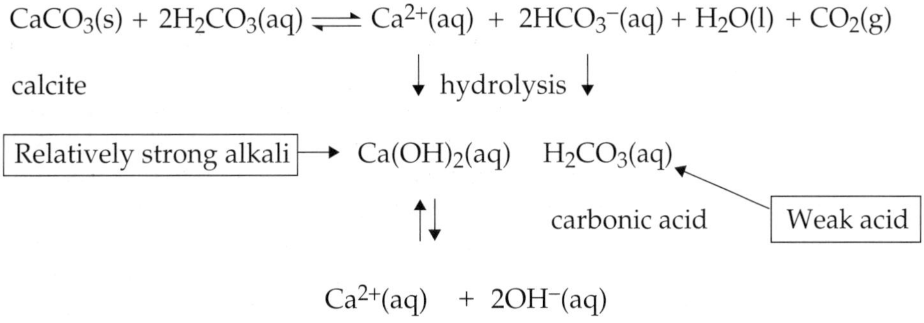

You can see that the products of the reactions shown in Example 4.16 are mainly alkaline.

The reactions shown in Examples 4.15 and 4.16 illustrate how the acidity of rainfall is effectively neutralized during the weathering of minerals. Gardeners and farmers use this principle to reduce soil acidity (often caused by acid precipitation or agricultural practices) by spreading finely ground lime (CaO), slaked lime ($Ca(OH)_2$) or limestone ($CaCO_3$) on the soil surface. In the soil, all of these materials extract exchangeable hydrogen ions from the soil and convert them to water, as follows:

$$CaCO_3(s) + {}^H_H\text{-soil} \rightleftharpoons \text{Ca-soil} + H_2O(l) + CO_2(g)$$

$$Ca(OH)_2(s) + {}^H_H\text{-soil} \rightleftharpoons \text{Ca-soil} + H_2O(l)$$

CaO reacts with soil water to form calcium hydroxide, which then neutralizes soil acidity, as shown above. The replacement of the exchangeable hydrogen ions tends to drive the reactions to the right, and the reactions will continue until the lime is consumed. Where soil cover is very thin and non-calcareous (e.g. in many parts of Scandinavia), the rainwater acidity may not be efficiently neutralized.

Other chemical weathering mechanisms include ion exchange and biotic extraction of minerals by lichens, although microbial weathering can also occur.

4.13 The Composition and Properties of Soil

As we have already seen, soil is formed by the weathering of rocks. It is a complex mixture of inorganic materials (clay, silt, gravel and sand), decaying organic matter, water, air and living organisms. Soils are open systems that undergo continual exchange with the atmosphere, hydrosphere and biosphere. Mature soils are arranged in layers called horizons, as shown in Figure 4.22. Each horizon has a distinct texture and composition that varies with different types of soils.

Soil is composed of particles of different sizes. The three main classes used to describe soil – sand, silt and clay – are identified by estimating the percentage of soil that passes through a 2 mm sieve. You will now have seen the term 'clay' used with three different meanings depending upon the context, i.e. as a soil mineral constituent fraction, a soil classification and a soil particle size fraction. The size distribution of soil particles is outlined in Table 4.11; the sand fraction is further sub-divided according to particle size. Soil material coarser than 2 mm is described as gravelly or stony.

A soil's texture depends upon the relative amounts of clay (very fine particles), silt (fine particles), sand (medium-sized particles) and gravel (coarse to very coarse particles). Figure 4.23 categorizes soils into textural classes according to clay, silt and sand content. The sum of the percentages of sand, silt and clay at any point in the triangle is 100. The dominant size fraction is used to describe the texture of the soil, e.g. point 'A' on Figure 4.23 represents a mixture of 15% clay, 20% silt and 65% sand, which is therefore known as a sandy loam. Definite lines of division separate the different classes of soil from each other, for example, soils containing a mixture of clay, sand and silt in which no fraction is dominant are called loams. The soil's

O-Horizon	The O-Horizon consists of surface litter, including organic debris, freshly fallen leaves and partially decomposed organic matter
A-Horizon	The A-Horizon is otherwise known as topsoil. It consists of humus, some inorganic minerals and living organisms
E-Horizon	The E-Horizon is a zone that separates the topsoil and the subsoil. Dissolved or suspended matter can move downward through this horizon, which is also known as the leaching zone
B-Horizon	The B-Horizon is otherwise known as subsoil. In this horizon, humic compounds, clay, iron and aluminium accumulate after leaching from the A- and E-Horizons
C-Horizon	The C-Horizon consists of the original parent material, usually partially weathered inorganic matter
R	Bedrock – unweathered rock below the parent material.

Figure 4.22 Schematic diagram of soil showing the different horizons

Table 4.11 Soil particle size categories (mm diameter)

Category	International system	US Department of agriculture system
Gravel	>2.0	>2.0
Very coarse sand	–	1.0–2.0
Coarse sand	0.2–2.0	0.5–1.0
Medium sand	–	0.1–0.5
Fine sand	0.02–0.2	0.05–0.1
Silt	0.002–0.02	0.002–0.05
Clay	<0.002	<0.002

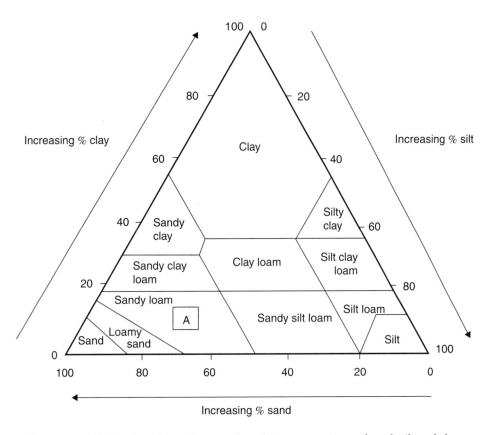

Figure 4.23 Soil texture depends upon the relative proportions of sand, silt and clay

texture determines the number of pores and therefore the amount of water it can hold – the higher the porosity, then the more water it can hold. The average pore size determines the permeability of the soil – the rate at which it can transmit water and air from upper to lower layers.

The pH of a soil determines the types of crops it can support. Acid soil can be treated with lime to raise its pH, while soil that is too alkaline can be treated with ammonium sulfate to lower its pH. A typical soil contains approximately 5% organic and 95% inorganic matter. Peat soils, however, may contain as much as 95% organic

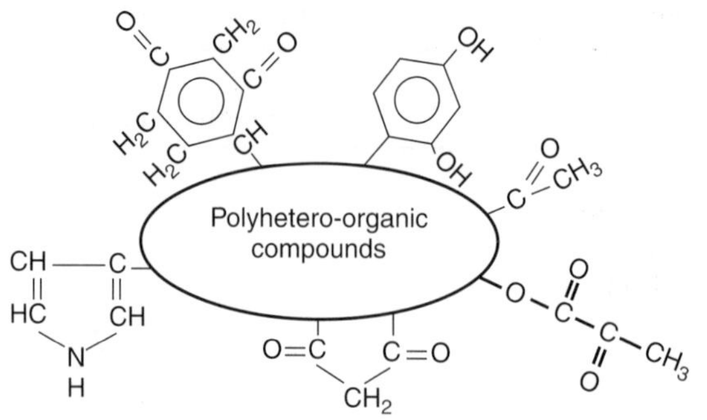

Figure 4.24 General structure of humic acids

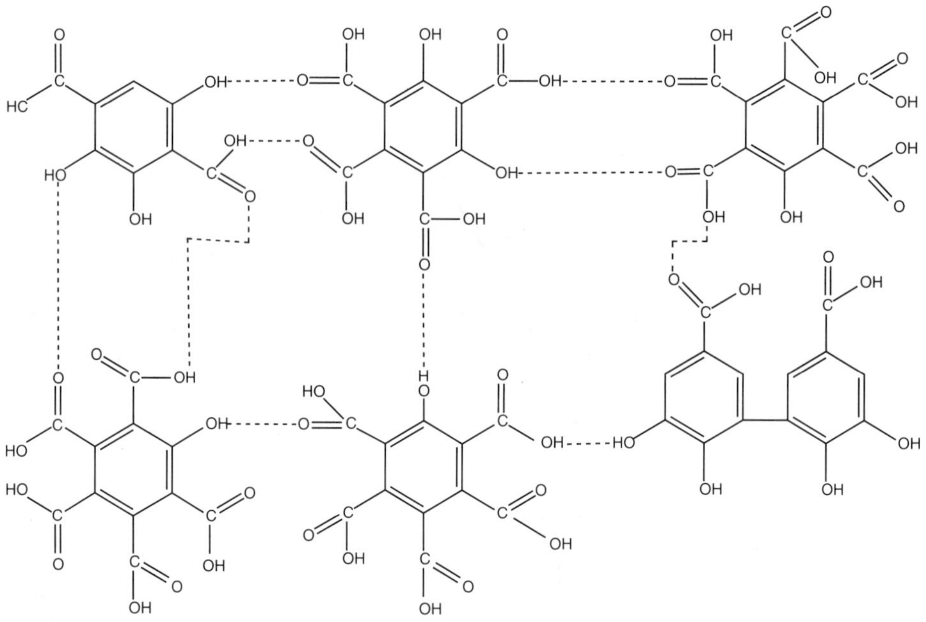

Figure 4.25 General structure of fulvic acids

material, most of it in the form of substances known as colloids that are composed of minute particles of one material evenly distributed in another material. Soil humus is the most important part of the organic matter. It is produced by partial decay of plants and consists of humic and fulvic acids which are high-molecular-weight macromolecules, as shown in Figures 4.24 and 4.25. These acids are particularly important because of their ability to complex metal ions and hence stabilize and improve the condition of soil. Humic substances are strongly coloured, and give water a colour ranging from yellow to brown, depending on the concentration. Soil organic matter also includes microorganisms such as bacteria, fungi, protozoa and algae. A typical topsoil, the top 10 cm of soil, usually contains approximately 1–3% of carbon in organic compounds, although this value may increase to 30% in peat soils.

The organic matter determines the soil's productivity. It provides food for microorganisms and also takes part in ion exchange. The accumulation of organic matter in soil depends on the temperature and availability of oxygen. In colder climates, degradation of organic matter is slow, and hence organic matter will build up. In soil saturated with water, decaying vegetation does not have easy access to oxygen and this also leads to the accumulation of organic matter.

About 35% of the volume of a typical soil is air, which fills the pores. However, the oxygen content of this air is much lower than that of a normal atmosphere. As organic matter decays, oxygen is consumed and carbon dioxide is produced. Consequently, the oxygen content may be as low as 15% and the CO_2 content may be as high as 3–5%.

Any chemical element or compound that an organism must take in to live, grow or reproduce is called a *nutrient*. Soil *macro-nutrients* – those required in large quantities – are carbon (C), hydrogen (H), oxygen (O), nitrogen (N), phosphorus (P), potassium (K), calcium (Ca), magnesium (Mg) and sulfur (S). Soil *micro-nutrients* – those required in small (trace) quantities – are boron (B), chloride (Cl^-), copper (Cu), iron (Fe), manganese (Mn), molybdenum (Mb), sodium (Na), vanadium (V) and zinc (Zn). Most of the micro-nutrients function as components of enzymes.

The order of occurrence of the twelve elements present in greatest concentration in a typical soil is as follows:

$$O > Si > Al > Fe = C = Ca > K > Na > Mg > Ti > N > S$$

All other elements are present in concentrations of less than 0.1%. Based on chemical composition, the main difference between soil and the rocks from which it is derived, is that soil contains a much higher concentration of carbon, nitrogen and sulfur from biological accumulation.

4.13.1 Soil solutions

Soil solutions contain a variety of dissolved matter that it transports to and from soil particles, as well as providing water and nutrients for plant growth via the roots. There is continuous interaction between clays and water in soil; water is adsorbed on to the surface of the clay particles and when it becomes saturated with water (waterlogged), there will be considerable changes in the physical and chemical properties of the soil. Under these conditions, microorganisms rapidly use up oxygen gas and the conditions become more reducing, thus causing the reduction of the next most easily reduced species, e.g.

$$MnO_2 + 4H^+ + 2e^- \rightleftharpoons Mn^{2+} + 2H_2O$$

$$Fe(OH)_3 + 3H^+ + e^- \rightleftharpoons Fe^{2+} + 3H_2O$$

These reactions lead to the mobilization of Fe(II) and Mn (II) which become toxic to plants at high concentrations.

One of the most important properties of the soil is to exchange cations. This is expressed as the *cation exchange capacity* (CEC), the quantity of monovalent cation that can be exchanged by 100 g of dry soil. Both the organic and the mineral part are involved in the ion exchange. The essential metals (K, Ca, Mg, etc.) and the trace metals (Zn, Cu) are made available to plants through cation-exchange processes. When plants take up metal ions, hydrogen ions are released, thus making the soil more acidic, as illustrated by the following reaction:

$$\text{Soil-Ca}^{2+} + 2CO_2 + 2H_2O \longrightarrow \text{soil-}(H^+)_2 + Ca^{2+}(\text{root})\ 2HCO_3^-$$

The pH of soil can be measured by using a suspension of soil in water or with an extracted soil solution (pH is discussed in Chapter 7). Soil can act as a *buffer* (see Section 7.9); the buffering capacity depends on the type of the soil present. For example, soils containing carbonates can neutralize added acids by forming soluble bicarbonates, which are leached out of the soil by increased precipitation, as follows:

$$CaCO_3(s) + H^+(aq) + HCO_3^-(aq) \rightleftharpoons Ca(HCO_3)_2(aq)$$

4.13.2 Soil erosion

Soil is generally regarded as a renewable resource because it is continuously regenerated by natural processes. However, in many areas of the world, the rate of soil erosion is faster than the rate of renewal, thus making it non-renewable. Soil erosion occurs when topsoil components are moved from one place to another. Soil may be worn away naturally by wind and rain, while anthropogenic erosion is due to a number of factors, including excessive farming, mining, logging, deforestation, off-road vehicles, overgrazing by livestock, etc. Erosion reduces the fertility and water-holding capacity of the soil. It is a slow process – it may take many decades for the effects to become apparent. Much of the soil removed from land ends up as sediment in rivers and water reservoirs and is harmful to fish and other forms of aquatic life. This process also increases the rate of water run-off and the risk of flooding as sediments accumulate in river channels and lakes. Soil erosion is widespread on the world's crop and animal grazing land and in deforested areas of Asia and South America.

4.13.3 Soil fertility

The term "soil fertility" is usually applied to land used for growing crops. A fertile soil is one that is capable of producing a particular crop with minimum (identified) yield and quality characteristics. Natural levels of soil fertility are rarely adequate to sustain long-term agricultural productivity due to nutrient removal when crops are harvested.

Agricultural practices such as crop rotation and intercropping may be used to retain soil fertility. Corn, tobacco and cotton remove large amounts of nutrients from the soil; when continuously grown, the soil becomes depleted. If, in the following year legumes, which add nitrogen to the soil as such species have nitrogen-fixing bacteria at their roots, are planted, the fertility may be restored. Intercropping is a technique that involves using mixtures of plant species to make more efficient use of the nutrients.

Soil fertility can also be improved by the use of natural or chemical fertilizers. The latter are used to remove the limitation to plant growth caused when there is an inadequate supply of nutrients in the soil. Natural fertilizers include animal manure, green manure and compost. Chemical fertilizers contain nitrogen (e.g. ammonium nitrate or sulfate and urea), phosphorus (e.g. single and triple superphosphate) and potassium (e.g. potassium chloride and sulfate). The disadvantage of chemical fertilizers is that they do not add humus or essential trace nutrients and therefore the organic content of the soil decreases. The porosity also decreases as well as the ability to retain water. Chemical fertilizers should therefore always be supplemented by organic ones, although a high organic content binds trace nutrients, thus reducing their bioavailability.

Water can reduce soil fertility as the nutrients are washed out into nearby surface waters. This often leads to excessive growth of algae in water bodies which prevents photosynthesis taking place, and therefore the water becomes depleted in oxygen. Further problems with chemical fertilizers occur as water seeps through the soil, leaching out nitrates and contaminating groundwater. High levels of nitrates may be toxic in drinking water, particularly to small children.

4.13.4 Soil contamination

Like air and water, soil is vulnerable to pollution from several sources. Air pollutants may be deposited on land and can settle in the soil. The deposition of long-lived radionuclides, such as ^{137}Cs, into the environment from the accident at the Chernobyl nuclear power station in 1986, caused considerable soil contamination in areas such as North Wales, the English Lake District and Scotland. Soil acidification due to acid precipitation is a worldwide problem that has caused nutrients to leach from soils and subsequent damage to trees and vegetation (see Section 7.10.2). Soil acidification is a natural process – most soils will become acidic if exposed to rainwater for a sufficient period of time – but the rate of acidification has increased markedly in the last forty years.

Hazardous waste buried in landfills or dumped in fields will damage soil and may have serious knock-on effects. The Love Canal waste tip near Niagara Falls, in the State of New York, USA, is one of the most notorious examples of contaminated land arising from the waste disposal of hazardous waste. The site was initially an abandoned excavation for a canal, but later became used as a site for the disposal of wastes from a nearby chemical works. Following its closure, the land was used for building a school and a housing estate. However, over 20 years later (in 1977–1978), and following severe weather, the dumped chemicals began to rise to the surface and contaminate land and groundwater. Eventually, the site was declared a disaster area and 239 families were then evacuated, followed by others later. Evaluation, monitoring and remediation activities at the Love Canal site have cost millions of dollars and even now, the site has not yet been fully cleaned.

Soil may also be damaged by agricultural practices such as irrigation and the use of pesticides. Irrigation helps to increase crop productivity in the short-term, but not in the long-term, as excessive salt builds up (salinization) and rising water tables occur (waterlogging). About 20% of the world's cropland is now irrigated in order to grow enough food. As a consequence, water dissolves minerals from the soil and the salinity of the water thus increases. When the water evaporates, the salt is left behind, so increasing salt concentration in the topsoil. Build-up of salt promotes excessive water use which leads to reduction in crop growth, thus making the land less productive. The removal of water for irrigation has caused the calamitous desertification of the Aral Sea between Kazakhstan and Uzbekistan.

Salinization is associated with waterlogging. In order to reduce the salinity, large amounts of irrigation water are applied. If the drainage is not adequate, water accumulates underground and the water table is raised closer to the surface. Salty water effects the roots of the plants and may kill them. The overall effect is the conversion of fertile land into infertile land and ultimately into desert. Soil reformation is very slow – it can take hundreds of years.

The use of pesticides temporarily increases crop productivity until insects develop genetic resistance to them, and they can also bioaccumulate in crops and run off land into watercourses.

4.14 Summary

In this chapter, we have hypothesized about how the Earth was formed and have built up a picture of the Earth's physical environment from its core to its atmosphere. We have studied important physical and chemical processes that occur within Earth's different environmental compartments, particularly those within the crust. We have discussed the structures and properties of the materials that make up the Earth, and described how they are influenced by the nature of chemical bonds. In the experiments that follow, we will further develop your laboratory skills by studying methods for the quantitative determination of frequently measured environmental indicators.

Experiment 4.1 Determination of Soil pH

Introduction

Soil pH is the most widely measured soil parameter. It influences a large number of environmental mechanisms, including the following:

- the leaching of nutrients from soil;

- the provision of nutrients and water to plants;

- the pH of water in rivers and lakes.

By definition, pH is the negative logarithm to the base ten of the numerical value of the H_3O^+ ion concentration (see Chapter 7). Consequently, it is not possible to measure the pH of a dry soil and we need to add water to soil prior to pH determination. It is not easy to decide how much water to add to soils since they have widely differing capacities to absorb water. Some soil scientists prefer to measure the pH of a thick paste of moist soil, since this provides more 'realistic' conditions and allows direct determination of moist soils in the field. However, this method reduces the precision of measurement and sufficient water must be present to allow liquid contact with the electrodes.

When distilled water (typical pH 5.6, although it can be lower) is added to a soil with a different pH, the pH of the water changes so that it is approximately the same as the solution in the moist soil. In acid soils, this process occurs due to the displacement of H^+ ions from exchange sites by ions such as Ca^{2+} and Al^{3+} as the soil solution is diluted. In alkaline soils, the pH of the soil solution is buffered primarily by the dissolution of $CaCO_3$. Thus, by using a standard procedure for the determination of soil pH (i.e. a set mass of soil and a set volume of water), comparisons between different soil types can be made.

Soil acidification is a natural process – most soils will become acidic if exposed to rainwater that causes thorough drainage for a sufficient period of time. Every continent in the world contains significant areas of acid soils. Alkaline soils usually contain significant amounts of calcium carbonate. Usually, soils of humid regions are acid while soils of arid regions are alkaline. It is rare to find a soil with a pH outside the range 2– 10.

Time required

One-two hours, depending upon the number of soil samples determined.

Equipment

Mortar and pestle
2 mm sieve and receiver
Spatula
Screw-cap bottle
25 cm^3 measuring cylinder
Laboratory shaker
Glass rod
pH meter
50 cm^3 beakers (for calibration of pH electrode)
Laboratory tissues
Balance accurate to ± 0.1 g

Reagents

pH 4 buffer solution
pH 9.2 buffer solution

De ionized water

Suitable air-dried soil sample(s) (ideally, at least two different samples)

Significant experimental hazards

- Students should be aware of the hazards associated with the use of all glassware (cuts) and electrical equipment (shock, burns).

Procedure

(1) Place a quantity of air-dried soil into a mortar and grind it with a pestle until the soil can pass through a 2 mm sieve. You may need to remove any stones present.

(2) Weigh accurately 10.0 ± 0.1 g of the homogenized soil into a bottle with a screw cap (the procedure for weighing chemicals is outlined in Chapter 1).

(3) Add 25 cm^3 of double-distilled, deionized water from a measuring cylinder to the soil in the bottle, and screw on the top.

(4) Shake the bottle using a laboratory shaker for 15 minutes. (If a laboratory shaker is not available, you can shake the bottle periodically by hand over a 15 minute period).

(5) While the bottle is shaking, calibrate the pH meter by using the two buffer solutions provided. You should initially wash the electrode with distilled water from a wash bottle and then gently dry the electrode with a laboratory tissue. Place the electrode in a pH 4 buffer solution and gently swirl the solution until a stable reading is obtained. If necessary, adjust the meter to pH 4 by using the controls. Remove the electrode, rinse with distilled water (as above), dry with a tissue and insert the electrode into the pH 7 buffer solution. Swirl the solution until a stable reading is obtained and adjust the meter as necessary. Repeat these readings until the pH meter is correctly calibrated for both buffer solutions.

(6) When shaking is completed, remove the bottle from the shaker and stir the resulting soil suspension with a glass rod.

(7) Insert the (cleaned and dried) pH electrode into the bottle, swirl the suspension gently until a stable reading is obtained, and record the pH value. A constant value should be obtained within about 1 min. Occasionally, the pH value will 'drift' i.e. slowly but continuously increase or decrease. If drifting occurs, record that the pH is drifting and note the pH limits over a certain (identified) period of time. You may also need to re-calibrate the electrodes.

(8) If you are determining the pH of more than one soil sample, it is most efficient to prepare and determine all of the samples together.

Issues to consider for your practical report

- What are the potential sources of error in this analytical determination? How could these be overcome?

- What are the alternative methods for determining the pH of soil? How do they compare with this method?

- What is the source of the unknown soil samples?

- What are typical pH values of soils in your local or national area? How do your data compare with these values?

- What are the main sources of (i) acidity and (ii) alkalinity in soils?

- What types of soils typically have a pH (i) < 7 and (ii) > 7?

- How can the acidity of soil be regulated?

- What are the potential environmental effects (if any) of extreme soil pH values?

Useful references

Hesse, P.R. (1971). *A Textbook of Soil Chemical Analysis*. John Murray Publishers Ltd, London. ISBN 0-7195-2061-4.

Petersen, L. (1986). 'Effects of acid deposition on soil and sensitivity of the soil to acidification'. *Experientia*, **42**, 340–344.

Experiment 4.2 Determination of the Organic Matter Content of Soil

Introduction

Soil is a complex mixture of organic and inorganic substances that are intimately mixed. As we have seen in Section 4.13, the composition of soil is constantly changing due to the chemical, physical and biological processes taking place. However, in spite of these changes, the organic content of soil remains fairly constant in soil that is not excessively cultivated; the continual input of plant and animal remains is balanced by decay processes generating carbon dioxide and other volatile substances.

The organic part of soil occurs in two main forms, as follows:

(a) undecomposed vegetable and animal remains;

(b) decomposed organic matter which is dark brown in colour, colloidal in form and generally referred to as humus. Humus is composed of a number of components, including humic and fulvic acids and humin.

In addition, carbon is found in soils in the form of:

(c)   carbonate minerals (e.g. $CaCO_3$, $MgCO_3$);

(d)   organic and elemental carbon (e.g. charcoal, graphite and coal, oil contaminants, pesticides, waste-derived polymers, etc.).

In agriculture, forms (a), (b) and (c) are the most important and when determining the organic matter content of a soil, the aim is to determine (a) and (b) only. The organic part of soil, although it only usually comprises 5–10% of the total, affects many of its most important properties, including fertility, water retention, the ability to bind ions into soil and the ability to withstand erosion. Determination of the organic content of soil therefore enables one to assess the quality of soil and its suitability for crop production.

Various methods are available for the determination of the organic content of soil. Some methods determine the total carbon content (all the forms listed above) and others the oxidizable organic matter (mainly forms (a) and (b)). No method measures only forms (a) and (b). In this experiment, two different methods will be used. The results obtained by using Method 1 will usually be higher than those obtained by using Method 2, and although both methods have their weaknesses, they do provide an indication of the relative amounts of organic matter present in soils.

Method 1 is based on the thermal decomposition of organic matter when heated to a high temperature. At 800°C, all the organic carbon is ignited (forms (a), (b) and (d)) and the carbonate minerals are decomposed, liberating carbon in the form of carbon dioxide. After ignition, the soil is treated with a solution of ammonium carbonate to reform the carbonate minerals that are destroyed in the furnace. Combined water in the soil colloids is lost and this results in errors.

In Method 2, the organic matter is oxidized by a strong oxidizing agent (potassium dichromate). Heat is required for the oxidation and this can be supplied externally by a heat source, or internally by the dilution of concentrated sulfuric acid. When using the latter method, less of the organic matter is oxidized and this is thought to be an advantage, since the less active organic matter is not measured (the latter does not play much part in agriculture). In order to determine the percentage organic carbon, a known amount of potassium dichromate is added to a given amount of soil and after the oxidation is complete, the amount of unused (excess) potassium dichromate is determined by titration with iron (II) ammonium sulfate. Thus, the amount of oxidizing agent used can be found and therefore the amount of organic carbon.

Time required

Method 1 typically requires up to two hours while Method 2 requires three hours. For a more in-depth study, you can extend the experiment to cover two or more experimental sessions by collecting the soil samples in the field, air-drying the soil and determining several samples from different locations.

Equipment

Method 1 – Determination by loss on ignition
Mortar and pestle
2 mm sieve and receiver
Spatula
Heat resistant gauntlets or gloves
High-temperature furnace
Desiccator
Silica crucible
Bunsen burner, tripod stand and metal gauze
Tongs

Method 2 – Determination by oxidation
500 cm^3 conical flasks
10 cm^3 pipettes and pipette filler
Tongs
50 cm^3 burette
50, 20 and 10 cm^3 measuring cylinders
Boss, clamp, retort stand and white tile

Reagents

Method 1 – Determination by loss on ignition
2% ammonium carbonate solution
Suitable air-dried soil sample(s)

Method 2 – Determination by oxidation
0.167 M potassium dichromate solution ($K_2Cr_2O_7$) (49.035 g $K_2Cr_2O_7$ in
   one litre of distilled water)
0.5 M iron (II) ammonium sulfate (otherwise known as ferrous ammonium
   sulfate (FAS), $FeSO_4(NH_4)_2SO_4$ (196 g of FAS dissolved in 0.2 M sulfuric
   acid and made up to one litre with distilled water)
Concentrated sulfuric acid (specific gravity 1.84 g cm^{-3})
Orthophosphoric acid (85 vol%)
Ferroin indicator solution (dissolve 0.25 g of sodium diphenylamine
   sulfonate in 100 cm^3 of distilled water)
Suitable air-dried soil sample(s)

Significant experimental hazards

Method 1 – Determination by loss on ignition
- Students should be aware of the hazards associated with the use of all glassware (cuts), Bunsen burners (burns, fire), tongs to move hot objects (burns, dropping hazards) and furnaces.

- Ammonium carbonate solution may be harmful if ingested in quantity and may cause severe burns to eyes and skin.
- Great care should be taken when removing the lid of the desiccator used during the experiment as the initial heating and subsequent cooling of the enclosed air will strengthen the seal between the lid and the base.

Method 2 – Determination by oxidation
- Students should be aware of the hazards associated with the use of all glassware (cuts) and fume cupboards.
- Potassium dichromate solution is harmful by ingestion, inhalation and skin contact, and may be corrosive to eyes and skin.
- Sulfuric acid is harmful by ingestion and may cause severe burns to eyes and skin.
- Orthophosphoric acid is harmful by ingestion and may cause severe burns to eyes and skin.
- FAS may be harmful if ingested in quantity and may irritate eyes and skin.
- Ferroin indicator may be harmful by ingestion and may irritate eyes and skin.

Procedure

Method 1 – Determination by loss on ignition
(1) Place a quantity of air-dried soil into a mortar and grind it with a pestle until the soil can pass through a 2 mm sieve. You may need to remove any stones present.

(2) Weigh accurately 5.00 ± 0.01 g of the ground soil into a previously ignited and weighed silica crucible (W_1 g) (the procedure for weighing chemicals is outlined in Chapter 1).

(3) Wearing heat-resistant gauntlets or gloves, place the silica crucible in a furnace at 800°C for 15–20 minutes. During this period, the organic carbon compounds will decompose together with the inorganic carbonates.

(4) Remove the silica crucible from the furnace by using the tongs and heat-resistant gauntlets and place immediately into a desiccator.

(5) Allow the crucible and contents to cool in the desiccator.

(6) After cooling, place the crucible on to a tripod stand and gauze and moisten the residue slowly and carefully with the 2% ammonium carbonate solution. This treatment will reform the inorganic carbonates.

(7) Heat the crucible gently by using a Bunsen burner in order to remove the excess carbonate solution. When the residue is dry, cool the crucible again in the desiccator and reweigh (W_2 g).

(8) Express the loss in weight of the soil (the organic matter content) as a percentage of the original soil weight.

Method 2 – Determination by oxidation

To save time when performing this determination, you could perform steps 1–3 of part (b) before starting part (a).

Part (a) – Standardization of Iron (II) Ammonium Sulfate (FAS)

(1) Pipette $10 \, cm^3$ of potassium dichromate solution into a $500 \, cm^3$ conical flask.

(2) In a fume cupboard, carefully and slowly add $20 \, cm^3$ of concentrated sulfuric acid to the contents of the conical flask by using a measuring cylinder. The addition should be accompanied by continuous swirling to mix the solution and to disperse the heat generated. When the addition is complete, cool the flask to room temperature in ice or by holding it (with tongs) under a tap of running cold water. (Make sure that the water does not splash into the flask.)

(3) After cooling, add $200 \, cm^3$ of double-distilled deionized water, $10 \, cm^3$ of orthophosphoric acid and $2 \, cm^3$ of indicator solution to the flask and shake thoroughly.

(4) Titrate the contents of the flask with the FAS solution, adding 0.5 cm^3 increments until the solution turns to an emerald-green colour. The colour changes observed should be from an initial red-black to a blue-black colour and finally to green at the end point. To determine an accurate end point, add a further $0.5 \, cm^3$ of potassium dichromate solution and note the change in colour from emerald green back to black. Add the FAS dropwise from the burette, mixing well between each addition, until the colour changes to green again. Record the volume of FAS added ($V_1 \, cm^3$) used for this *blank* determination.

(5) Repeat this procedure to obtain a second value of V_1.

Part (b) – Oxidation of Organic Matter in Soil

(1) Weigh between 0.2 and 0.3 g of air-dried soil accurately (accurate to 0.001 g) into a dry $500 \, cm^3$ conical flask.

(2) Pipette $10 \, cm^3$ of potassium dichromate solution into the flask.

(3) In a fume cupboard, carefully and slowly add $20 \, cm^3$ of concentrated sulfuric acid to the contents of the conical flask by using a measuring cylinder. The addition should be accompanied by continuous swirling to mix the solution and to disperse the heat generated. Mix well and allow the flask to stand for approximately 30 minutes to allow the oxidation of the organic matter to proceed to completion.

(4) After standing, add 200 cm^3 of double-distilled deionized water, 10 cm^3 of orthophosphoric acid and 2 cm^3 of indicator solution to the flask and shake thoroughly.

(5) Titrate the contents of the flask with the FAS solution until the solution turns to an emerald-green colour, using the method described in Step 4 of part (a). Record the volume of FAS (V_2) required to neutralize the potassium dichromate solution that has not reacted with the organic matter in the soil.

(6) Repeat this procedure with another accurately weighed sample of soil to obtain a second value of V_2.

Calculations

Method 1 – Determination by loss on ignition
The difference between the initial and final weight of the soil corresponds to the amount of organic carbon ($W_1 - W_2$ g). The percentage of organic carbon in the soil is given by the following:

$$\% \text{ Organic matter} = \frac{W_1 - W_2}{W_1} \times 100$$

Method 2 – Determination by oxidation
The difference between the volume of FAS used in the blank titration (standardization) and that used in the titration containing the soil sample corresponds to the amount of organic carbon which was oxidized ($V_1 - V_2$). The reaction between potassium dichromate and FAS is an example of a redox reaction and two separate equations can be written to represent these processes, as follows:

$$Fe^{2+}(aq) \longrightarrow Fe^{3+}(aq) + e^-$$

$$Cr_2O_7{}^{2-}(aq) + 14H^+(aq) + 6e^- \longrightarrow 2Cr^{3+}(aq) + 7H_2O(l)$$

Combining these two equations then gives the overall reaction:

$$Cr_2O_7{}^{2-}(aq) + 6Fe^{2+}(aq) + 14H^+(aq) \longrightarrow 2Cr^{3+}(aq) + 6Fe^{3+}(aq)$$
$$+ 7H_2O(l)$$

and therefore:

$$6 \text{ moles of } Fe^{2+} \equiv 1 \text{ mole of } Cr_2O_7{}^{2-}$$

$$1 \text{ mole of } Fe^{2+} \equiv 1/6 \text{ mole of } Cr_2O_7{}^{2-} \equiv 0.167 \text{ mole of } Cr_2O_7{}^{2-}$$

V_1 is the volume of FAS used in the standardization reaction.
$V_1 \text{ cm}^3$ of FAS solution reacts with 10.5 cm^3 of $0.167 \text{ M } Cr_2O_7{}^{2-}$

$$= 10.5 \times 10^{-3}1 \times 0.167 \text{ moles } Cr_2O_7{}^{2-}$$

$$= 10.5 \times 0.167 \text{ millimoles } Cr_2O_7{}^{2-}$$

Thus, 1 cm^3 of FAS solution reacts with:

$$\frac{10.5 \times 0.167}{V_1} \text{ millimoles } Cr_2O_7{}^{2-}$$

V_2 is the volume of FAS required to neutralize the potassium dichromate solution that has not reacted with the organic carbon in the soil. V_2 cm^3 of FAS solution is equivalent to:

$$\frac{10.5 \times 0.167 \times V_2}{V_1} \text{ millimoles } Cr_2O_7{}^{2-}$$

The amount of soil organic carbon oxidized by the dichromate solution is equivalent to $(V_1 - V_2)$ cm^3 of FAS, and hence (by substitution), the number of millimoles of $Cr_2O_7{}^{2-}$ equivalent to organic carbon is given by:

$$(10.5 \times 0.167) - \frac{10.5 \times 0.167 \times V_1}{V_2} = (10.5 \times 0.167) \left(1 - \frac{V_2}{V_1}\right)$$

Thus, the volume of 0.167 millimoles of $Cr_2O_7{}^{2-}$ required in order to oxidize the organic carbon in the soil sample:

$$= 10.5 \left(1 - \frac{V_2}{V_1}\right) = V$$

The oxidation of organic carbon to carbon dioxide is a four-electron process, while the reduction of $Cr_2O_7{}^{2-}$ is a six-electron process. Therefore:

$$3 \text{ moles of C} \equiv 2 \text{ moles of } Cr_2O_7{}^{2-} \text{ or (cross-multiplying)}$$

$$3/2 \text{ moles of C} \equiv 1 \text{ mole of } Cr_2O_7{}^{2-}$$

$$\Rightarrow 1 \text{ millimole of } Cr_2O_7{}^{2-} \equiv 3/2 \text{ millimoles of C}$$

Therefore:

$$1/6 \text{ millimole of } Cr_2O_7{}^{2-} \equiv 3/2 \times 1/6 \text{ millimoles of C}$$

$$\equiv 3/12 \text{ millimoles of C}$$

$$= 1/4 \text{ millimoles of C} = 3 \text{ mg of C}$$

and so:

$$\% \text{ oxidizable carbon} = \frac{V \times 3 \times 100}{M}$$

where M $=$ the weight of soil in mg.

However, we need to include some correction factors in our calculation. This is because only 3/4 of the organic carbon is oxidized by the dichromate solution and it is assumed that the soil organic matter contains 58% of organic carbon. By incorporating these two correction factors (1.33 and 1.724), the percentage of oxidizable carbon can be calculated as follows:

$$\% \text{ oxidizable carbon} = \frac{V \times 3 \times 100 \times 1.33 \times 1.724}{M} = \frac{688 \times V}{M}$$

NB – If your soil contains a known percentage of organic matter, you should replace the assumed correction factor, 1.724 (100/58 = 1.724), with the known correction factor.

Issues to consider for your practical report

- What are the potential sources of error in these analytical determinations? How could they be overcome?
- How many significant figures should you use when reporting the % oxidizable carbon?
- What are the alternative methods for determining the organic matter content of soil? How do they compare to Method 1 and Method 2 in terms of accuracy and ease of determination?
- What is the source of the soil sample(s)?
- Identify the factors that can affect the organic matter content of soil. How do your data compare with previously published data for similar soil types?

Useful references

Hesse, P.R. (1971). *A Textbook of Soil Chemical Analysis*. John Murray Publishers Ltd, London. ISBN 0-7195-2061-4.

Johnston, A.E. (1986). 'Soil organic matter, effects on soil and crops'. *Soil Use and Management*, **2**, 97–104.

Experiment 4.3 Study of the Cation Exchange Capacity of Soil

Introduction

Soils are complex mixtures of inorganic materials (clay, silt and sand), decaying organic matter, water, air and living organisms. The composition of soil is constantly changing due to the various biological and chemical processes taking place. One of the most important functions of soil is to deliver nutrients to the plant root. This is done through the exchange of cations between the solid soil and the soil solution with which it is in contact. This property is expressed as the *cation exchange capacity* (CEC) of soil – the quantity of monovalent cation that can be exchanged by 100 g of dry soil. The value of CEC is usually expressed in terms of milli-equivalent (meq).

$$1 \text{ meq} = 1 \text{ millimole of a monovalent ion (Na}^+)$$

$$= 0.5 \text{ millimole of a divalent cation (Ca}^{2+})$$

Both the organic and mineral parts of soil are involved in the exchange. Essential metals (K, Ca and Mg) and trace metals (Zn, Fe, Cu, etc.) are made available to plants through cation-exchange processes. When metal ions are taken up by a plant, hydrogen ions are released, with the result that the soil is made more acidic. Anions are not retained by soil; they are mobile and are easily leached away (with the exception of phosphate, which tends to bind to aluminium and iron oxides).

Knowledge of a soil's CEC is important for agricultural purposes as it can indicate the degree of leaching of essential nutrients like potassium and its replacement by sodium (due to irrigation of land). The exchange mechanism can also play a part in removing a variety of pollutants, including ionized herbicides and pesticides, as well as metals. Typical CEC values of soil vary between 10–30 meq, although the CEC of soil with a high organic content may be several hundred meq.

The cation-exchange capacity of soil can be determined by various methods. One simple method is based on measuring the amount of sodium ion that is taken up by a given weight of dry soil. A flame photometer is then used for the determination of sodium that has been replaced in the soil by ammonium ions.

Time required

Three hours, depending upon the number of soil samples determined. (A large-group experiment can take longer, especially if a limited number of centrifuges are available.)

Equipment

Mortar and pestle
2 mm sieve and receiver
Weighing boat
Spatula
Centrifuge
50 cm^3 centrifuge tubes
50 cm^3 measuring cylinders
Ultrasonic bath
Flame photometer
100 cm^3 volumetric flasks
1, 2, 5 and 10 cm^3 pipettes and pipette filler

Reagents

1.00 M sodium ethanoate (CH_3COONa) solution
1.00 M ammonium ethanoate (CH_3COONH_4) solution
Ethanol (CH_3CH_2OH)
Standard solution of sodium (100 mg l^{-1})
Suitable air-dried soil sample(s)

Significant experimental hazards

- Students should be aware of the hazards associated with the use of all glassware (cuts), electrical equipment (shock, burns) and flame photometers (burns).

- Ethanol is highly flammable, intoxicating if inhaled or ingested, and may irritate eyes and skin.

- Ammonium ethanoate may be harmful if ingested in quantity, and may irritate eyes and skin.

- Sodium ethanoate may be harmful if ingested in quantity, and may irritate eyes and skin.

- The standard solution of sodium may be harmful if ingested, may irritate eyes and skin, and may contain a small amount of nitric acid as a "preservative".

Procedure

(1) Place a quantity of air-dried soil into a mortar and grind it with a pestle until the soil can pass through a 2 mm sieve. You may need to remove any stones present.

(2) Weigh 5.000 g of the ground soil accurately (accurate to 0.001 g) into a 50 cm^3 centrifuge tube (the procedure for weighing chemicals is outlined in Chapter 1).

(3) Add 30 cm^3 of 1.00 M sodium ethanoate to the soil in the centrifuge tube, place it into an ultrasonic bath and agitate for 5 minutes.

(4) Centrifuge the tube at about 200 rpm for 5 minutes until the supernatant liquid (liquid above the solid) is clear. (You will need to "balance" the centrifuge for it to operate properly). Decant and discard the supernatant liquid.

(5) Repeat Steps 3–4 twice more with fresh sodium ethanoate. (During Steps 3–4, the sodium ions will replace the exchangeable cations in the soil.)

(6) Add 30 cm^3 of ethanol to the soil in the centrifuge tube. Place it again into an ultrasonic bath and agitate for 5 minutes.

(7) Centrifuge the tube at about 200 rpm for 5 minutes until the supernatant liquid (liquid above the solid) is clear. Decant and discard the supernatant liquid.

(8) Repeat Steps 6 and 7 twice more with fresh ethanol. (This procedure will remove any excess of sodium ethanoate.)

(9) Add 20 cm^3 of ammonium ethanoate to the soil in the centrifuge tube. Place it into an ultrasonic bath and agitate for 5 minutes.

(10) Centrifuge the tube at about 200 rpm for 5 minutes until the supernatant liquid (liquid above the solid) is clear. Decant and collect the extract (supernatant liquid) in a 100 cm^3 volumetric flask.

(11) Repeat Steps 9 and 10 twice more with fresh ammonium ethanoate. (The sodium ions will be replaced by ammonium ions.) Collect the extracts (containing the sodium ions) in the same 100 cm^3 volumetric flask and make up to the mark with double-distilled deionized water. It is occasionally necessary to filter the extracts to remove suspended particles.

(12) Determine the sodium content of the combined extracts by using a flame photometer that has been previously calibrated (dilute the solution to within the range of the photometer if necessary). The principle of operation of a flame photometer is outlined in Box 4.2. For the calibration of the flame photometer, prepare standard solutions containing 10.0, 8.0, 6.0, 4.0 and 2.0 mg cm^{-1} of Na from the stock solution using the 100 cm^3 volumetric flasks (the procedure for diluting solutions is outlined in Chapter 1). You should also prepare a blank solution. Make sure you label the flasks clearly. Calibrate the instrument by adjusting the photometer reading to the concentration of each of the above standards.

Box 4.2 Flame Photometry

Flame photometry may be used for the quantitative determination of metals that produce characteristic flame emissions when their solutions are introduced in the form of sprays into a steady flame. This process enables atomization of the metals to occur. Furthermore, some of the metal atoms acquire sufficient energy to promote their electrons from the ground state to a higher energy level. The higher level is not stable; the electrons return to their ground state and emit radiation, the frequency of which corresponds to the difference in energy between the higher level and the ground state. This process may be represented by the Planck equation, $E = h\nu$, where E is the energy corresponding to the electronic transition, h is the Planck constant and ν is the frequency of the radiation emitted.

Although only some of the atoms are promoted to a higher level it can be assumed that their number is proportional to the total number of atoms, and hence to the concentration of the element in solution.

In the flame photometer, the intensity of the emitted light is measured by a photocell. As the intensity of the emitted light is proportional to the number of atoms promoted, measurement of the intensity gives information about the concentration of ions in solution. Compressed air is used to atomize the solution. The fuel and air carrying the sample enter the base of the burner and the characteristic emission of the element is obtained in the flame. The radiation is focused with the help of a lens. The optical filter absorbs unwanted wavelengths (those not due to emission of the element being studied) of radiation. The photoelectric cell produces a current which is proportional to the intensity of the radiation and hence to the concentration of the analyte. The current is amplified and measured by the galvanometer. The galvanometer reading can be adjusted with standard solutions to show their concentration expressed as appropriate units, provided that the calibration is within the linear range or that any curvature can be electronically corrected.

You should be aware that the calibration curve for a flame photometric determination is not always linear due a process called *self-absorption*. The latter is a phenomenon that occurs due to re-absorption of emitted radiation by the analyte as it moves away from the centre of the flame. The amount of self-absorption increases with the concentration of the analyte.

Calculation

Plot a graph of flame photometer reading (y-axis) versus concentration of sodium (x-axis) and from this determine the concentration of the solution tested. This will correspond to the amount of sodium in the 5.000 g sample of soil. Multiply this value by the appropriate dilution factor and by 20 in order to obtain the concentration of sodium present in 100 g of soil. Dividing this value by 23.0 (A_R of sodium) gives the required CEC of the soil investigated.

Issues to consider for your practical report

- What are the potential sources of error in this analytical determination? How could they be overcome?

- Are there alternative methods for determining the CEC of a soil? If so, how do they compare to this method?

- What is the source of the soil sample(s)?

- What are typical CEC values for soils in your local area? How do your data compare with these values?

- Explain why the presence of exchangeable cations in soil is important for the growth of crops. What factors can affect the presence of exchangeable cations in soil?

- What are the potential human health and environmental effects (if any) of excess trace metals in soils?

Useful references

Hesse, P.R. (1971). *A Textbook of Soil Chemical Analysis''*. John Murray Publishers Ltd, London. ISBN 0-7195-2061-4.

Gillingham, J.T. (1965). *Canadian Journal of Soil Science*, **45**, 102.

Legg, W.T. (1963). *Soil Science*, **95**, 214.

Self-Study Exercises

Structure of the Earth

4.1 Explain the difference between the continental and oceanic crusts.

4.2 Identify the ten most abundant elements in the Earth's crust.

4.3 What is the chemical composition of each of the six zones that comprise the upper mantle?

4.4 Define the following terms:

   (a) hydrosphere;

   (b) mantle;

   (c) siderophile;

   (d) sima;

   (e) differentiation;

   (f) core;

   (g) asthenosphere.

4.5 Explain, in detail, (i) why the Earth has a molten inner core and (ii) why the Earth has an atmosphere comprised predominantly of gaseous nitrogen.

4.6 Explain how the oceans play a major role in regulating the physical and chemical nature of the Earth's surface.

4.7 Why do we believe that air is a mixture rather than a compound?

Chemical bonds

4.8 Define the following terms:

   (a) co-ordination number

   (b) dipole;

   (c) delocalized electron;

    (d)   chemical bond;

    (e)   ligand;

    (f)   co-ordinate bond;

    (g)   resonance structure;

    (h)   complex ion;

    (i)   crystal.

4.9   Distinguish between ionic and covalent bonding.

4.10 Show how the transfer of electrons forms ionic bonds in the following compounds:

    (a)   KCl;

    (b)   MgO;

    (c) Li_2O

4.11 Draw Lewis formulae for the following covalent compounds:

    (a)   HBr;

    (b) Cl_2;

    (c) CH_3F;

    (d) CH_3COOH;

    (e) $(NH_2)_2CO$.

4.12 Give two examples for each of the following:

    (a)   monodentate ligands;

    (b)   polydentate ligands;

    (c)   neutral complex ions;

    (d)   ionic complex ions.

4.13 Explain why metallic bonding is different from co-ordination and ionic bonding.

4.14 Outline how the theory of metallic bonding accounts for the special properties of metals.

Electronegativity

4.15 Place the following elements in ascending order of electronegativity:

    (a)   I, F, Br, Cl;

    (b)   Si, Al, P, Na, Cl, S, Mg;

    (c)   Ca, Be, Sr, Mg;

    (d)   O, Li, B, F, C, Be, N.

4.16 Identify the following compounds as ionic, covalent or polar covalent:

(a) NaCl;

(b) N_2;

(c) NO_2;

(d) OF_2;

(e) SF_6;

(f) $AlCl_3$;

(g) Mg_2O.

4.17 Identify the limitations to the use of electronegativity values.

Shapes of molecules, ions and metals

4.18 Using VSEPR theory, suggest shapes for the following molecules and ions about their central atoms:

(a) H_2S;

(b) H_3O^+;

(c) NO;

(d) NO_3^-;

(e) PCl_4F;

(f) ClO_4^-;

(g) TeF_6;

(h) O_3.

4.19 Explain the difference between hexagonal and cubic close-packed structures.

4.20 Define the following terms:

(a) unit cell;

(b) lone pair;

(c) radius ratio;

(d) close-packed structure;

(e) isomorphism;

(f) tetrahedral gap.

4.21 Predict the shapes of the following oxides by using the radius ratio rule:

(a) $(SiO_4)^{4-}$;

(b) $(AlO_4)^{5-}$;

(c) MgO;

(d) K_2O;

(e) B_2O_3.

Silicates, aluminates and aluminosilicates

4.22 Distinguish between an aluminate and an aluminosilicate.

4.23 What feature distinguishes silicates and aluminosilicates from other complex anions?

4.24 Sketch the structures of the different classes of silicate minerals.

4.25 Give two examples of each of the seven different silicate types.

4.26 Explain the difference between the structures of talc and mica.

4.27 Look up "Mohs scale of hardness" and assign hardness numbers to quartz, talc and topaz.

4.28 List the uses of silicates, both natural and man-made.

4.29 Decide whether the following statements are true or false, giving reason(s) for your answer:

    (a) All rocks are composed wholly, or in part, of silicates.

    (b) Pyroxenes contain a characteristic plane of weakness along which the rock crystals prefer to break.

    (c) The mica group of minerals have good thermal and electrical insulation properties.

    (d) Forsterite and fayalite are examples of disilicates.

    (e) No hydrogen bonding exists in 2:1 clay mineral structures.

    (f) Layer silicates form giant, two-dimensional honeycomb sheets.

    (g) Cyclosilicates may be represented by the formula $(SiO_3)_n^{3n-}$.

Weathering processes

4.30 Define the following terms:

    (a) weathering;

    (b) physical weathering;

    (c) chemical weathering.

4.31 Identify ten factors that contribute to the rate and extent of physical weathering.

4.32 Identify the four main chemical weathering mechanisms.

4.33 Explain how the acidity of rainfall is effectively neutralized during the weathering of minerals.

Composition and properties of soil

4.34 Sketch a diagram of the different horizons in a mature soil.

4.35 List and discuss methods of soil conservation.

4.36 Using Figure 4.22, identify the textural class of each of the following soils:

(a) 15% silt, 10% sand, 75% clay;

(b) 52% silt, 13% sand, 35% clay;

(c) 22% silt, 48% sand, 30% clay;

(d) 40% silt, 40% sand, 20% clay;

(e) 5% silt, 45% sand, 50% clay.

4.37 Discuss the links between soil erosion, food production and overpopulation.

4.38 List ten examples of soil contamination.

4.39 Decide whether the following statements are true or false, giving reason(s) for your answer:

(a) A typical soil contains 5% inorganic and 95% organic matter.

(b) Soil is generally regarded as a renewable resource.

(c) Cation exchange is more important than anion exchange in soils.

(d) Soil irrigation increases long-term crop productivity.

(e) The higher the porosity of a soil, then the less water it can hold.

(f) Chemical fertilizers frequently contain nitrogen, phosphorus and potassium.

(g) About 35% of the volume of a typical soil is air.

Challenging Exercises

Structure of the Earth

4.40 Outline the evidence available to support the "Big Bang" theory.

4.41 Identify the advantages and disadvantages of living on a differentiated planet.

Electronegativity

4.42 Discuss the following statement: In practice, all ionic bonds have some covalent character.

Shapes of molecules, ions and metals

4.43 Write the Lewis structures and predict the shapes of the following molecules and ions about each central atom:

(a) $S_2O_3^{2-}$;

(b) SO_3^{2-};

(c) NH_2^-;

    (d) NH_4^+;

    (e) $OCCl_2$.

Weathering Processes

4.44   Specify the difference between weathering and erosion.

4.45   Explain how physical weathering is different from, but contributes to, chemical weathering.

4.46   If we continue to burn fossil fuels at an increasing rate, we may eventually double the carbon dioxide in our atmosphere. How might this affect the weathering of rocks and minerals?

Composition and properties of soil

4.47   Discuss the economic, environmental, political and social consequences of the rapid erosion of fertile soil in (i) your country, and (ii) your continent.

4.48   Explain in detail, using equations as appropriate, why soil acidification is a natural process.

4.49   Using your research skills to search the scientific literature, identify the optimum soil pH ranges of typical field and vegetable crops.

4.50   Explain why the percentage base saturation is more useful than the cation-exchange capacity when analysing acidic soils that have a pH-dependent charge.

4.51   Discuss in detail, using specific examples, the effects of eroded soil (sediment) on water quality.

4.52   The moisture content of a given volume of soil is defined as the ratio of the weight of water to the weight of solid particles, expressed as a percentage. Devise a method to quantitatively determine the moisture content of soil, outlining all reagent(s), equipment and procedures used.

Further Sources of Information

Further reading

Aide, M.T. and Cwick, G.J. (1998). 'Chemical weathering in soils from the glacial Lake Agassiz region of Manitoba, Canada'. *Environmental Geology*, **33**(2–3), 115–121.

Camuffo, D. (1995). 'Physical weathering of stones'. *Science of the Total Environment*, **167**, 1–14.

Clow, D.W. and Drever, J.I. (1996). 'Weathering rates as a function of flow through an alpine soil'. *Chemical Geology*, **132**(1–4), 131–141.

Deegan, J. (1987). 'Looking back at Love Canal'. *Environmental Science and Technology*, **21**(5), 421–426.

Drever, J.I. (1994). 'The effect of land plants on weathering rates of silicate minerals'. *Geochimica et Cosmochimica Acta*, **58**(10), 2325–2332.

Hoffman, A.J. (1995). 'An uneasy rebirth at Love Canal'. *Environment*, **37**(2), 4.

Langan, S.J. Hodson, M.E., Bain, D.C., Skeffington, R.A. and Wilson, M.J. (1995). 'A preliminary review of weathering rates in relation to their method of calculation for acid sensitive soil parent materials'. *Water, Air and Soil Pollution*, **85**(3), 1075–1081.

Lasaga, A.C., Soler, J.M., Ganor, J., Burch, T.E., and Nagy, K.L. (1994). 'Chemical weathering rate laws and global geochemical cycles'. *Geochimica et Cosmochimica Acta*, **58**(10), 2361–2386.

Press, F. and Siever, R. (1986). *Earth* (4th Edn). W.H. Freeman and Company, New York. ISBN 0-7167-1776-X.

Rowell, D.L. (1994). *Soil Science: Methods and Applications*. Longman Publishers, Harlow, UK. ISBN 0-582-08784-8.

White, A.F. and Blum, A.E. (1995). 'Effects of climate on chemical weathering in watersheds'. *Geochimica et Cosmaochimica Acta*, **59**(9), 1729–1747.

Wild, A. (1993). *Soils and the Environment: An Introduction*. Cambridge University Press, Cambridge, UK. ISBN 0-521-43859-4.

Useful reference books

Deer, W.A., Howie, R.A. and Zussman, J. (1982). *Rock-Forming Minerals* (2nd Edn). Longman Publishers, Harlow, UK. There are five volumes of this excellent book, each covering a different class of mineral: Volume 1 (*Ortho*- and Ring Silicates), Volume 2 (Chain Silicates), Volume 3 (Sheet Silicates), Volume 4 (Framework Silicates), and Volume 5 (Non-Silicates).

Carmichael, R.S. (1989). *CRC Practical Handbook of Physical Properties of Rock and Minerals*. CRC Press Inc., Boca Raton, FL, USA. ISBN 0–8493-3703–8.

Boulding, J.R. (1995). *Practical Handbook of Soil, Vadose Zone and Groundwater Contamination: Assessment, Prevention and Remediation*. Lewis Publishers, Boca Raton, FL, USA. ISBN 1–56670-051–5.

Useful websites

Soil

Association for the Environmental Health of Soils
http://www.aehs.com/

Canadian Journal of Soil Science
http://nrc.ca/aic-journals/cjss.html

Soil Survey and Land Research Centre
http://www.silsoe.cranfield.ac.uk/sslrc/

5 Earth's Vital Resources – Minerals, Metals and Fossil Fuels

Built in Spain; owned by a Norwegian; registered in Cyprus; managed from Glasgow; chartered by the French; crewed by Russians; flying a Liberian flag; carrying an American cargo; and pouring oil on to the Welsh coast.

Headline from *The Independent*, 22 February 1996, after the *Sea Empress* tanker spill.

5.1 Introduction

As the population of the Earth has grown, so has the demand for goods and energy. Although the Earth's crust contains huge quantities of exploitable natural resources, many of these vital assets are non-renewable and depleting rapidly. Consequently, it is essential that we understand the formation processes and physico-chemical properties of existing key resources, so that we can develop alternative, renewable and sustainable sources of energy and materials for utilization by future generations.

The term *mineral* may be used to describe natural resources that can be obtained by mining. However, in a geological sense, minerals are the natural inorganic, crystalline materials that make up the earth and most of its rock. Many minerals have a constant general formula (e.g. quartz, SiO_2) but some have a range of compositions (e.g. olivine, $(Mg,Fe)_2SiO_4$, which may contain only magnesium, only iron, or a combination of both, in addition to silicon and oxygen). Minerals have characteristic physical properties, including hardness, colour, density and cleavage planes, which are determined by their composition and structure. The rocks that make up the solid material of the Earth's crust are any natural material, hard or soft (e.g. clay), that consist of one or more minerals. They may be divided into 3 types depending upon how they are formed.

- *Igneous rocks* result from the crystallization of magma or lava, or from the accumulation and consolidation of volcanic material, such as ash. Magma tends to be rich in silicon, oxygen, aluminium, sodium, potassium, calcium, iron and magnesium. Examples of igneous rocks include granite, micas, feldspars and quartz.

- *Sedimentary rocks* are formed by the accumulation and consolidation of rock deposits, weathered and eroded rock particles, precipitation of matter from solution or compaction of plant and animal remains. Examples of sedimentary rocks include sandstone, limestone, rock salt and gypsum.

- *Metamorphic rocks* result from the alteration of any rocks by pressure, heat or chemically active fluids after their original formation. Marble is a classic example of a metamorphic rock – it is formed by the metamorphosis of limestone.

Over 3500 minerals have been identified on Earth but only about 25 are commonly occurring. In order to simplify their classification, the anion of a mineral is used as a means of dividing minerals into chemical classes. The main chemical classes of minerals are sulfides (S^{2-}), sulfates (SO_4^{2-}), oxides (O^{2-}) and hydroxides (OH^-), halides (Cl^-, F^-, Br^-, I^-), carbonates (CO_3^{2-}), phosphates (PO_4^{3-}), silicates (SiO_4^{4-}) and native (free) elements.

Metals are a class of chemical elements with particular chemical characteristics and physical properties. About 75% of all elements are metals. Their chemical properties are largely determined by the extent to which their atoms can lose one or more electrons to form cations. Elemental metals are good conductors of heat and electricity, and are malleable and ductile. Sixteen metals can occur in nature in the elemental form (free or native metals) – bismuth, cobalt, copper, gold, iridium, iron, lead, mercury, nickel, osmium, palladium, platinum, ruthenium, rhodium, tin and silver. However, most metals occur naturally as compounds and mineral *ores*, mainly in combination with O (as oxides) or S (as sulfides). An ore is a mineral from which a metal can be profitably extracted (although in general use, the term has been extended to non-metallic rocks such as coal). The science of extracting metals from mineral ores and preparing them for use is called *extractive metallurgy*. Metals have been used for structural and decorative purposes since prehistoric times, and may be classified according to their usage:

- *Precious metals*, such as gold, silver, platinum and mercury, which have a high economic value and are often used for jewellery.

- *Heavy metals*, such as iron, copper, lead, tin and zinc, commonly used in engineering.

- *Rarer heavy metals*, such as chromium, manganese, tungsten and bismuth, frequently alloyed with heavy metals to make substances with special properties.

- *Light metals*, such as aluminium and magnesium, used in lightweight alloys for aircraft and other construction purposes.

- *Alkali* (sodium, potassium and lithium) and *alkaline earth metals* (calcium, magnesium, barium and strontium), used mainly for chemical purposes.

Fossil fuels are non-renewable sources of energy that include peat, coal, oil (petroleum) and natural gas, formed from the remains of plants and animals that lived hundreds of millions of years ago. *Coal* is a black or blackish mineral substance that has been used as a fuel since the early inhabitants of Wales used it for funeral pyres 3–4 thousand years ago. It was mined on a small scale from Roman times in the UK, but became the mainstay of the industrial revolution in the nineteenth and twentieth centuries, being used in the steel industry, the development of the railway engine, general domestic heating and finally the generation of electricity. The extraction and combustion of coal has lead to considerable environmental problems, including land subsidence, acid waste seepage, acid precipitation and global warming.

Petroleum (or crude oil) is a natural mineral oil found underground in porous rocks. It consists predominantly of hydrogen and carbon, together with elements such as oxygen, nitrogen and sulfur, in varying proportions. The occurrence of petroleum was known in ancient times but it was not extracted commercially until the 1850s. Products of petroleum are used in huge quantities as fuels for the generation of power in vehicles and power stations and for the manufacture of plastics, fertilizers, pharmaceuticals, toiletries and detergents. *Natural gas*, a mixture of flammable gases, is often found together with petroleum. Like coal, the extraction and usage of petroleum (and its products) has brought enormous benefits and wealth to society, as well as causing enormous environmental problems.

This chapter aims to provide an understanding of the following:

- the chemistry of metal ore formation, the extraction techniques necessary to remove metals from their mineral ores, the properties and utilities of selected metals and methods of protecting metals from environmental corrosion;
- the formation and utility of fossil fuels;
- the environmental implications of mining, transporting and utilizing Earth's vital resources;
- the techniques that may be used to determine the environmental concentrations of selected metals and anthropogenic substances.

5.2 Mineral, Metal and Fuel Resources

A *resource* may be defined as a concentration of naturally occurring material (solid, liquid or gas) in or on the Earth's crust in such a form and amount that its extraction is currently or potentially feasible. Thus, resources include undiscovered, as well as discovered, material. Resources may be divided into three categories, as follows:

- *metallic resources* (see Table 5.1 for a summary of the Earth's metallic resources);
- *non-metallic resources* (includes sand, gravel, stone and sulfur);
- *energy resources* (includes coal, petroleum, natural gas and uranium).

Reserves are that part of the resource base which may be extracted economically (i.e. profitably). The Earth's resources have been steadily and increasingly used and industrialized societies are totally dependent upon their availability. Supplies of some resources (e.g. sand) will be available for indefinite periods, but other (non-renewable) resources are being used at a rate faster than they are being formed (e.g. natural gas and oil). Current estimates of the world's non-renewable energy reserves are:

- $>1 \times 10^{12}$ tonnes of coal;
- >2330 billion barrels of petroleum;
- $>140\,000$ billion m^3 of natural gas.

Table 5.1   Summary of the Earth's metallic resources

Element	Mineral	Formula	Main occurrence
Lithium	Spodumene	$LiAl(SiO_3)_2$	Found in most igneous rocks
Beryllium	Beryl	$Be_3Al_2(SiO_3)_6$	Brazil, Colombia, Siberia, USA
Boron	Borax	$Na_2B_4O_7.10H_2O$	USA, Turkey, China, Chile, Italy
Sodium	Rock salt	$NaCl$	Resources unlimited – found
	Chile	$NaNO_3$	extensively in the Earth's crust
	saltpetre	Na_3AlF_6	and in seawater
	Cryolite		
Magnesium	Magnesite	$MgCO_3$	Found extensively in the Earth's
	Dolomite	$MgCO_3.CaCO_3$	crust and in seawater
	Seawater	Mg^{2+} ions	
Aluminium	Bauxite	Al_2O_3	Australia, Guinea, Jamaica
Potassium	Carnallite	$KCl.MgCl_2.6H_2O$	Found extensively in the Earth's
	Saltpetre	KNO_3	crust and in seawater
	Sylvite	KCl	
Calcium	Limestone	$CaCO_3$	Found extensively in the Earth's
	Gypsum	$CaSO_4.2H_2O$	crust and in seawater
	Fluorspar	CaF_2	
	Fluorapatite	$CaF_2.3Ca_3(PO_4)_2$	
Scandium	Thortveitite	$(Sc,Y)_2Si_2O_7$	Scandinavia
Titanium	Ilmenite	$FeO.TiO_2$	Scandinavia, Canada, India,
	Rutile	TiO_2	Australia, Russia
Vanadium	Vanadite	$(PbO)_9(V_2O_5)_3PbCl_2$	USA, Mexico, Russia, Argentina, Scotland
Chromium	Chromite	$FeCr_2O_4$	Russia, South Africa, Turkey,
	Crocite	$PbCrO_4$	Philippines, Zimbabwe, India
Manganese	Pyrolusite	MnO_2	Germany, Russia, India, Brazil,
	Nodules		Cuba, USA, Gabon, South Africa
Iron	Haematite	Fe_2O_3	Russia, Switzerland, Italy, UK,
	Magnetite	Fe_3O_4	Brazil, China, Canada
	Siderite	$FeCO_3$	
	Pyrite	FeS_2	
Cobalt	Cobaltite	$(Co,Fe)AsS$	Russia, Poland, UK, Germany, India, Australia
Nickel	Pentlandite	$(Ni,Fe)_9S_8$	USA, Canada, Norway, Cuba,
	Millerite	NiS	Japan, Russia
Copper	Native	Cu	USA, Russia, UK, Australia, Bolivia,
	Chalcocite	CuS_2	Chile, Peru, Mexico, Zambia,
	Chalcopyrite	$CuFeS_2$	Zaire, Zimbabwe, Canada
	Bornite	Cu_5FeS_4	
	Azurite	$Cu_3(CO_3)_2(OH)_2$	
	Malachite	$Cu_2CO_3(OH)_2$	
	Chrysocolla	$CuSiO_3.2H_2O$	
Zinc	Sphalerite	ZnS	USA, Mexico, Canada, Australia,
	Smithsonite	$ZnCO_3$	China, Peru, Russia
Strontium	Celestite	$SrSO_4$	Scotland, Mexico, Spain, Turkey,
	Strontianite	$SrCO_3$	Iran, China, USA, Canada
Palladium	Native	Pa	Canada
Silver	Native	Ag	Mexico, USA, Peru, Canada, Russia
	Argentite	Ag_2S	
Tin	Tinstone	SnO_2	China, Brazil, Indonesia, Malaysia

Table 5.1 *(Continued)*

Element	Mineral	Formula	Main occurrence
Barium	Barite	$BaSO_4$	USA
	Witherite	$BaCO_3$	
Platinum	Native	Pt	Canada, South Africa
Gold	Native	Ag	South Africa, USA, Russia, China, Australia, Canada
Mercury	Cinnabar	HgS	USA, Spain, Peru, Italy, Slovenia
Lead	Galena	PbS	Russia, USA, Mexico, Australia, Serbia, Canada, Peru
Uranium	Uraninite	UO_2	Canada, USA, Australia, South Africa

These estimates may seem huge, but as our potential for energy consumption is almost limitless and many of the world's countries are underdeveloped, these resources could be depleted very rapidly.

5.3 The Chemistry of Metal Ore Formation

Before we look at how metals are extracted, it is worth considering how the original ores were formed.

5.3.1 Cooling and heating processes

Many of the processes taking place in and under the Earth's crust occur at high temperature and pressure, and as these decrease, metal ores may be formed. Mineral deposits may be formed from cooling (solidifying) magma, a process known as *magmatic concentration*. Individual minerals crystallize from magma at different temperatures, forming deposits that have characteristic physical and chemical properties. Minerals with low and high densities, respectively, relative to the molten magma will rise or sink to form a layer (band) that is enriched in the elements forming the mineral. For example, chromite $((Mg,Fe)_2CrO_4)$ deposits are formed as the dense, crystalline oxide settles in the magma to form a layer enriched in chromium. Important ore deposits of this type have been found in Canada, the United States and South Africa.

The hot, aqueous solutions associated with molten magma may also form mineral deposits by *hydrothermal processes*. Hydrothermal fluid is superheated aqueous solution that ascends through the Earth's crust. As the temperature and pressure of the fluid decreases, the solubility of its dissolved substances decreases and these may be deposited in rock cavities or substitute for materials already present. Similarly, surface or groundwater may dissolve minerals from hot magma and then deposit them as the solution cools. Such processes deposit so-called *hydrothermal minerals* such as lead, copper, mercury and zinc sulfides, for example:

$$Pb^{2+}(aq) + S^{2-}(aq) \longrightarrow PbS(s)$$

The bed of the Red Sea is particularly rich in hydrothermal minerals.

Evaporites are minerals formed from the direct evaporation of seawater. Halite (NaCl) is the most obvious example, but others include gypsum ($CaSO_4 \cdot 2H_2O$), Chile saltpetre ($NaNO_3$), potassium (e.g. sylvite, KCl) and magnesium salts (e.g. epsomite, $MgSO_4 \cdot 7H_2O$).

$$NaCl(aq) \xrightarrow{-H_2O(g)} NaCl(s)$$

5.2.2 Weathering processes

A number of weathering processes may give rise to mineral deposits. Minerals may be removed from rocks by weathering processes and then transported downstream by flowing water. These minerals may then be deposited in a particular location by sedimentation because of their size and density, while lighter impurities are dissolved or kept suspended in the flowing water and washed away. These ores, known as *placer deposits*, tend to concentrate on river bottoms and beaches, e.g. gold deposits in California, and iron-sands in New Zealand and Canada.

Minerals resistant to weathering can also be moved around by water and concentrated in specific regions by *mechanical concentration*, e.g. alluvial gold and ilmenite, $FeTiO_3$. Some mineral deposits are formed by *residual concentration* when the material remaining after weathering, and resistant to further changes, becomes an ore, e.g. bauxite deposits at Lake Superior. Bauxite deposits often form in hot, tropical countries after weathering and heavy rainfall has removed silicates and other, more soluble minerals. This type of mineral is known as a *laterite*.

5.2.3 Oxidation and reduction processes

Following surface weathering, certain ions, e.g. Cu ions, leach into the Earth's crust and produce ore enrichment, a process known as *supergene enrichment*. These replacement reactions are important for copper sulfide deposits in Arizona because of the low solubility of such sulfides:

$$Cu^{2+}(aq) + ZnS(s) \longrightarrow CuS(s) + Zn^{2+}(aq)$$

i.e. a replacement reaction, following which Zn ions are transported away.

Reduction may also occur as follows:

$$ZnS(s) \longrightarrow Zn^{2+}(aq) + S^{2-}(aq)$$

$$S^{2-}(aq) + 4H_2O \longrightarrow SO_4^{2-}(aq) + 8H^+ + 8e^-$$

$$Cu^{2+}(aq) + e^- \longrightarrow Cu^+(aq)$$

$$2Cu^+(aq) + S^{2-}(aq) \longrightarrow Cu_2S(s)$$

Redox reactions also take place during microbiological processes (such reactions are introduced in Section 1.9.2). *Heterotrophic bacteria* are dependent on organic material for carbon. They encompass a wide range of taxa, including sulfate-reducing bacteria, which produce hydrogen sulfide gas that reacts to form insoluble sulfides with metals such as Cu, Pb, Zn and As:

$$SO_4{}^{2-}(aq) + 2(CH_2O) + 2H^+(aq) \xrightarrow{\text{bacteria}} H_2S(g) + 2CO_2(g) + 2H_2O(l)$$

and then:

$$Pb^{2+}(aq) + H_2S(g) \longrightarrow PbS(s) + 2H^+(aq)$$
$$\text{Galena}$$

Autotrophic bacteria, which are not dependent on organic material for their carbon, are also important, e.g. bacteria such as *Ferrobacillus* catalyse the oxidation of Fe^{2+} to Fe^{3+} and the energy released is then utilized by the bacteria:

$$4FeCO_3 + O_2 + 6H_2O \xrightarrow{\text{bacteria}} 4Fe(OH)_3 + 4CO_2$$

5.4 The Chemistry of Metal Extraction Processes

All metals in ores occur in an oxidized state, M^{n+}, and therefore some form of reduction (i.e. gain of electrons) is required to obtain the pure metal. The reduction may occur directly on the solid metal oxide ($M^{n+}(s)$) or in the aqueous form ($M^{n+}(aq)$). This process may be represented diagrammatically by using an enthalpy diagram, as shown in Figure 5.1 (enthalpy is discussed in detail in Box 5.1). We can learn a great deal about the chemistry and energy processes involved in metal extraction processes by studying these reactions in detail.

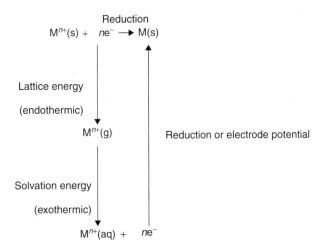

Figure 5.1 Enthalpy diagram for the extraction of a metal from its ore

> ## Box 5.1 Enthalpy
>
> The term *enthalpy* is a thermodynamic state function used in chemistry to describe the energy released or absorbed during a chemical reaction and is denoted by the letter H. The *enthalpy change*, ΔH, of a system is the heat absorbed or released by the system when the reaction occurs at constant pressure. When heat is released during a chemical reaction – an *exothermic* reaction – ΔH is negative. When heat is absorbed during a chemical reaction – an *endothermic* reaction – ΔH is positive. The *standard molar enthalpy* of a reaction, $\Delta H_{r,m}^{\ominus}$, is the enthalpy change that occurs under standard conditions per mole of a reaction as specified by a balanced chemical equation. Standard molar enthalpies can be defined for all types of chemical reactions, including formation, combustion, vaporization, fusion, solution and hydration processes.
>
> Many reaction enthalpies cannot be determined experimentally and have to be calculated from the enthalpies of other reactions. We can illustrate this process by using a simple example. The enthalpy of freezing for water is the change in enthalpy per mole when liquid water turns into solid ice. The enthalpy of freezing for water at 0°C is -6.0 kJ mol^{-1} since 6.0 kJ of energy are released when one mole of water freezes. The enthalpy of fusion for water – the change in enthalpy per mole when solid ice turns into liquid water – is $+6.0 \text{ kJ mol}^{-1}$ at 0°C (melting is an endothermic process). It is quite clear that the enthalpy of water must be the same after being frozen and then melted as it was before it was frozen. Consequently, the enthalpy of freezing for water is the negative of its enthalpy of fusion since freezing and melting are reverse processes. In fact, because enthalpy is a state property, it is generally true that the enthalpy change for a reverse reaction is the negative of the enthalpy change for a forward reaction.
>
> We can extend this example by considering the disappearance of frost on dry, cold mornings. The frost (solid ice) sublimes directly into water vapour. The same effect would be achieved if the frost first melted into liquid water before evaporating into water vapour. Consequently, we can calculate the enthalpy of sublimation of frost by adding the enthalpy change for melting and vaporization of water at the same temperature, as follows:
>
> $$\Delta H_{\text{Sublimation}} = \Delta H_{\text{Fusion}} + \Delta H_{\text{Vaporisation}}$$
>
> This is an example of *Hess's Law* of constant heat summation, which states that the enthalpy change for an overall reaction is the sum of the reaction enthalpies for each of the steps into which the reaction can be divided.

First, let us consider reducing the solid metal oxide. The lattice enthalpy (or energy) is the heat required to vaporize the solid ore to a gas of ions. Conversely, a heat equal to the lattice enthalpy is released when the gaseous ions form a solid ore. The lattice enthalpy provides a measure of the strength of ionic bonds. The *standard molar lattice enthalpy* (symbol $\Delta H^{\ominus}$) is the enthalpy change that accompanies the formation of one mole of the solid ionic compound from its gaseous ions under standard conditions. Lattice enthalpies cannot be determined directly, but can be calculated from experimental data by using Hess's Law, as illustrated in Example 5.1.

Example 5.1

Let us consider the formation of magnesium oxide from its elements in their standard states. The overall reaction is exothermic:

$$Mg(s) + \tfrac{1}{2}O_2(g) \longrightarrow Mg^{2+}O^{2-}(s) \qquad \Delta H_{r,m}^{\ominus} = -604 \text{ kJ mol}^{-1} \qquad (5.1)$$

Equation 5.1 represents the reverse of the extraction process, which must therefore be endothermic. In theory, this reaction can be broken down into a series of five steps.

Step 1 The sublimation (atomization) of magnesium:

$$Mg(s) \longrightarrow Mg(g) \qquad\qquad \Delta H_{s1,m}^{\ominus} = +153 \text{ kJ mol}^{-1}$$

Step 2 The atomization of oxygen:

$$\tfrac{1}{2}O_2(g) \longrightarrow O(g) \qquad\qquad \Delta H_{s2,m}^{\ominus} = +248 \text{ kJ mol}^{-1}$$

Step 3 The ionization of gaseous magnesium:

$$Mg(g) \longrightarrow Mg^{2+}(g) + 2e^- \qquad \Delta H_{s3,m}^{\ominus} = +2180 \text{ kJ mol}^{-1}$$

Step 4 The ionization of oxygen:

$$O(g) + 2e^- \longrightarrow O^{2-}(g) \qquad\qquad \Delta H_{s4,m}^{\ominus} = +745 \text{ kJ mol}^{-1}$$

Step 5 The formation of magnesium oxide from its gaseous ions:

$$Mg^{2+}(g) + O^{2-}(g) \longrightarrow Mg^{2+}O^{2-}(s) \qquad \Delta H_{s5,m}^{\ominus} = ? \text{ kJ mol}^{-1}$$

$\Delta H_{s5,m}^{\ominus}$ is the standard molar lattice enthalpy of magnesium oxide. We can determine this value by applying Hess's Law, since:

$$\Delta H_{r,m}^{\ominus} = \Delta H_{s1,m}^{\ominus} + \Delta H_{s2,m}^{\ominus} + \Delta H_{s3,m}^{\ominus} + \Delta H_{s4,m}^{\ominus} + \Delta H_{s5,m}^{\ominus}$$
$$-604 = 153 + 248 + 2180 + 745 + \Delta H_{s5,m}^{\ominus}$$

Thus, $\Delta H_{s5,m}^{\ominus} = -3930 \text{ kJ mol}^{-1}$.

The information used in Example 5.1 can be summarized in an enthalpy diagram known as a *Born–Haber Cycle*, as shown in Figure 5.2. The reason why the formation of the ionic lattice of magnesium oxide is strongly exothermic is that both the ions, Mg^{2+} and O^{2-}, are very small and both carry two units of charge. Consequently, when these two opposite charges are brought close together, there is a big release of energy. It is clear from this example that large amounts of energy would be required to obtain pure magnesium by reducing magnesium oxide.

Now let us consider extracting a metal from its ore by dissolving it in aqueous solution. When one mole of an ionic solute is dissolved in a solvent to form an infinitely dilute solution, the enthalpy change that occurs is called the *enthalpy of solution* (or the solvation energy), $\Delta H_{solution,m}^{\ominus}$. In aqueous solution, this process can be broken down into two steps. First, energy is required to separate the ions (bond

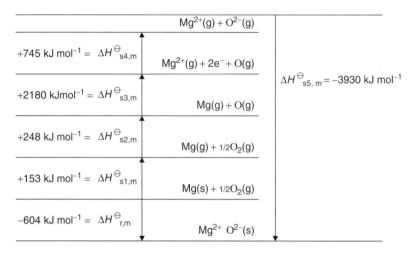

Figure 5.2   Born–Haber cycle for magnesium oxide

breaking), so that the ionic lattice breaks down. The breaking down of the lattice is obviously the reverse of its formation, i.e. $-\Delta H^{\ominus}_{\text{lattice,m}}$. Secondly, the separated ions enter the solvent – they become solvated. When the solvent is water, the second step is called *hydration*. The water molecules will form hydrogen bonds with anions and the negative partial charge on the oxygen atoms will attract cations. Such powerful interactions always result in exothermic reactions when ionic compounds are hydrated. Values for hydration enthalpies cannot be found experimentally, but may be determined by applying Hess's Law, as follows:

$$\Delta H^{\ominus}_{\text{solution,m}} = -\Delta H^{\ominus}_{\text{lattice,m}} + \Delta H^{\ominus}_{\text{hydration,m}} \tag{5.2}$$

and hence:

$$\Delta H^{\ominus}_{\text{hydration,m}} = \Delta H^{\ominus}_{\text{lattice,m}} + \Delta H^{\ominus}_{\text{solution,m}}$$

The breaking of the lattice and the hydration of the ions occur simultaneously.

Example 5.2

Consider the dissolution of sodium chloride at 298.15 K:

$$NaCl(s) \longrightarrow Na^+(aq) + Cl^-(aq)$$

In theory, this reaction can be broken down into two steps.

Step 1   The dissociation of solid sodium chloride into its gaseous ions:

$$NaCl(s) \longrightarrow Na^+(g) + Cl^-(g) \qquad \Delta H^{\ominus}_{\text{lattice,m}} = +788 \text{ kJ mol}^{-1}$$

Step 2   The hydration of the dissociated ions:

$$Na^+(g) + Cl^-(g) \longrightarrow Na^+(aq) + Cl^-(aq) \qquad \Delta H^{\ominus}_{\text{hydration,m}} = +784 \text{ kJ mol}^{-1}$$

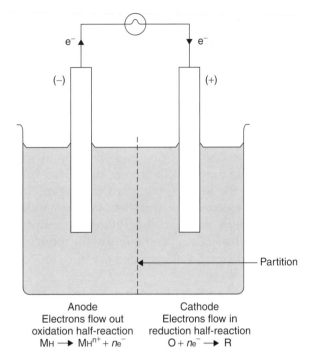

Figure 5.3 Schematic diagram of an electrochemical cell; an electromotive force is generated by the cell

From Equation 5.2:

$$\Delta H^{\ominus}_{\text{solution,m}} = -788 + 784 = -4 \text{ kJ mol}^{-1}$$

Thus the overall reaction is (just) exothermic.

The lattice energy and the solvation energy are linked by the value of the *electrode potential* (see Figure 5.1), which therefore provides a useful indication of the ease of reduction of metal ores. The electrode potential of a system may be determined by using an *electrochemical cell* – a device for converting chemical energy into electrical energy. An electrochemical cell consists of two half-cells, as shown in Figure 5.3. In one half-cell, an oxidation half-reaction takes place, while in the other, a reduction half-reaction occurs.

Consider an electrode made of a metal, M, which dips into a solution containing ions of the same metal, M^{n+}. Metal atoms from the solid electrode will pass into solution to form hydrated ions, leaving the electrons on the electrode. This process, the oxidation process, results in a negative charge on the electrode, as follows:

$$M(s) \longrightarrow M^{n+}(aq) + ne^- \qquad (5.3)$$

However, hydrated metal ions in solution will simultaneously gain electrons from the electrode to form metal atoms that immediately become part of the metal electrode. This process, the reduction process, results in a positive charge on the electrode, as follows:

$$M^{n+}(aq) + ne^- \longrightarrow M(s) \qquad (5.4)$$

Eventually, equilibrium is established:

$$M^{n+}(aq) + ne^- \rightleftharpoons M(s) \qquad (5.5)$$

The overall charge on the electrode depends upon which of the two processes occurs most readily, and thus on the position of equilibrium. If the equilibrium lies to the right of Equation 5.5, reduction predominates and the electrode acquires a positive charge; if the equilibrium lies to the left, oxidation predominates and the electrode acquires a negative charge. Either way, the result is a potential difference between the electrode and the ions in solution. The *standard electrode potential, $E^\ominus$*, is the potential difference measured in an electrochemical cell when the electrode on the left is a standard hydrogen electrode and the electrode on the right is the electrode in question under standard conditions (i.e., 298 K, 1 atmosphere pressure and a 1 M solution concentration). The standard electrode potential is written as $E^\ominus_{M^{n+}/M}$ and refers to the reduction reaction taking place at the electrode, i.e. Equation 5.4. Electrode potentials for half-cells are often called redox potentials.

The $E^\ominus$ values for a series of metals are listed in Table 5.2. When metals are placed in order of their standard electrode potentials, the *electrochemical series of metals* is obtained. This series closely approximates to the order of reactivity of metals. Those metals with the most negative $E^\ominus$ values (i.e. Li, K, Na, Ca) are the most difficult ions to reduce. In general, metal ions with $E^\ominus <= -1.66$ V require electrolytic reduction to remove them from their ores, whereas metal ions with less negative potentials

Table 5.2 Standard electrode potentials for selected elements at 25°C

	Half-reaction Oxidized species	+	ze^-	⇌	Reduced species	$E^\ominus$, (V)	
↑	$Li^+(aq)$	+	e^-	⇌	$Li(s)$	−3.05	
	$K^+(aq)$	+	e^-	⇌	$K(s)$	−2.93	
	$Ca^{2+}(aq)$	+	$2e^-$	⇌	$Ca(s)$	−2.87	
	$Na^+(aq)$	+	e^-	⇌	$Na(s)$	−2.71	
	$Mg^{2+}(aq)$	+	$2e^-$	⇌	$Mg(s)$	−2.36	
Increasing:	$Al^{3+}(aq)$	+	$3e^-$	⇌	$Al(s)$	−1.66	Increasing:
(1) tendency for	$Zn^{2+}(aq)$	+	$2e^-$	⇌	$Zn(s)$	−0.76	(1) tendency for
reverse reaction	$Fe^{2+}(aq)$	+	$2e^-$	⇌	$Fe(s)$	−0.44	forward reaction
(oxidation) to	$Sn^{2+}(aq)$	+	$2e^-$	⇌	$Sn(s)$	−0.14	(reduction) to
occur	$Pb^{2+}(aq)$	+	$2e^-$	⇌	$Pb(s)$	−0.13	occur
(2) tendency to lose	$Fe^{3+}(aq)$	+	$3e^-$	⇌	$Fe(s)$	−0.04	(2) tendency to gain
electrons	$2H^+(aq)$	+	$2e^-$	⇌	$H_2(g)$	0.00[a]	electrons
(3) power as a	$Cu^{2+}(aq)$	+	$2e^-$	⇌	$Cu(s)$	+0.34	(3) power as an
reducing agent	$I_2(g)$	+	$2e^-$	⇌	$2I^-(aq)$	+0.54	oxidizing agent
	$Fe^{3+}(aq)$	+	e^-	⇌	$Fe^{2+}(aq)$	+0.77	
	$Ag^+(aq)$	+	e^-	⇌	$Ag(s)$	+0.80	
	$Br_2(l)$	+	$2e^-$	⇌	$2Br^-(aq)$	+1.09	
	$Cl_2(l)$	+	$2e^-$	⇌	$2Cl^-(aq)$	+1.36	
	$Au^+(aq)$	+	e^-	⇌	$Au(s)$	+1.69	
	$F_2(g)$	+	$2e^-$	⇌	$2F^-(aq)$	+2.87	↓

[a] By definition.

require chemical reducing agents such as carbon, dihydrogen gas and aluminium. Metal ions having positive $E^{\ominus}$ values, (e.g. Au and Ag), may occur free in nature. However, they can also be obtained by heating the naturally occurring ore *without* the presence of an added reducing agent, for example:

$$2HgS(s) + 3O_2(g) \longrightarrow 2HgO(s) + 2SO_2(g)$$

$$2HgO(s) \xrightarrow{500°\,C} 2Hg(l) + O_2(g)\,(\text{i.e. } Hg^{2+} + 2e^- \longrightarrow Hg(l))$$

Metals high in the electrochemical series of metals will reduce the oxidized forms of metals lower in the series from their oxides or solutions of their salts, for example:

$$Mg(s) + Cu^{2+}(aq) \longrightarrow Mg^{2+}(aq) + Cu(s)$$

5.4.1　Chemical reduction

Before reduction processes occur, metal ores often have to be crushed and sorted to a specific size. Magnetite (Fe_3O_4) can be concentrated magnetically. Froth floatation is used to separate many minerals; a basic description of this process is that ores attach themselves to air bubbles produced in a bath of water and then float off in the froth. Once they have been washed and concentrated, metals may be extracted from their ores by chemical reduction. Reducing agents used for this purpose include carbon or other metals. When the metal ore occurs as a metal oxide, reduction can occur directly, but ores existing as sulfides or carbonates must first be converted to the oxides by roasting in air:

$$2MS(s) + 3O_2(g) \longrightarrow 2MO(s) + 2SO_2(g)$$

$$4FeCO_3(s) + O_2(g) \longrightarrow 2Fe_2O_3(s) + 4CO_2(g)$$

It is important to understand the chemistry of reduction reactions in order to identify a suitable reducing agent for an extraction process. Consider the following reactions:

R1　　　metal + oxide of carbon $\longrightarrow$ metal oxide + carbon

In R1, the metal acts as the reducing agent.

R2　　　metal oxide + carbon $\longrightarrow$ metal + oxide of carbon

Here the carbon acts as the reducing agent, so reducing the metal oxide.

It is clear that R2 is the reverse of R1. For some metals R1 applies, while for others R2 applies. How can we predict which reaction will apply for a given metal? The answer is to use *thermodynamics*.

The *Gibbs Free Energy* (G) of a compound is a measure of its stability. The free energy change of a reaction (ΔG) is a measure of the change in the total entropy of a system and its surroundings at constant pressure. It is a measure of the spontaneity (or feasibility) of a reaction since only reactions with negative ΔG values (i.e. $\Delta G < 0$) can occur spontaneously.

Let us consider the chemical reduction reactions occurring in Case 2 reaction (R2). The individual reactions are:

$$MO \longrightarrow M + \tfrac{1}{2}O_2 \qquad \Delta G_1^\circ$$

$$\tfrac{1}{2}O_2 + R \longrightarrow RO \quad \Delta G_2^\circ$$

The overall reaction equation is:

$$MO + R \longrightarrow M + RO \qquad \Delta G_{RO}^\circ = \Delta G_1^\circ + \Delta G_2^\circ$$

$$\uparrow$$

Standard free energy change for the formation of RO

For this reaction to proceed spontaneously (i.e. for $\Delta G_{RO}^\circ < 0$), the free energy change during the formation of the product oxide (RO) must be more negative than the free energy change during the formation of the reactant oxide (MO) (i.e. $\Delta G_{RO}^\circ < \Delta G_{MO}^\circ$).

An example of a Case 2 reaction could be the proposed reduction of CaO by C. Calcium oxide is formed by a spontaneous reaction:

$$2Ca(s) + O_2(g) \longrightarrow 2CaO(s) \qquad \Delta G_{CaO}^\circ = -1208.4 \text{ kJ mol}^{-1}$$

Thus, for the reverse reaction:

$$2CaO(s) \longrightarrow 2Ca(s) + O_2(g) \qquad \Delta G_{Ca}^\circ = +1208.4 \text{ kJ mol}^{-1}$$

Carbon would be oxidized during the reaction of calcium oxide:

$$C(s) + O_2(g) \longrightarrow CO_2(g) \qquad \Delta G_{CO_2}^\circ = -394.5 \text{ kJ mol}^{-1}$$

Hence, for the proposed reduction of CaO by C, the overall equation is:

$$2CaO(s) + C(s) \longrightarrow 2Ca(s) + CO_2(g) \qquad \Delta G_r^\circ = \Delta G_{Ca}^\circ + \Delta G_{CO_2}^\circ$$

$$\Delta G_r^\circ = 1208.4 - 394.5$$

$$\Delta G_r^\circ = +813.9 \text{ kJ mol}^{-1}$$

This clearly represents a very unfavourable reaction because of the large positive ΔG_r°. Alternatively, we can say that the free energy of the product oxide $(\Delta G_{CO_2}^\circ)$ must be more negative than the free energy of the reactant oxide (ΔG_{CaO}°) for the reaction to occur spontaneously. This is clearly not the case (i.e. -394.5 kJ mol^{-1} is more positive than -1208.4 kJ mol^{-1}); calcium oxide will not be reduced by carbon since calcium oxide is more stable than carbon dioxide.

Example 5.3

Let us consider the reduction of haematite, Fe_2O_3, by carbon at 25°C:

$$Fe_2O_3(s) + 3C(s) \longrightarrow 2Fe(s) + 3CO(g) \quad \Delta G^\circ = +330 \text{ kJ mol}^{-1}$$

This is clearly an unfavourable reaction at 25°C since $\Delta G°$ is positive. However, what will happen at higher temperatures, i.e. how does ΔG vary with increasing temperatures? We need to consider the following relationship:

$$\Delta G = \Delta H - T\Delta S \tag{5.6}$$

where ΔG is the free energy change, ΔH is the enthalpy change, T is the temperature and ΔS is the entropy change.

Changes in enthalpy and entropy can be calculated over a wide range of temperatures, and hence it is possible to calculate free energy changes over a similar temperature range.

We can use Equation 5.6 to demonstrate the factors that govern the direction of a spontaneous change at constant pressure by determining the values of ΔH, T and ΔS that result in a negative value for ΔG. These factors are summarized in Table 5.3. Figure 5.4 illustrates the variation of $\Delta G°$, $\Delta H°$ and $T\Delta S°$ with temperature for Example 5.3.

The variation of $\Delta G°$, $\Delta H°$ and $T\Delta S°$ with temperature for the formation of CO_2 from carbon and oxygen gas is shown in Figure 5.4(a). There is very little change of entropy in this reaction since we start and finish with one mole of gas. Consequently, from Equation 5.6, we can see that there is very little change in $\Delta G°$ and $\Delta H°$.

$$C(s) + O_2(g) \longrightarrow CO_2(g)$$

$$1\,\text{mole} \longrightarrow 1 \text{ mole}$$

Figure 5.4(b) shows the situation if carbon is the reducing agent and is oxidized to carbon monoxide, CO:

$$2C(s) + O_2(g) \longrightarrow 2CO(g)$$

$$1 \text{ mole} \longrightarrow 2 \text{ moles}$$

In this case, the entropy increases (ΔS is positive) since we start with one mole of gas and finish with two. This results in a large reduction in $\Delta G°$ as the temperature increases.

The formation of ferric oxide by the oxidation of iron is shown in Figure 5.4(c). You can see that the entropy decreases (ΔS is negative) since no gaseous molecules are produced. As $\Delta H°$ remains relatively constant, $\Delta G°$ increases as the temperature

Table 5.3 Factors that govern the occurrence of a spontaneous change

Enthalpy change	Entropy change	Outcome of reaction[a]
Exothermic ($\Delta H < 0$)	Increase ($\Delta S > 0$)	Spontaneous reaction occurs, as $\Delta G < 0$
Exothermic ($\Delta H < 0$)	Decrease ($\Delta S < 0$)	Spontaneous reaction occurs if $|T\Delta S| < |\Delta H|$, as $\Delta G < 0$
Endothermic ($\Delta H > 0$)	Increase ($\Delta S > 0$)	Spontaneous reaction occurs if $T\Delta S > \Delta H$, as $\Delta G < 0$
Endothermic ($\Delta H > 0$)	Decrease ($\Delta S < 0$)	Spontaneous reaction does not occur, as $\Delta G > 0$

[a]The symbol $|x|$ represents the absolute value of x, its value disregarding its sign – this is known as the magnitude.

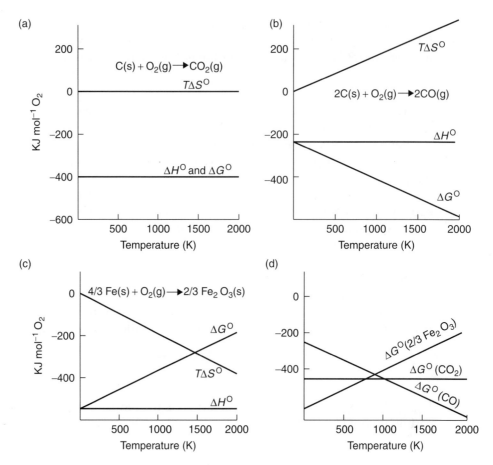

Figure 5.4 The variation of $\Delta G°$, $\Delta H°$ and $T\Delta S°$ with temperature for: (a) CO_2; (b) CO; (c) $2/3Fe_2O_3$; (d) combined reaction (see Example 5.3)

increases.

$$\tfrac{4}{3}Fe(s) + O_2(g) \longrightarrow \tfrac{2}{3}Fe_2O_3(s)$$

$$1 \text{ mole}$$

Figure 5.4(d) shows all three $\Delta G°$ lines plotted on the same graph and it can be seen that there is a temperature at which the free energy of formation of CO, $\Delta G°(CO)$, becomes more negative than the free energy of formation of Fe_2O_3, $\Delta G°(\tfrac{2}{3}Fe_2O_3)$. This is at approximately 900 K and therefore at this temperature the overall free energy for the reaction:

$$\tfrac{2}{3}Fe_2O_3(s) + 2C(s) \longrightarrow \tfrac{4}{3}Fe(s) + 2CO(g)$$

becomes negative and thus the reduction process will occur. The alternative way to consider this relationship is that above 900 K, CO(g) becomes more stable than $Fe_2O_3(s)$ and therefore C will reduce the metal oxide.

The behaviour for Fe_2O_3 formation is typical of all metal oxide formations in that the free energies become less negative because of the large negative entropy

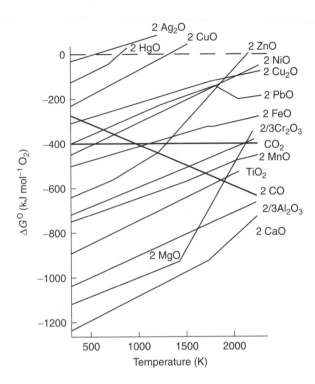

Figure 5.5 Ellingham diagram for oxide formation

change as a result of the loss of a mole of oxygen. A range of these lines for different metal oxides is plotted in Figure 5.5 – an *Ellingham Diagram*. The latter is simply a graph showing the variation of standard molar free energy of formation (ΔG_f°) with temperature (T). The discontinuities on the graph are due to phase changes – the melting points and boiling points of metals and/or oxides. These result in increases in entropy and a more negative value for ΔG_f°.

In an Ellingham diagram, the higher the lines, then the less stable the oxide. Thus, carbon monoxide is less stable than nickel (II) oxide at temperatures below 680 K whereas at higher temperatures it is more stable. Consequently, nickel (II) oxide will be reduced by carbon monoxide at temperatures above 680 K. In principle, any oxide whose Ellingham plot is lower than that of another metal oxide will reduce it, e.g. the aluminium oxide line is below the chromium oxide line at all temperatures. This means that the value of ΔG_r° for the reaction:

$$2Al(s) + Cr_2O_3(s) \longrightarrow 2Cr(s) + Al_2O_3(s)$$

is always negative over the temperature range shown (i.e. $\Delta G_f^\circ(Al_2O_3(s))$) is always lower than $\Delta G_f^\circ(Cr_2O_3(s))$). Aluminium will therefore reduce chromium oxide at any temperature within this range. At 1000 K, ΔG° for this reaction is −530 kJ; this temperature is required in order to overcome an energy barrier, the *activation energy*. The latter is the energy which colliding molecules must possess before a collision will result in a reaction.

Some metal oxides, e.g. silver and mercury, have lines which start close to $\Delta G^\circ = 0$, and which cross it at quite low temperatures. Silver (I) oxide (Ag_2O) becomes unstable at about 500 K and therefore by heating to this temperature the oxide will decompose into the metal without requiring the presence of a reducing agent.

The utility of carbon as a reducing agent is clearly illustrated by the data shown in Figure 5.5, as at some specific temperature it is able to reduce most metal oxides, particularly when carbon is oxidized to carbon monoxide. Hence, carbon would reduce zinc oxide to zinc at 1275 K. Note that this is above the boiling point of zinc metal, which therefore has to be condensed from the vapour. In theory, carbon would reduce aluminium oxide above 2300 K, but this temperature is so high that it is impractical to construct a smelter to withstand the heat. Consequently, electrolytic reduction is used for the extraction of aluminium from its oxide. A similar situation applies to the reduction of calcium oxide.

The utility of hydrogen (H_2) as a reducing agent can also be considered via Ellingham Diagrams:

$$H_2(g) + O_2(g) \longrightarrow 2H_2O(l)$$

1 mole 1 mole

The entropy decreases during this reaction (ΔS is negative) since no gaseous molecules are produced; $\Delta G°$ runs parallel to many metal oxide lines. Only metal oxides whose formation lines lie above the H_2 line will be reduced. H_2 may remain dissolved within the produced metal, thus affecting its properties.

5.5 Extraction Technology for Selected Metals

5.5.1 Extraction of s-block metals

Metals in Groups I and II of the Periodic Table are known as the s-block metals because they all have one or two electrons in their outer shells. The s-block metals are too reactive to occur free in nature – both Group 1 and Group II metals have high electrode potentials and are powerful reducing agents (see Table 5.2). Indeed, Group I metals are the strongest reducing agents known and consequently cannot be extracted by reduction of their oxides or electrolysis of an aqueous solution. They can only be extracted commercially by the high-temperature electrolysis of their fused halides, although potassium is extracted by exposing molten potassium chloride to sodium vapour:

$$KCl(l) + Na(g) \xrightarrow{750°C} NaCl(s) + K(g)$$

The potassium vapour is collected by condensation in a cooled collecting tank. Group II metals are also strong reducing agents and are either extracted by electrolysis of their fused halides or by high-temperature thermal reduction using aluminium or silicon as the reducing agent.

Extraction of sodium

Sodium occurs extensively in the Earth's crust in a variety of minerals, including rock salt (NaCl), Chile saltpetre ($NaNO_3$) and cryolite (Na_3AlF_6). Both sodium

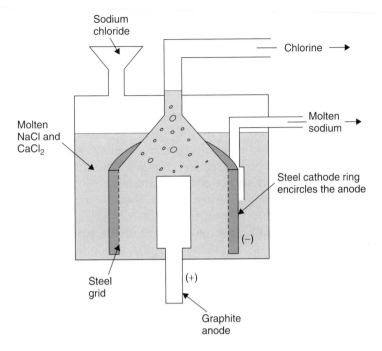

Figure 5.6 Schematic diagram of a downs cell

and chlorine are extracted commercially from fused (molten) sodium chloride by *electrolysis* using the *Downs Cell*. (The chemistry of electrochemical reactions is discussed in Section 5.6.1). The sodium chloride is obtained from conventional rock salt mining, solution mining and solar evaporation of seawater.

A schematic diagram of the Downs Cell is shown in Figure 5.6. This consists of a central graphite anode surrounded by a circular steel cathode in a cylindrical steel cell. Sodium chloride melts at 800°C, which causes two serious problems for the extraction process; the steel cell corrodes rapidly at this temperature and some of the sodium produced by the electrolysis dissolves in the molten sodium chloride, so halting electrolytic decomposition. The temperature in the cell has to be reduced, and this is achieved by using an electrolyte containing 60% calcium chloride and 40% sodium chloride. This electrolyte reduces the melting point to approximately 600°C.

A typical Downs Cell operates at a high current (30 000 A) and a low voltage (7 V). At the cathode, the molten sodium ions gain electrons to form molten metal, while at the anode, chlorine gas is formed:

Cathodic reaction $\qquad Na^+(l) + e^- \longrightarrow Na(s)$

Anodic reaction $\qquad 2Cl^-(l) \longrightarrow Cl_2(g) + 2e^-$

The chlorine gas rises to the surface of the electrolyte where it is collected in a hood and the liquid sodium is drawn off from the cell by using a trap. A steel gauze to prevent them from reacting together separates the pure products.

Sodium is a highly reactive, soft, wax-like metal that has a very low density and is silver-white in colour. The metal has a range of uses – in street lighting, as a coolant in nuclear reactors, in metallurgical and manufacturing processes and in spectroscopy.

Extraction of magnesium

Like sodium, magnesium occurs extensively in the Earth's crust in a variety of minerals, including magnesite ($MgCO_3$), dolomite ($MgCa(CO_3)_2$), magnesium chloride ($MgCl_2$) and in a variety of magnesium silicates (e.g. forsterite, Mg_2SiO_4). Magnesium may be extracted commercially from its ores by two methods, depending upon the source of raw material.

Dolomite and magnesite are used as the raw materials for the thermal reduction of magnesium oxide during the *Pidgeon Process*, named after Lloyd M. Pidgeon, the Canadian metallurgist who invented the process. Initially, the minerals are calcinated by heating. A calcinated mineral is one from which carbon dioxide has been removed by heating:

$$MgCO_3(s) \xrightarrow{\text{heat}} MgO(s) + CO_2(g)$$

Magnesite

$$MgCa(CO_3)_2(s) \xrightarrow{\text{heat}} MgO.CaO(s) + 2CO_2(g)$$

Dolomite

The calcinated minerals are then heated with an alloy of iron and silicon. The silicon reduces the magnesium oxide to magnesium vapour that can be trapped and removed. The reaction also produces silicon oxide, which reacts to remove the calcium oxide from dolomite as a calcium silicate slag.

$$2MgO(s) + Si(s) \longrightarrow SiO_2(s) + 2Mg(g)$$

$$CaO(s) + SiO_2(s) \longrightarrow CaSiO_3(l)$$

Magnesium is also extracted from seawater or a slurry of calcinated dolomite in seawater. The extraction is a three-stage process; initially, the magnesium is precipitated as magnesium hydroxide. Precipitation is achieved in seawater by adding hydrated lime ($Ca(OH)_2$); if a slurry is used, calcium hydroxide is already present due to the reaction between calcium oxide (from dolomite) and water:

$$Mg^{2+}(aq) + Ca(OH)_2(s) \longrightarrow Mg(OH)_2(s) + Ca^{2+}(aq)$$

The precipitated magnesium hydroxide is converted (using hydrochloric acid, followed by evaporation in a furnace) to hydrated magnesium chloride, $MgCl_2.6H_2O$, which is then dehydrated to anhydrous magnesium chloride, $MgCl_2$:

$$Mg(OH)_2(s) + 2HCl(aq) \longrightarrow MgCl_2(aq) + 2H_2O(l)$$

Molten magnesium chloride is then electrolysed in a cylindrical cell using carbon anodes and steel cathodes. The electrolyte is actually a fused mixture containing

25% $MgCl_2$, 15% $CaCl_2$ and 60% NaCl. The additional salts are used to lower the melting point and to increase the density of the electrolyte so that the molten magnesium can float to the surface of the electrolyte for removal.

Cathodic reaction　　$Mg^{2+}(l) + 2e^- \longrightarrow Mg(l)$

Anodic reaction　　　$2Cl^-(l) \longrightarrow Cl_2(g) + 2e^-$

Magnesium is a silvery-white metal, which is very lightweight, ductile and malleable. It is widely used in alloys and flash photography.

5.5.2　Extraction of d-block elements

The d-block elements are characterized by inner building of d-sub-shells – the s-orbital in the outer shell is usually filled before the d-orbitals in the next to outermost shell. The d-block metals are also known as the main transition metals, with the exception of zinc, which has a filled d-shell. The most important sources of d-block elements are listed in Table 5.1.

Extraction of iron

Iron is the second most abundant metal in the Earth's crust, and commonly occurs with calcium, silicon, oxygen, phosphorous and sulfur. Its most important ores include *haematite* (Fe_2O_3), *magnetite* (Fe_3O_4), *ilmenite* ($FeTiO_4$), *chromite* ($FeCr_2O_4$) and *siderite* ($FeCO_3$). *Pyrite*, FeS_2, is also widely available, but is not used as a raw material in the extraction of iron since the sulfur is difficult to remove. Large quantities of the world's iron are extracted in Russia, Kazakhstan and the Ukraine, while other important producers include Australia, Brazil, Canada, France and the USA. Iron ore is typically drilled and blasted in open pits. High-grade ore (iron content 50–70%) is crushed and sieved to produce lump ore (6.3–31.5 mm) and fine ore (< 6.3 mm). Low-grade ore is upgraded in a concentrator.

　　Iron is extracted by reduction of its ores in a blast furnace (Figure 5.7). Part of the 70 m high blast furnace is a tall (30 m) steel cylinder standing on a base (hearth). Cylinders are lined with heat-resistant bricks that eventually require replacement. The whole process (known as a campaign) goes on continuously for ten years or more. This is because if the process is stopped and the furnace is allowed to cool, damage could be caused to the furnace lining by the cracking of the refractory bricks as they cool.

　　Iron manufacture is essentially a counter-current process, in which the charge (iron ore, coke and limestone) is added at the top of the furnace through a pair of valves and pre-heated air (600–700°C) is forced through blowpipes (tuyeres) at the bottom of the furnace. Coke is produced from a mixture of coals that are crushed and ground into a powder. The powdered coal is heated in an oven for 18–24 hours to remove volatile matter before being cooled and screened into 2–10 cm pieces which typically contain 90–93% carbon.

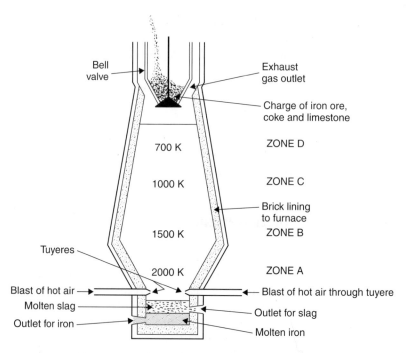

Figure 5.7 Schematic diagram of a blast furnace

The process is continuous and a complex series of reactions take place. Different reactions occur in different temperature zones in the furnace. In the lower section of the furnace, Zone A, the blast of air passing up the furnace reacts with some of the carbon in the coke in a very exothermic reaction, forming CO_2, and thus heating up the charge.

Zone A $C(s) + O_2(g) \longrightarrow CO_2(g)$ $\Delta H^\ominus_{r,m} = -392 \text{ kJ mol}^{-1}$

As the carbon dioxide passes up the furnace (into Zone B), it is reduced to carbon monoxide (CO). This reaction is endothermic, lowering the furnace temperature to 1300°C.

Zone B $C(s) + CO_2(g) \longrightarrow 2CO(g)$ $\Delta H^\ominus_{r,m} = +172 \text{ kJ mol}^{-1}$

The reduction of the iron ore takes place in stages. The CO produced by the above reaction rises up the furnace to Zone C, where it acts as a reducing agent for the iron ore.

Zone C $Fe_2O_3(s) + 3CO(g) \longrightarrow 2Fe(l) + 3CO_2(g)$

          $Fe_3O_4(s) + 4CO \longrightarrow 3Fe(l) + 4CO_2(g)$

The excess O in higher oxides, such as Fe_2O_3 and Fe_3O_4, is also removed in a number of reactions with CO. This occurs in the cooler (upper) regions of the furnace (Zone D) because of the lesser stability of Fe(III) towards reduction. The ores are partially reduced to FeO, which is reduced to iron lower in the furnace.

Zone D $\quad 3Fe_2O_3(s) + CO(g) \longrightarrow 2Fe_3O_4(s) + CO_2(g)$

$\qquad\qquad Fe_2O_3(s) + CO(g) \longrightarrow 2FeO(s) + CO_2(g)$

$\qquad\qquad Fe_3O_4(s) + CO(g) \longrightarrow 3FeO(s) + CO_2(g)$

Zone C $\quad FeO(s) + CO(g) \longrightarrow Fe(l) + CO_2(g)$

At the higher temperatures lower in the furnace, direct reduction by C occurs, since at temperatures above 1000 K, CO is more stable than CO_2.

Zone B $\quad FeO(s) + C(s) \longrightarrow Fe(l) + CO(g)$

Crushed and ground limestone is added to the charge in order to remove impurities. At the top of the furnace (Zone D), the limestone is thermally decomposed to lime (calcium oxide) and carbon dioxide.

Zone D $\quad CaCO_3(s) \longrightarrow CaO(s) + CO_2(g)$

This process serves to lower the melting points of the silica (SiO_2), alumina (Al_2O_3) and phosphate impurities present in the iron ore, which helps to reduce the consumption (and therefore the cost) of the coke fuel. The basic character of CaO serves to neutralize these acidic impurities as a mixture of products known as slag.

Zones A and B $\quad CaO(s) + SiO_2(s) \longrightarrow CaSiO_3(l)$

$\qquad\qquad\quad CaO(s) + Al_2O_3(s) \longrightarrow Ca(AlO_2)_2(l)$

$\qquad\qquad\quad 6CaO(s) + P_4O_{10}(s) \longrightarrow 2Ca_3(PO_4)_2(l)$

The raw materials require 6–8 hours to descend to the bottom of the furnace where they become final products. The molten iron and a layer of floating slag collects in the hearth at the bottom of the furnace. The slag melts in the hot part of the stack, flows rapidly down the furnace and floats on the molten iron, thus preventing oxidation of the latter by the air blast. The slag is drawn off as necessary and is used to make building materials for the construction industry. Other materials in the ore undergo various reactions in the blast furnace, as follows:

- Above 700°C, Fe_3C forms, which dissolves in the iron.
- Above 1200°C, SiO_2 is reduced to Si, which also dissolves in the iron.
- Manganese oxides reduce to the pure metal, which then alloys with the iron.
- Sulfur, mainly from the coke, reacts with iron to produce FeS, which is then removed from the iron:

$$FeS(s) + CaO(s) + C(s) \longrightarrow Fe(l) + CO(g) + CaS(l)(\text{appears in the slag})$$

Typically, the molten iron is 'tapped off' about four times a day and transferred to a storage furnace where it is kept molten prior to transfer to the steel-making process. In smaller, less modern furnaces, molten iron may be solidified into 'pig' or cast iron by cooling in sand moulds. The so-called *pig iron* produced is hard, but brittle because the impurities cause defects in the crystal lattice. A typical composition for pig iron is 93.5–95.0% iron, 4.1–4.4% carbon, 0.55–0.75% manganese, 0.3–0.9% silicon, 0.03–0.09% phosphorus and 0.025–0.050% sulfur.

Cast iron is pig iron that has been melted with scrap steel and cooled in a mould of some definite shape. Heating the pig iron under a current of air to remove most of the impurities forms material that is used to manufacture *wrought iron*. Pig iron is converted into *steel* by oxidation of these impurities in preference to iron.

The production of each tonne of iron requires 2–3 tonnes of ore, about 1 tonne of coke and nearly 6 tonnes of air. The process is quite efficient as only about 2–3% of the original iron is lost during this process.

Steel making

Steel is an alloy of iron with carbon and other elements, with the amounts being strictly controlled to give the required properties. The carbon content of pig iron is reduced and other impurities are removed by oxidation using techniques such as the oxygen furnace (also known as the basic oxygen process or the Linz–Donawitz process), the electric arc furnace (EAF) and the open hearth furnace.

The oxygen furnace is an efficient process for steel manufacture, using pure O_2 as the oxidizing agent. Modern furnaces can take up to 350 tonnes of charge and convert it to steel in less than 40 minutes. The furnace is charged with scrap metal before molten iron is added (70% molten iron, 30% scrap metal). An *oxygen lance*, cooled by circulating water, is then lowered into the furnace and pure oxygen is blown on to the metal. In the oxidation, Si and Mn form SiO_2 and MnO_2, respectively, which float on the surface as a slag; oxides of S and P react with lime that is added in the amounts necessary to form a slag.

The main raw materials for the electric arc furnace are iron and steel scrap. In recent times, the increased availability of scrap has stimulated the growth of electric arc steel making and led to its use in the production of many 'bulk' steel products. The main features of the EAF are shown in Figure 5.8. The hearth of the furnace is typically constructed of magnesite bricks on to which a layer of granular magnesite or dolomite is rammed. The lower part of the furnace walls and the area of the roof around the electrodes are also lined with refractory material, while the remainder of the roof and side wall, not in contact with molten metal, is water-cooled. The actual sequence of furnace operations is shown in Figure 5.9. A modern EAF melts about 100–120 tonnes per heat and operates at a power level of 70–90 megawatts(MW). When operating on full power, such a furnace consumes as much electricity as a small town.

Steels have various hardnesses, tensile strength and ductilities, depending upon their composition. In general, the higher the carbon content of steel, then the greater its hardness and brittleness. Alloying the steel improves its corrosion resistance and creates a variety of steels with special properties, for example:

- 2.5% Si makes steel elastic (for use in springs);
- 1% Mn gives steel a high tensile strength;

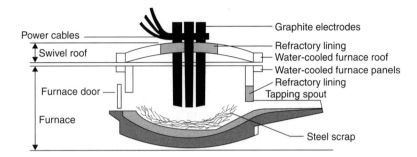

Figure 5.8 Schematic diagram of an electric arc furnace

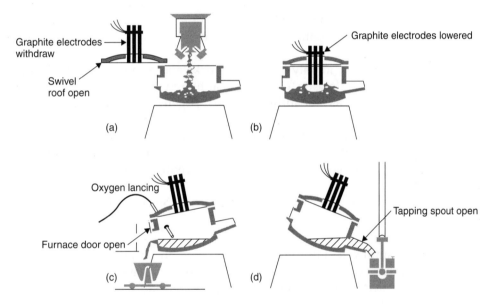

Figure 5.9 Schematic representation of the sequence of operations in an electric arc furnace: (a) charging scrap; (b) melting; (c) slagging and decarburizing; (d) tapping

- 12–15% Cr gives stainless steel, which is extremely corrosion resistant;

- W provides steel with a strong cutting edge.

Extraction of titanium, nickel and copper

Titanium is the ninth most abundant element in the Earth's crust and is present in practically all igneous rocks and their sedimentary deposits. It was first discovered over 200 years ago and named after Titan, one of the powerful giants of Greek mythology. The main ores of titanium, a mixed oxide of iron and titanium called ilmenite, $FeTiO_3$, and rutile, TiO_2, are primarily found in North and South America, India, China, Russia and Australia. The metal is most commonly extracted from its oxide by the *Kroll Process*. Titanium oxide is separated magnetically from ilmenite and then converted to titanium pentachloride by heating with coke (or tar) in a stream of chlorine gas.

$$TiO_2(s) + C(s) + 2Cl_2(g) \xrightarrow{800-1000°C} TiCl_4(g) + CO_2(g)$$

The titanium pentachloride gas is cooled to a colourless liquid which is purified by continuous fractional distillation before reduction by molten magnesium (or sodium) in an atmosphere of argon gas.

$$TiCl_4(s) + 2Mg(l) \longrightarrow Ti(s) + 2MgCl_2(l)$$

The molten magnesium chloride is removed (for reprocessing and recycling) and the titanium is melted with an electric arc before being moulded into ingots. The ingot is not poured, but solidifies under vacuum in the melting furnace. Titanium is very strong, lightweight and resistant to corrosion, and consequently is used to make alloys for use in the construction of high-speed aircraft, spacecraft and heart pacemakers.

Nickel occurs mainly in igneous rocks and as a free metal. The commonest ore of nickel is pentlandite, $(Ni, Fe)_9S_8$, although other ores include millerite (NiS) and arsenides such as chloanthite $(NiAs_2)$. Together with iron, it is a component of the Earth's core.

Nickel sulfides are roasted with coke, sand and limestone and the resultant iron impurities are removed as a slag. The nickel (II) sulfide is then converted to nickel (II) oxide by heating in a furnace and then reduced to the metal using carbon as the reducing agent.

$$NiS(s) + 3O_2(g) \longrightarrow 2NiO(s) + 2SO_2(g)$$

$$2NiO(s) + C(s) \longrightarrow 2Ni(s) + CO_2(g)$$

The resulting metal is impure and needs to be purified. This is achieved in an electrolysis cell by using the impure nickel as an anode, pure nickel as the cathode and an electrolyte of nickel (II) sulfate or chloride.

Nickel is hard, malleable and ductile, and has a high melting point and low electrical and thermal conductivities. It does not tarnish, and consequently is widely used for alloys, electroplating and coinage.

Copper is present in over 360 ores and also occurs naturally in its native form. The main copper ores include chalcocite, Cu_2S, chalcopyrite, $CuFeS_2$, bornite, Cu_5FeS_4, azurite, $Cu_3(CO_3)_2(OH)_2$, malachite, $Cu_2CO_3(OH)_2$, and chrysocolla, $CuSiO_3.2H_2O$. Native copper and the copper sulfides are usually associated with igneous rocks, while chrysocolla and the carbonates are the products of weathering of copper-bearing rocks. Like iron, copper is mainly mined in open pits before being crushed, ground and concentrated. The sulfide ores are then converted to iron (II) oxide by roasting the ore in a limited supply of air.

$$2CuFeS_2(s) + 4O_2(g) \longrightarrow CuS_2(s) + 3SO_2(g) + 2FeO(s)$$

The iron oxide is removed as a slag of iron (II) silicate by adding sand. The copper (I) sulfide is then further reduced to molten copper by heating in a limited supply of air. This molten, impure copper is moulded and cooled before electrolytic refinement.

$$Cu_2S(s) + O_2(g) \longrightarrow 2Cu(l) + SO_2(g)$$

Copper is widely used for its high thermal and electrical conductivities, durability, and resistance to corrosion and pliability.

5.5.3 Extraction of aluminium

Aluminium occurs abundantly as aluminosilicates in rocks and clays, but it is not economically recoverable from them. *Bauxite* ($Al_2O_3.xH_2O$, where $x = 1-3$) is the principal ore used for the extraction of aluminium, but it is only economically viable if the ore contains >45% aluminium and <5% silica. The reduction reactions to produce aluminium by using carbon alone requires temperatures of >2000°C, and is therefore very expensive (see Section 5.4). In order to reduce the temperature required (and thus the cost), aluminium is extracted by the electrolysis of bauxite and some sodium hexafluoroaluminate (*cryolite*), which lowers the melting point of aluminium oxide (Al_2O_3) from 2050 to just under 1000°C. Cryolite is a suitable solvent because of its solubility powers for aluminium oxide (also known as alumina), its non-reactivity with other components, and its stability under the applied potentials.

There are two main stages in the extraction.

Bauxite purification

The principal impurities in bauxite are iron (III) oxide, silica and titanium (IV) oxide. These are removed by mixing the crushed, powdered bauxite with 10% sodium hydroxide solution and heating the resulting solution under high pressure at around 150°C. The bauxite and silica oxide dissolve, but iron (III) oxide and other insoluble minerals remain insoluble and are removed by filtration.

$$Al_2O_3(s) + 2OH^-(aq) + 3H_2O(l) \longrightarrow 2[Al(OH)_4]^-(aq)$$

$$SiO_2(s) + 2OH^-(aq) \longrightarrow SiO_3^{2-}(aq) + H_2O(l)$$

Aluminium hydroxide is precipitated from solution by adding carbon dioxide ($CO_2(g)$) and seed crystals of $Al_2O_3.3H_2O$ and then dehydrated in a rotary kiln at 1000°C to alumina.

$$2[Al(OH)_4]^-(aq) + CO_2(g) \longrightarrow 2Al(OH)_3(s) + CO_3^{2-}(aq) + H_2O(l)$$

$$2Al(OH)_3(s) \longrightarrow Al_2O_3(s) + 3H_2O(l)$$

Electrolytic reduction of aluminium oxide to aluminium

The reduction cell for Al_2O_3 is a shallow steel container lined with carbon made by baking anthracite and pitch (see Figure 5.10). The lining acts as the cathode, while the anode is a consumable block of carbon suspended in the cell. The cells are connected in series, operate at a potential of about 5 V, a current of 100 000 A and a temperature of 850°C.

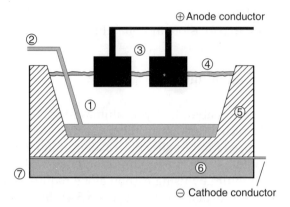

Figure 5.10 Schematic diagram of the Hall–Heroult cell for the production of aluminium; (1) electrolyte of molten Na_3AlF_6 and Al_2O_3; (2) removal of molten aluminium; (3) graphite anode blocks (periodically replaced); (4) crust of solid Al_2O_3 protecting molten aluminium from oxidation; (5) graphite cathode; (6) insulation block (7) steel case

The electrolyte is mainly cryolite, with some calcium fluoride to lower its melting point, and contains about 5% of alumina. The cell reactions are somewhat uncertain, but a possible explanation is that the alumina oxidizes as follows:

$$Al_2O_3(l) \longrightarrow Al^{3+}(l) + AlO_3^{3-}(l)$$

and redox reactions occur at the electrodes:

Cathodic reaction $Al^{3+}(l) + 3e^- \longrightarrow Al(s)$

Anodic reaction $2AlO_3^{3-}(l) \longrightarrow Al_2O_3(l) + 3O(g) + 6e^-$

$$O(g) + O(g) \longrightarrow O_2(g)$$

The oxygen produced at the anodes reacts with the carbon and energy is produced.

$$2C(s) + O_2(g) \longrightarrow 2CO(g)$$

This energy production compensates some of the energy required for the electrolysis process:

$$\tfrac{2}{3}Al_2O_3(l) \longrightarrow \tfrac{4}{3}Al(l) + O_2(g)$$

so that the overall reaction becomes:

$$\tfrac{2}{3}Al_2O_3(l) + 2C(s) \longrightarrow \tfrac{4}{3}Al(l) + 2CO(g)$$

Thus, indirectly, carbon assists in the reduction of Al_2O_3 and only a little more than one half of the expected energy is required for the overall reduction process. There is, of course, an energy requirement to produce the C electrodes, but this is less than the above energy saving in power.

Aluminium is a versatile metal that is ductile, malleable, lightweight and an excellent conductor of electricity. It also oxidizes easily, forming a layer of oxide on its surface that makes it highly resistant to tarnishing. It is widely used in the shipbuilding and aircraft industries, in the manufacture of cooking utensils, beer and soft-drinks cans, foil and overhead cables. A plastic form of aluminium, developed in the 1970s, is widely used in the electronics, car and construction industries, because it moulds to any shape and can extend to several times its original length. Aluminium sulfate is used to help in the coagulation and precipitation of suspended matter in the purification of water for drinking.

5.5.4 Extraction of manganese

Manganese ores may accumulate in metamorphic rocks or as sedimentary deposits, frequently forming nodules on the sea floor. *Manganese nodules* are found on or near sediment surfaces in nearly all deep-sea areas, as well as in coastal waters and some freshwater lakes. They are generally dull black ovoid objects, several centimetres in diameter and weighing approximately 1 kg (although weights up to 850 kg have been recorded). Manganese nodules consist of the fine-grained oxides of iron (Fe_2O_3) and manganese (MnO_2), as well as small amounts of clays, $CaCO_3$, SiO_2 and organic matter. A typical composition by weight of a Pacific Ocean nodule would be:

8–50% Mn (average 24%);
2–27% Fe (average 14%);
0.2–2% Ni (average 1.0%);
0.03–1.6% Cu (average 0.5%);
0.01–2.3% Co (average 0.4%).

Such ores are highly enriched by any standard and it may eventually become economic to harvest manganese nodules for extraction of these trace metals. The development of a technology for sea-floor mining which will be non-polluting and sufficiently cheap has been carried out by American, Japanese and German companies since the early 1970s.

Cross-sectional investigations of manganese nodules show that they grow by deposition of ferromanganese minerals around a foreign nucleus, e.g. a rock fragment, clay conglomerate, volcanic glass or even a shark's tooth (see Figure 5.11). The rates of growth of the nodules differ between deep-sea nodules (very slow) and those in coastal waters (relatively rapid). The slow formation of deep-sea nodules would be expected to result in burial by sediment deposition but the activities of

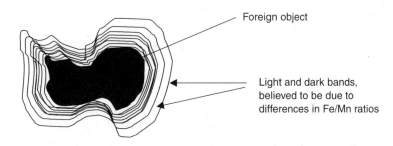

Foreign object

Light and dark bands, believed to be due to differences in Fe/Mn ratios

Figure 5.11 Cross-section of a typical manganese nodule

burrowing organisms and fish are believed to counteract this and keep the nodules on the surface. The action of strong bottom water currents would also assist this.

The mechanism of nodule formation is generally believed to be by deposition of manganese and iron by precipitation reactions caused by a change in oxidation state. Mn^{2+} is soluble in seawater, but on oxidation to Mn^{4+} this solubility is greatly reduced, causing precipitation and hence deposition on the outermost layers of the nodule. The inclusion of trace metals such as Cu, Co and Ni is by either:

- co-precipitation, or;
- adsorption on the freshly formed ferromanganese surface.

Current research involves the investigation of the role of biological processes in nodule formation, e.g. bacterial and plankton activity.

There are a number of possible sources of manganese for the formation of manganese nodules, as follows:

- Rivers contain substantial amounts of dissolved and particulate manganese and inflows to coastal waters are obvious sources for nodule formation.
- Submarine volcanicity injects many metals, e.g. Mn, Fe, Co, Ni and Cu, into seawater.
- Diffusion from the pore waters of sediments. These waters can contain higher amounts of dissolved manganese (compared to overlying seawater) due to conversion of insoluble Mn^{4+} to soluble Mn^{2+} at the lower redox potential that exists due to degradation of organic material. Hence, there is a concentration gradient allowing soluble Mn^{2+} to pass from pore waters to overlying seawater.

However, since it is still technically difficult to extract manganese from seabed nodules, manganese is extracted from *pyrolusite*, MnO_2, by thermal reduction.

$$3MnO_2(s) + 4Al(s) \longrightarrow 3Mn(l) + 2AlO_3(s)$$

Manganese is a hard, brittle, grey-white metal that resembles iron. It is mainly used to make alloys with iron, aluminium and copper, and in fertilizers, industrial chemicals and paints.

5.6 Environmental Warfare – the Corrosion of Metals

Corrosion is usually regarded as the chemical attack of a metal by the environment, (although environmental corrosion of stone and other materials also occurs). Corrosion cannot be stopped but it may be controlled. The most common type of corrosion is due to the action of the atmosphere, together with moisture and substances dissolved in it, e.g. CO_2, SO_2, etc. Corrosion reduces the lifetimes of steel products such as cars, bridges and industrial equipment and costs huge sums of money in corrosion protection, replacement of corroded material and lost production time. The costs of corrosion are estimated as costing the UK 3.5% of its gross national product.

5.6.1 Corrosion chemistry

The aqueous corrosion of a metal is an electrochemical reaction that involves the type of electron transfer which is characteristic of electrochemical cells (see Section 5.2.3). For metal corrosion to occur, an oxidation reaction (often the dissolution of the metal or the formation of an oxide) and a reduction reaction (often proton or oxygen reduction) must occur simultaneously. The corrosion process may be written as two separate reactions, occurring at two distinct sites on the same metal surface. The two reactions are coupled to form an electrochemical cell.

The oxidation (or anodic) reaction occurs at the site known as the *anode*. The latter is the site (electrode) of an electrochemical cell where oxidation occurs as the principal reaction. Electrons (e) flow away from the anode.

Anodic reaction $\quad$ Me $\longrightarrow$ M^{n+} $-$ ne^-

The reduction (or cathodic) reaction occurs at the site known as the *cathode*. The latter is the electrode of an electrochemical cell where reduction is the principal reaction. Electrons flow towards the cathode.

Cathodic reaction $\quad$ O (oxidizing reagent) $+$ ne^- $\longrightarrow$ R (reduced species)

Thus, the cathode consumes the electrons liberated by the anode. Since corrosion involves the movement of electrons, it may be expressed in terms of an electrochemical current, the corrosion current, i_{corr}. The total current flowing in the direction of the cathodic reaction (Σi_c) must be equal, and opposite in sign to, the total current flowing out of the anodic reaction ($-\Sigma i_a$).

$$i_{corr} = -\Sigma i_a = \Sigma i_c$$

The overall reaction for these two processes, shown schematically in Figure 5.12, may be represented as follows:

Overall reaction $\quad$ M $+$ O $\longrightarrow$ M^{n+} $+$ R

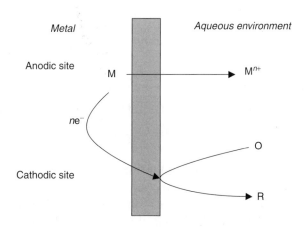

Figure 5.12 The corrosion process M $+$ O $\longrightarrow$ M^{n+} $+$ R

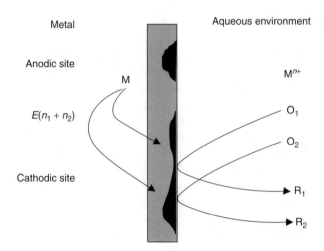

Figure 5.13   Corrosion process involving two cathodic reactions

The corrosion of a metal by anodic oxidation may be supported by several simultaneous cathodic reactions, as illustrated in Figure 5.13. The area of the anodic and cathodic sites (A_a and A_c, respectively) may be very different providing that the anodic and cathodic currents are equal. Examples 5.4 and 5.5 illustrate the corrosion of iron in two different aqueous environments.

Example 5.4

The corrosion of iron in acid solution may be written as two separate reactions:

Anodic reaction	$Fe(s) \longrightarrow Fe^{2+}(aq) - 2e^-$
Cathodic reaction	$2H^+(aq) + 2e^- \longrightarrow H_2(g)$
Overall reaction	$Fe(s) + 2H^+(aq) \longrightarrow Fe^+(aq) + H_2(g)$

Example 5.5

The corrosion of iron in a solution containing dissolved oxygen may be written as follows:

Anodic reaction	$Fe(s) \longrightarrow Fe^{2+}(aq) + 2e^-$
Cathodic reaction	$O_2(aq) + 4H^+(aq) + 4e^- \longrightarrow 2H_2O(l)$
Overall reaction	$2Fe(s) + O_2(aq) + 4H^+(aq) \longrightarrow 2Fe^{2+}(aq) + 2H_2O(l)$

Corrosion chemistry is complex but it can be simplified. In general, any metal that has a more positive electrode potential than water can be oxidized under standard conditions. At pH 7, water has a standard electrode potential, $E^\ominus$, of -0.42 V (pH is discussed in Chapter 7). This is almost identical to the value for iron ($E^\ominus = -0.44$V for $Fe^{2+}(aq) + 2e^- \longrightarrow Fe(s)$) and hence iron has only got a slight tendency to be oxidized by pure water. However, when iron is exposed to damp air (i.e. to both oxygen and water simultaneously), the following half-reaction occurs:

$$O_2(g) + 4e^- + 4H^+(aq) \longrightarrow 2H_2O(l) \qquad E^\ominus = +0.81V \text{ at pH7}$$

Because the $E^\ominus$ for this reaction is more positive than the value for iron, oxygen and water can jointly oxidize Fe to Fe^{2+}, which can subsequently be oxidized to Fe^{3+}. We saw in Section 5.4 that metals placed in order of their standard electrode potentials give rise to the electrochemical series of metals (see Table 5.2). Very active metals (those with the most negative $E^\ominus$ values) will be oxidized even in the absence of moisture (Na, K, etc.) Less active metals (Mn, Cr, etc.) are oxidized only in the presence of moisture. The noble metals, also known as the platinum group metals (Au, Pt, Pd) are the only metals which resist corrosion.

However, the electrode potential of a metal is not the only factor that affects corrosion. For example, aluminium and magnesium are very active metals, but the thin layer of oxide that is formed on their surfaces gives them corrosion protection. They corrode less in air than iron does – this is opposite to expectation on the basis of their electrode potential values. Oxide, more commonly known as rust, is formed on the surface of iron, but it is so flaky and porous that corroding chemicals can pass through it easily and attack the underlying metal. The rusting process is described schematically in Figure 5.14, and can be described as follows:

$$4Fe(s) + 3O_2(g) + 2H_2O(l) \longrightarrow 2Fe_2O_3 \cdot H_2O(s)$$

In the absence of water – in dry, heated buildings – steel will only rust superficially and the rate of corrosion will be slow. Steel deeply buried in soil will also corrode very slowly because of the limited amount of oxygen. On the other hand, on ships, offshore structures and other marine structures, steel corrosion rates will be high due to the presence of high concentrations of dissolved salts. If salts are dissolved in water, oxidation is speeded up because electrolytes are a good medium for electrochemical oxidation–reduction reactions. Oxygen dissolved in water, variations in temperature, dissolved salts and the presence of microorganisms (iron bacteria) also influence corrosion. The most important atmospheric contaminants affecting corrosion are chlorides (present in marine environments) and sulfates (produced by the burning of fossil fuels).

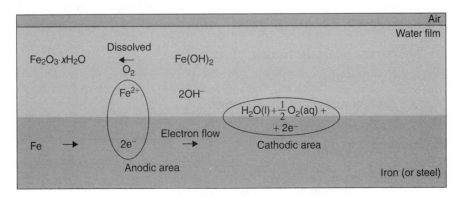

Figure 5.14 Chemical reactions involved in the process of rusting

Thus, in practice, metals and alloys are listed according to their relative electrode (or in this context, corrosion) potentials *in a given environment*. Such ordered lists are known as a *galvanic series*. The position of metals and alloys in a galvanic series and their corrosion behaviour depends upon the specific environment to which the materials are exposed. There are as many galvanic series as there are aqueous environments.

Biological as well as chemical processes may initiate corrosion. Corrosion in sewers is the destruction of pipe materials by chemicals generated from the biological production of sulfuric acid from industrial or domestic wastes that are not treated prior to discharge. Biological activity in wastewater in sewers can produce anaerobic conditions leading to the formation of hydrogen sulfide. When oxygen becomes available in the presence of bacteria, sulfuric acid will be produced as follows:

$$H_2S(aq) + 2O_2(g) \longrightarrow H_2SO_4(aq)$$

In concrete sewers, the sulfuric acid combines with lime to produce a very soft mineral, calcium sulfate (gypsum); hence, the concrete is weakened and the pipe can collapse.

It is clear therefore, that aqueous corrosion is a complicated process that can occur in various forms and is affected by many variables, including the following.

- the composition and metallurgical properties of the metal or alloy (or other material that can undergo corrosion);

- the chemical, physical and microbiological properties of the environment;

- the presence or absence of surface films;

- the properties of surface films (e.g. thickness and resistivity).

5.6.2 Corrosion protection

There are two general methods of corrosion protection – the application of a protective layer (barrier protection) and the application of a sacrificial metal, otherwise known as cathodic protection. The simplest way to prevent corrosion is to coat the metal with a protective barrier that resists penetration by environmental constituents. The main types of barrier protection are as follows:

- the use of anodic oxides;

- metal coatings;

- ceramic and inorganic coatings;

- use of corrosion inhibitors;

- organic coatings;

- conversion coatings.

We will briefly study each of these different types of corrosion protection.

Anodic oxides

Anodic oxides are widely used to protect aluminium from corrosion. Aluminium anodizing is an electrochemical method of converting aluminium into a film of protective aluminium oxide (Al_2O_3) at the surface of the item being coated. The most common acids used to produce anodic coatings are sulfuric (H_2SO_4) and chromic (CrO_3) acids. The metal piece requiring protection is used as the anode in a suitable electrolytic cell – for example, immersing the aluminium work piece into an aqueous solution of CrO_3 forms chromic anodize.

Metal coatings

Metal coatings are often used to provide corrosion resistance but are also used to provide decorative effects, enhancement of material properties such as wear resistance, electrical and magnetic properties. The coatings may be applied by electroplating or dipping the material to be protected in a bath of molten metal.

Electroplating is the electrodeposition of an adherent metal coating upon an electrode (made of the material to be protected) to create a surface with different properties or dimensions from the original. Coatings may be anodic or cathodic to the metal substrate. Only a rough idea of which situation applies can be obtained from the electrochemical series of metals, because coatings that are anodic to a substrate in one environment, may be cathodic to the same substrate in another. The most desirable coatings are anodic with respect to the material they protect and are resistant to corrosion because of the formation of a passive, oxide film on the material's surface.

Food cans made from steel are coated with tin to provide protection from corrosion. Tinplate is the name given to a low-carbon steel strip coated on both sides with a thin layer of tin. It has been used for 200 years to make containers – tin cans – for the long-term storage of food. In theory, if the tin coating is broken, corrosion should be rapid as the iron is oxidized in preference to the tin; iron is higher in the electrochemical series of metals and thus has a greater tendency to donate its electrons. However, tin is able to form a variety of chemical complexes with the organic liquids present in foodstuffs, converting it into a sacrificial anode and thus diminishing the ability of iron to dissolve and rust. In addition, the tinplate industry has become very sophisticated in its use of lacquers to provide a further inert barrier against metal dissolution.

Hot dip coating is a process in which a protective coating is applied to a metal by immersing it in a molten bath of the coating metal. Galvanized (zinc-coated) iron and steel have been in use for over 100 years. Coating iron with an unbroken film of zinc provides protection from corrosion by two methods. First, in many environments, zinc corrodes much more slowly than iron and effectively forms a barrier between iron and the environment. This is because, although zinc is more reactive than iron, when it reacts with oxygen in the presence of moisture and carbon dioxide, it forms a tough film of $Zn(OH)_2 \cdot ZnCO_3$ which resists further attack. Secondly, zinc provides electrochemical protection. When zinc is coupled to steel, the steel is polarized to such a potential that it becomes the cathode. In practice, this means that steel exposed to the environment will not rust until nearby zinc is consumed.

Ceramic and inorganic coatings

Ceramic coatings are generally used to protect steel from corrosion. They act as a physical barrier to the external environment and also maintain an alkaline environment at the steel/ceramic interface in which the corrosion rate is low. Porcelain enamels are perhaps the best known type of ceramic coating. They offer only barrier protection to the metal substrate and hence must be free from defects and discontinuities.

Porcelain enamels are distinguished from paint by their inorganic composition and the fusion of the coating matrix to the coated metal. They are composed of a mixture of melted metal oxides (e.g. SiO_2, B_2O_2, B_2O_3, Na_2O, K_2O, CaO, TiO_2, MnO_2) which may be applied to the metal substrate by dipping, spraying and electrodeposition. In addition to corrosion resistance, they are used to provide chemical resistance, weather resistance, specific mechanical or electrical properties, appearance or colour needs, cleanability and thermal shock capability. Porcelain enamels have a large number of applications, including the following:

- Industrial products – chemical reactors and, food processing vessels;
- Household appliances – refrigerators, freezers, dishwashers and space heaters;
- Plumbing fixtures – laundry tubs, kitchen sinks, baths and lavatories;
- Signs- street, highway and advertising signs;
- General products – camping equipment, hospital ware, venetian blinds, table-tops, etc.

Use of corrosion inhibitors

Inhibitors or pH regulators (neutralizers) may also be used for corrosion protection. Inhibitors act in the opposite way to catalysts in that they slow down the rate of chemical reactions. The role of the inhibitor is often to form a surface coating on the substrate that may be one or several molecular layers thick. Neutralizers are widely used in large industrial plants to lessen the corrosivity of the environment by reducing the H^+ ion concentration.

Many effective organic inhibitors consist of a reactive group attached to a hydrocarbon. The reactive group interacts with the metal surface while the hydrocarbon portion of the inhibitor is in contact with the environment.

Organic coatings

Organic coatings and linings act as a protective film to isolate the material surface from the environment. They often act as a barrier to water, oxygen and ions in order to prevent (i) cathodic reactions occurring at the surface of a metal and/or (ii) general wear and weathering reactions on building, metal or stone surfaces. Organic coatings may be painted, sprayed or powder-coated on to the material surface and many different types are available, depending upon the type of protection required. Examples of organic coatings include the following:

- alkyds (primers for metal and steel surfaces);

- acrylics (primers for concrete and masonry surfaces);

- epoxy esters (used for atmospheric resistance on tank exteriors);

- vinyls (used on portable water tanks and sanitary equipment);

- phenolics (used to coat tanks used for alcohol storage);

- chlorinated rubbers (used in concrete and masonry paints);

- coal tar pitch (used on pavement sections);

- asphalt pitch (used for roof coatings);

- polyamides and polyurethanes (used to enhance moisture resistance);

- water emulsion latex (used as a concrete and masonry sealant);

- organic zinc resins (used to repair galvanized coatings and as a primer on bridges, offshore structures and steel in chemical processing industries).

Conversion coatings

Organic coatings are only really effective if they can protect the entire surface of the metal substrate. However, many appliances require corrosion protection if the organic coating becomes damaged and is not repaired – motor vehicles are a prime example. Conversion coatings – phosphate, chromate or mixed-oxide coatings applied before the organic coating – are often used to provide corrosion resistance in these circumstances. Such coatings derive their name from the conversion of the base metal to a coating in which ions of the base metal are a component.

For example, phosphates may be coated on to a metal surface by the precipitation of divalent metal and phosphate ions (PO_4^{3-}). Phosphate salts are soluble in acid solutions and insoluble in neutral or basic solutions. Metal objects requiring protection are placed into a bath containing an acidified phosphate solution and the acid attacks the metal (M) surface on contact. Two changes occur in the boundary layer of solution directly adjacent to the metal surface, namely:

- the acid is neutralized and the pH rises;

- the concentration of metal ions increases.

The metal ions react with the phosphate ions to precipitate a phosphate coating on the surface of the metal. The chemical reactions involved in the phosphating process may be represented as follows:

$$2H^+(aq) + M(s) \longrightarrow H_2(g) + M^{2+}(aq)$$

$$2H_2PO_4^-(aq) + 3M^{2+}(aq) \longrightarrow M_3(PO_4)_2(s) + 4H^+(aq)$$

$$4H_2PO_4^-(aq) + 5M^{2+}(aq) \longrightarrow M_5H_2(PO_4)_4(s) + 6H^+(aq)$$

Iron phosphates are usually used when the primary consideration is good paint adhesion and low cost, e.g. in car manufacturing processes. Chromate conversion coatings are applied to:

- enhance corrosion resistance;
- improve the adhesion of paint or other organic finishes;
- provide the metallic surface with a decorative finish.

Cathodic protection

Large structures cannot be adequately protected by using physical barriers or galvanizing, and so another technique is required. *Cathodic protection* is an electrochemical means of corrosion protection in which the oxidation reaction of a galvanic cell is concentrated at the anode and suppresses corrosion of the cathode in the same cell. The are two methods of cathodic protection.

- *Passive systems*, in which *sacrificial anodes*, made from active (more reducing) metals like Mg or Zn, are connected at intervals by wire to the metal to be protected.
- *Active systems*, in which a low-voltage DC current is applied to the metal requiring protection.

During cathodic protection, the object to be protected is the cathode, as shown in Figure 5.15. If the iron pipe and the magnesium block were not connected, the following reactions would occur in the presence of acidic groundwater:

Oxidation reaction	$Fe(s) \longrightarrow Fe^{2+}(aq) + 2e^-$
Reduction reaction	$2H^+(aq) + 2e^- \longrightarrow H_2(g)$
Net reaction	$2H^+(aq) + Fe(s) \longrightarrow Fe^{2+}(aq) + H_2(g)$
Oxidation reaction	$Mg(s) \longrightarrow Mg^{2+}(aq) + 2e^-$
Reduction reaction	$2H^+(aq) + 2e^- \longrightarrow H_2(g)$
Net reaction	$2H^+(aq) + Mg(s) \longrightarrow Mg^{2+}(aq) + H_2(g)$

Both metals will corrode in these circumstances. However, if the two metals are immersed in the same electrolyte (acidic groundwater) and electrically connected, the reaction for the oxidation of magnesium (the more active metal) would dominate,

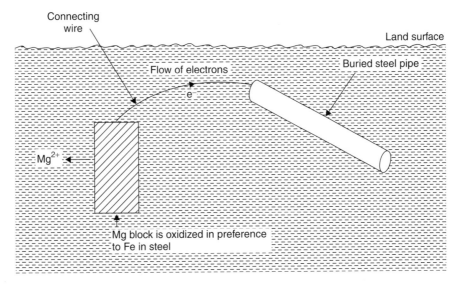

Figure 5.15 Cathodic protection of a buried pipeline using a 'buried' magnesium anode

resulting in the corrosion of magnesium at the anode. The iron pipeline will become the cathode and the site for the reduction reaction (hydrogen liberation). Thus the magnesium anode is used to protect the iron cathode.

Some corrosion of the iron may still occur, depending upon the relative sizes of the magnesium and iron electrodes. (Remember that the anode is the electrode at which a *net oxidation reaction* occurs, whereas cathodes are electrodes at *which net reduction reactions* occur). By careful design of the galvanic cell, the oxidation rate at the cathode may be suppressed to the point at which it becomes negligible, and at this point cathodic protection is achieved. The sacrificial anode will obviously have to be replaced periodically.

The use of a sacrificial anode is commonly used to prevent corrosion of the interior surfaces of steel tanks by suspending anodes in contaminated water. It is also useful for the protection of buried pipelines, bridges, ships' hulls etc. However, this type of cathodic protection does not work well on structures exposed to air, as air is a poor electrolyte and prevents current from flowing from the anode to the cathode.

In active systems, the application of an impressed current from the negative terminal of a DC source provides electrons to the cathodic metal requiring protection and hence suppresses the tendency of the metal to lose electrons during corrosion. The positive terminal of the DC source is connected to a non-sacrificial, inert anode that is usually made of graphite.

5.7 Fossil Fuels

5.7.1 The formation and classification of coal

Coal is a mixture of organic compounds of high molecular weight and complex structure, (see Figure 5.16) containing a large percentage of carbon with small

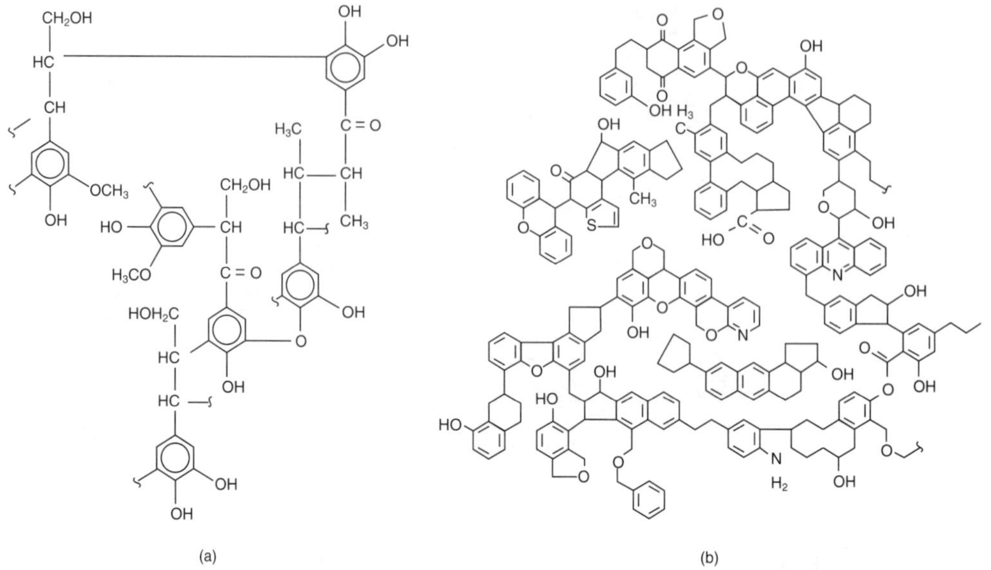

Figure 5.16 Schematic representations of two types of coal: (a) lignite coal (reprinted from *Organic Chemistry*, Volume 16, P. G. Hatcher, 'Structural Model of Lignite Coal', p. 959. Copyright 1990, with permission from elsevier science); (b) bituminous coal. (Reprinted from *fuel*, **63**, J. H. Shinn, 'Structural Model of Bituminous Coal', p. 1187. Copyright 1984, with permission from elsevier science)

amounts of hydrogen, oxygen, nitrogen and sulfur. The burning of coal can lead to severe air pollution and acid precipitation because most coals contain 2–6% sulfur. Unlike petroleum, which being a liquid may have migrated large distances from its place of origin, coal betrays its origin by fossils and by its inter-bedding with sedimentary rocks. Coal is a metamorphic rock produced by the partial anaerobic decomposition of buried terrestrial vegetation, typically in a near-shore swamp environment, whereas oil originates from organisms buried in offshore marine and lacustrine environments. However, the processes of coal formation are less well understood.

Coal is classified by its purity (carbon content), type (composition of the original vegetation) and rank (degree of metamorphism). *Anthracite*, or hard coal, contains the highest carbon content (> 90%), followed by *bituminous (soft) coal* (> 80% carbon), *lignite* (> 70% carbon), and *peat*. There are two general types of coal. *Sapropelic coals* are derived primarily from spores, pollen and algal remains. The most common type of coal, *humic coals*, originated from forest peats and therefore from wood and bark substances, leaves and the roots of swamp vegetation.

The formation of coal may be divided into two stages. The first stage – the formation of peat – is known as the *biochemical stage*, since microorganisms play a major role in the process. The vegetation initially consists of carbohydrates, proteins and lignin, which after dying undergoes decomposition by anaerobic reactions as the vegetation is usually protected from aerobic decomposition by water and rapid burial by further inputs of organic matter. In this process, the carbohydrates are broken down to CO_2 and water and the residues are concentrated and begin their transformation to humic substances (organic material with high content of carboxylic and phenolic groups). The overall composition and properties of a given peat depend on the environmental conditions that existed during the accumulation and initial burial of the plant material. For example:

- The access of air (oxygen), high temperatures and an alkaline environment enhance the formation of humic substances.

- Plants found in freshwater environments have a different biochemistry to those found in marine environments and decay to different products.

In swampy environments, anaerobic conditions prevail and depending upon the conditions, sulfate reduction or fermentation reactions may occur. During this stage in maturation, purely microbiological processes reduce the amounts of nitrogen and sulfur by the evolution of ammonia and hydrogen sulfide, respectively. The result is the formation of peat that has an average content of 90% of adsorbed water.

The second stage in the formation of coal – the conversion of peat to coal – is poorly understood and only occurs when layers of sediment impervious to air cover the peat. This second stage, known as the *metamorphic stage*, involves physical and chemical changes in peat due to the passage of time and increases in temperature and pressure. These factors cause changes in elemental composition, a decrease in moisture content and an increase in hardness. The ratio of carbon to volatile constituents (O, H, N) increases in a sequence passing from peat to lignite (brown coal), through bituminous coals to anthracite (see Figures 5.17 and 5.18). This is because as the temperature increases, gaseous compounds (carbon dioxide and methane) are liberated and the coal becomes relatively richer in carbon. The position of a coal within this series is called its *rank*. The rank generally increases with burial

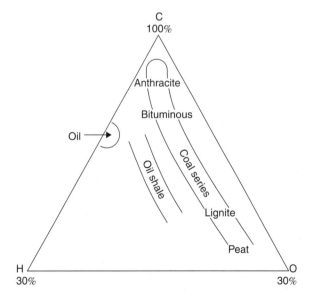

Figure 5.17 Schematic diagram of the compositional relationship between different ranks of coal, oil shale and crude oil

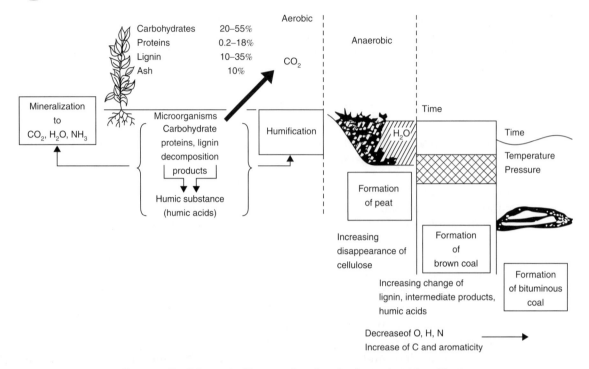

Figure 5.18 Schematic diagram showing the formation of coal beds

depth due mainly to increases in temperature rather than the increased pressure. The same rank of coal can be produced by short burial times at high temperatures or by long burial times at lower temperatures.

5.7.2 Formation and accumulation of oil (petroleum)

Petroleum is a biological product derived from the organic debris of former life that has been buried, transformed and preserved. Petroleum (from the Greek words meaning 'rock oil') is a complex mixture of hydrocarbons – alkanes, cycloalkanes and aromatics – as well as oxygen-, nitrogen- and sulfur-containing organic and inorganic compounds and trace amounts of metallic substances.

The hydrocarbon compounds of crude oils usually make up 90–95% of the total liquid. The elemental composition varies from deposit to deposit but usually ranges from about 78–90% carbon and from about 8–15% hydrogen. The abundances of typical hydrocarbon groups found in petroleum are shown in Table 5.4. In fact, over 600 different hydrocarbon compounds have been detected in a single sample of petroleum.

The non-hydrocarbon compounds of crude oils usually make up 5–10% of the total liquid, but may be more abundant in heavy crude oils. Oxygen occurs mainly in organic acids and phenols, nitrogen occurs in a wide range of compounds, and sulfur occurs in thiols, thiophenes, sulfides and disulfides. The sulfur content of crude oils is very variable, and may be as high as 10% by weight. The main metals found in oil are vanadium, nickel and iron.

Table 5.4 Typical hydrocarbon composition obtained by refining a crude oil

Fraction	Number of carbon atoms in molecule	Boiling point range (°C)	Vol%
Gasoline	5–10	30–200	27
Kerosene	11–13	180–200	13
Diesel fuel	14–18	250–380	12
Heavy gas oil	19–25	>350	10
Lubricating oil	26–40	>350	20
Residue	>40	>500	18

It is generally accepted that oil and natural gas were formed from the remains of microscopic plants and animals that lived over 50 million years ago. Upon dying, these organisms sank to sea or lake beds, where they were buried beneath layers of sand and silt. Over time, the weight and pressure of the beds combined together with chemical and biological processes to convert the organic remains into oil and gas. Later, the fossil fuels were squeezed by earth movements into trap structures, e.g. anticlinal traps and faults (see Figure 5.19), where they collect, ensuring their survival to the present day. The evidence that petroleum originated from biogenic (living) matter seems fairly convincing, namely:

- petroleum is not found in rocks that formed prior to the development of life on Earth;

- petroleum contains biomarkers, compounds either formed by living organisms or resulting from minor alteration of these compounds;

- petroleum shows the same property as all organic matter relative to inorganic matter, in that it is enriched in ^{12}C relative to ^{13}C.

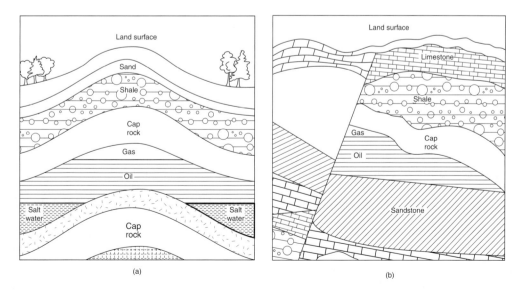

Figure 5.19 Oil traps in the Earth: (a) an anticlinal trap; (b) a fault

However, the majority of the constituents of crude oil have low molecular weights and simple structures compared to the complex organic compounds found in living organisms. In addition:

- light hydrocarbons (C_3–C_4) are almost entirely absent from modern organisms but constitute a significant proportion of a typical crude oil;

- hydrocarbons in modern organisms show a preference for odd-numbered carbon chains, whereas in crude oils the odd- and even-numbered chains are present in equal proportions;

- alkenes are the most abundant hydrocarbons in organisms but the least abundant in petroleum.

These observations, and other evidence, suggest that the original organic material did not simply accumulate over time. In fact, the original organic material begins to form petroleum only at depths of 1–2 km, when temperatures become sufficiently high for the organic matter to become thermally unstable and break down to hydrocarbons. Maximum oil generation occurs when the source material has been buried to about 3 km. The resulting oil becomes more mobile and is able to move from its source beds into the overlying reservoir rocks, e.g. porous sandstones, from which it may eventually be extracted. It is important to appreciate that these crude oil reservoirs are *not* giant underground lakes of oil-like water reservoirs, but are held in pores of sandstone, rather like water held in a bath sponge.

5.7.3 Formation and accumulation of natural gas

Natural gas is formed by a similar process to that of oil and is often found above crude oil reservoirs. Gas is formed from the cracking (thermal decomposition) of crude oil and the original organic material at depths greater than about 4 km.

Natural gas typically consists of 50–90% methane, depending on its source. The other components of natural gas are ethane, propane and butane, along with nitrogen, carbon dioxide and hydrogen sulfide (which must be removed before use). Natural gas found in the Texas panhandle and in Oklahoma is also a source of helium. *Conventional natural gas* lies above most reservoirs of crude oil, whereas *unconventional natural gas* is found by itself in other underground sources, e.g. coal seams and shale rock. It is not currently economically feasible to remove natural gas from unconventional sources, although the extraction technology is being developed rapidly.

5.7.4 Uses of coal, oil and natural gas

Coal is predominantly used as a fuel for heating and power generation but it may also be destructively distilled to produce a range of chemicals. These chemicals are purified or further reacted to create a huge range of products, including pharmaceuticals, coal tar soap and shampoo, dyestuffs, furniture oils, bitumen and linoleum. Coal may also be gasified to produce carbon monoxide, carbon dioxide, methane, hydrogen and nitrogen. These gases are used as fuels or raw materials for the production of chemicals and fertilisers.

Crude oil has no uses in its raw form and must be refined by primary distillation to provide useful products. It is the most important of all modern raw materials and is the source of 70% of the UK's organic chemicals. Other products include petrol, diesel or gas oil, kerosene, lubricating oils, waxes and bitumen (asphalt).

Compared to other fossil fuels, natural gas is relatively non-polluting and easily transportable, and is consequently known as a "premium fuel". It is used to generate electricity in power stations, combined heat and power systems in industry, domestic space heating and cooking, and to manufacture chemicals such as methanol. More recently, natural gas has been used to power some experimental motor vehicles and tumble dryers.

5.7.5 Crude oil and the environment

Oil pollution at sea may occur as a result of deliberate discharges from ships, ships colliding or running aground, blow out and explosions on oil rigs and fractures in underwater pipelines. Some of the worst marine oil disasters are outlined in Box 5.2.

Box 5.2 Major Marine Oil Spills

1955	*Gerd Maersk*	Grounded near the mouth of the river Elbe, losing 9 000 tonnes of crude oil. It is estimated that 50 000–500 000 birds were killed.
1957	*Tampico Maru*	A small tanker, which was wrecked off the Mexican coat, losing its cargo of diesel fuel. Two species of sea urchin were either killed or left the affected area.
1966	*Seestern*	Oil from this tanker overflowed and was carried over mud-flats in the Medway Estuary, Kent, where waders and gulls were roosting – 5 000 birds were killed.
1967	*Torrey Canyon*	This ship ran aground off Land's End in Cornwall. The publicity from the incident drew public attention to the environmental effects of a large oil spill.
1972	*Sea Star*	115 000 tonnes of oil spilled in the Gulf of Oman.
1973	*Zoe Colocotroni*	Contaminated 20 acres of mangrove and shoreline when it ran aground on the coast of Puerto Rico. In the court case that followed, the judge estimated that 92 109 720 animals were killed!
1977	*Ekofisk blow out*	An offshore well in the North Sea which discharged 20 000–30 000 tonnes of oil into the sea in seven days.
1978	*Amoco Cadiz*	Ran aground on the coast of Brittany, killing 5 000 seabirds and thousands of tonnes of fish and shellfish.

1979	*Atlantic Express*	The biggest oil tanker spill in history. Over 250 000 tonnes of crude oil was released
	Aegean Captain	when these two supertankers collided off the coast of Tobago.
1989	*Exxon Valdez*	Hit underground rocks in Prince William Sound, Alaska. The oil eventually fetched up on 1 700 km of coastline. 400 000 birds and 4 000 otters estimated dead as a result of the spill, $15 billion damage caused.
1991	*Gulf War*	Analysts estimated that 816 000 tonnes of oil were spilled or deliberately released as a result of this conflict.
1992	*Aegean Sea*	70 000 tonnes of oil lost from this tanker near La Coruna, Spain.
1993	*Braer*	Ran aground on Fitful Head, Quendale Bay, Shetland. Over 1 500 sea birds were found dead.
1996	*Sea Empress*	On 15 February 1996, the *Sea Empress*, a 147 000 tonne supertanker, ran aground at the entrance to Milford Haven harbour. The vessel lost 75 000–80 000 tonnes of oil before it was salvaged, causing massive environmental damage to the area.
1999	*Erika*	The *Erika* sank off the coast of Brittany during gale force winds on 12 December, spilling over 20 000 tonnes of heavy fuel oil. Pollution damage to the French coast was severe, particularly to tourist regions, effecting over 400 km of coastline. More than 50 000 birds were killed, and inter-tidal and sub-tidal communities, including commercial oyster and mussel beds, were badly contaminated.

There are a variety of techniques available to scientists when they need to clean up oil from coastal environments. These range from cheap and simple techniques to more complicated and expensive methods; some examples are described below.

Manual Recovery of oil on land

This often involves simply digging the oil off the beach and placing it in containers such as skips for later removal and disposal. During the Sea Empress clean-up operation in Wales during 1996, high-pressure hoses were used to clean the oil off rocks and beaches into specially created pits. The collected oil was then pumped from the pits into tankers. At the tourist resort of Tenby, large digger trucks were used to mechanically scrape oil off the beaches.

Chemical processes

The heaviest areas of the oil slick at sea can be sprayed with a dispersant from an aircraft. Dispersants are chemicals that speed up the natural process of emulsification of oil in the water. However, dispersants are not very effective against oil that has been on the sea for more than 48 hours and can cause chemical damage to the environment.

Barriers to the movement of oil

Floating booms can be laid in the path of a slick to control the movement of oil. These barriers can be used to deflect the oil away from sensitive areas such as beaches or salt marshes or to contain the oil using a V-shaped boom downwind of the spill. The oil will collect at the bottom of the boom and can be pumped out when appropriate.

Booms can also be towed in formation by ships, thus allowing barriers to be deployed rapidly over a large area. However, there is a limit to the effectiveness of such barriers as wind, waves and sea current can allow oil to escape under or over the boom.

Mechanical recovery of oil at sea

Oil can be removed from the surface of water by skimming the slick using a moving drum, belt or disc. Skimmers can be stand-alone devices or can be built into booms. Ships also operate which can separate the oil from the water by using large suction pumps and settling tanks.

Other techniques

Alternative methods that can be used to remove oil include the following:

- burning (can only be used at specific locations under carefully controlled conditions);

- sinking (increasing the density of the oil by adding chalk or sand and allow it to settle to the seabed);

- gelling (increasing the thickness of the oil to reduce the speed at which the slick spreads);

- bioremediation (use of microorganisms to degrade the oil).

5.8 Summary

After carefully studying this chapter, you will have an understanding of the formation of metal ores and fossil fuels, as well as their properties and utilities. You will have studied the extraction techniques necessary to remove metals from their mineral ores, the methods of protecting metals from environmental corrosion and the environmental implications of mining, transporting and utilizing these primary resources. In the following experiments, we will further develop your practical skills by studying techniques that may be used to determine the environmental concentrations of selected metals and anthropogenic substances.

Experiment 5.1 Spectrophotometric Determination of Iron in Water

Introduction

A sensitive method for the determination of the iron content of unknown water samples, e.g. from a river or lake, is based on the formation of the red-orange iron (II) orthophenanthroline complex. As orthophenanthroline is a weak base in acid solution, it exists as the phenanthrolium ion, here given the symbol PhH^+ for simplicity. The complex can be described as follows:

$$Fe^{2+}3PhH^+ \rightleftharpoons Fe(Ph)_3{}^{2+} + 3H^+$$

Quantitative formation of the complex occurs in the pH range 3 to 9, although a pH of between 2.9 to 3.5 is recommended to prevent the precipitation of a number of iron salts.

An excess of reducing agent is necessary to prevent the oxidation of Fe(II) to Fe(III). Hydroxylamine hydrochloride solution is used for this purpose. Once the complex is formed, it is stable and its absorbance can be measured with a spectrophotometer at 508 nm or with a photometer equipped with a green filter.

Time Required

Two–three hours, depending upon the number of water samples determined.

Equipment

Spectrophotometer or photometer with green filter
1.00 and 10.0 cm^3 pipettes and pipette filler
100 cm^3 volumetric flasks
10.0 cm^3 measuring cylinder

Reagents

Standard iron solution, 0.1 mg cm^{-3} (100 mg dm^{-3} of ferrous ammonium sulfate)

Hydroxylamine hydrochloride solution (dissolve 10 g of H_3NOHCl in 100 cm^3 of distilled water)

Orthophenanthroline solution (Dissolve 0.3 g of orthophenanthroline monohydrate in 100 cm^3 of distilled water. Warm gently, if necessary to effect solution, and stir well. Store in a dark place. The solution must be discarded if it darkens.)

1.2 M Sodium ethanoate ($NaCH_3COO$)

Double-distilled, de ionized water

Suitable water sample(s)

Significant Experimental Hazards

- Students should be aware of the hazards associated with the use of all glassware (cuts) and electrical equipment (burns, shock).

- Ferrous ammonium sulfate is harmful by ingestion in quantity, irritating to eyes and skin and may cause burns if contact is prolonged.

- Hydroxylamine hydrochloride solution is harmful if ingested or inhaled and may irritate eyes and skin. Prolonged skin contact may cause dermatitis.

- Orthophenanthroline solution is toxic by ingestion and may irritate eyes and skin.

- Sodium ethanoate solution is harmful by ingestion in quantity and may irritate eyes and skin.

Procedure

(1) First, you need to prepare a calibration graph for iron (the procedure for preparing a calibration graph is outlined in Chapter 1). Pipette 1.00 cm^3 of the standard iron solution in to a calibrated 100 cm^3 volumetric flask. Add about 1 cm^3 of double distilled, deionized water to a second 100 cm^3 volumetric flask (this is your blank). Make sure you label the flasks clearly. To each flask, add 1.0 cm^3 of hydroxylamine solution, 10.0 cm^3 of sodium ethanoate and 10.0 cm^3 of orthophenanthroline. Allow the mixtures to stand for about 5 minutes and then dilute each to the mark with double de-ionised distilled, deionised water.

(2) Clean a pair of matched cells for the spectrophotometer (or photometer). Wash one cell at least three times with small portions of the blank and the other with the standard.

(3) Determine the absorbance of the standard with respect to the blank and make a note of your readings.

(4) Repeat stages 1–3 with three other volumes of iron, such that an absorbance range between about 0.1–1.0 is covered. Plot your readings as a calibration curve.

(5) To determine the iron concentrations of your unknown water samples using a least-squares line fit, pipette 10.0 cm^3 of the unknown into a 100 cm^3 volumetric flask and treat in exactly the same way as for the standards. Measure the absorbance of the unknown with respect to the blank. Obtain replicate measurements after adjusting the sample volume so that the absorbance is between 0.1–1.0.

(6) Report your data as mg dm^{-3} of iron in the sample.

Issues to Consider for your Practical Report

- What are the potential sources of error in this experiment? How could they be overcome?

- How many significant figures should you use when reporting the iron concentration in the water sample?

- Are there alternative methods for determining iron in water? How do they compare to this method?

- What is the source of the unknown river and lake-water samples?

- What are typical iron concentrations in rivers, streams, lakes and drinking water in your local area? How do your data compare with these values?

- What are the potential sources of iron in environmental waters?

- What are the legal limits (if any) of iron in drinking water? Do your values exceed such legal limits?

- What are the potential human health and environmental effects (if any) of excess iron in potable and river waters?

Useful References

Mehlig, R.P. and Hulett, R.H. (1942). 'Spectrophotometric determination of iron with *o*-phenanthroline and with nitro-*o*-phenanthroline'. *Industrial Engineering Chemistry*, **14**, 869.

Caldwell, D.H. and Adams, R.B. (1946). 'Colorimetric determination of iron in water with *o*-phenanthroline'. *Journal of the American Water Works Association*, **38**, 727.

Greenberg, A.E., Clesceri, L.S. and Eaton, A.D. (Eds), (1992). *Standard Methods for the Examination of Water and Wastewater*. American Public Health Association, Washington, DC, USA. ISBN 0–87553–207–1.

Experiment 5.2 Determination of Sodium and Potassium in Water Samples by Flame Photometry

Introduction

Sodium and potassium are essential elements for life and are widely dispersed in the environment. They may be quantitatively determined in environmental samples by using a flame photometer. A *flame photometer* is an instrument used to determine Group I and Group II metals (see Box 4.2). When metal ions in solution are aspirated into a low temperature flame in an aerosol form, the electrons of the ions are excited to higher energy states. They are unstable in this excited state and rapidly return to the ground state, losing the excitation energy as a discrete wavelength of visible light. The emission lines of alkali metals atoms are well separated and consequently the required emission line (wavelength) can be isolated from other wavelengths by an optical filter. The amount of light emitted can be detected with a suitable photo-detector (a simple photo-voltaic cell). For low concentrations of metal ions, the amount of light emitted is proportional to the number of ions in solution. In the instrument, the electrical signal from the photo-detector is amplified and displayed on a digital read-out. The flame photometer has a number of advantages over other techniques, principally low cost and freedom from spectral interferences.

Time Required

Three hours.

Equipment

Flame photometer
  Glass rod
  100 cm^3 beakers
  100 and 250 cm^3 volumetric flasks
  5.00, 2.00 and 1.00 cm^3 pipettes and pipette filler

Reagents

AnalaR grade sodium chloride (NaCl)[†]
  AnalaR grade potassium chloride (KCl)
  Double-distilled, deionized water
  Suitable water sample(s) (water samples should be stored in plastic bottles to eliminate possible contamination due to leaching of metals from glass containers)

[†] *ANALAR* is a trade name of BDH Chemicals Ltd., used to describe a range of high-purity chemicals suitable for use in analytical determinations.

Significant Experimental Hazards

- Students should be aware of the hazards associated with the use of all glassware (cuts), electrical equipment (shocks, burns) and the flame photometer. (burns, fire).

Procedure

(1) Weigh 0.318 g of *Analar* sodium chloride accurately into a 100 cm^3 beaker (the procedure for weighing chemicals is outlined in Chapter 1).

(2) Dissolve the sodium chloride in about 40–50 cm^3 of double-distilled deionized water. Transfer this solution quantitatively into a 250 cm^3 volumetric flask and dilute to the mark. This is your stock solution.

(3) Dilute the stock solution 1:25 to give a 20 mg dm^{-3} Na standard for flame photometry (the procedure for diluting solutions is outlined in Chapter 1). This is your top standard.

(4) From your top standard, prepare standard solutions containing 15.0, 10.0, 5.0 and 2.5 mg dm^{-3} of Na using the 100 cm^3 volumetric flasks. Make sure you label the flasks clearly.

(5) Prepare a stock solution for potassium by using the procedure outlined above and the calculation below.

(6) From your stock solution, prepare a top standard of 10.0 mg l^{-1} K, and from this standard, prepare standard solutions containing 8.0, 6.0, 4.0 and 2.0 mg l^{-1} of K by using the 100 cm^3 volumetric flasks

(7) Get a (laboratory) demonstrator to show you how to use the flame photometer safely and correctly. Use the flame photometer to prepare calibration graphs for Na and K (the procedure for preparing a calibration graph is outlined in Chapter 1).

(8) Determine the Na and K concentrations in the unknown water samples provided (you may need to dilute the samples in order to achieve an appropriate reading on the flame photometer). Obtain replicate measurements for each sample.

(9) Report your data as mg dm^{-3} of Na and K in the samples.

Calculation

Molar mass (NaCl) $= 58.46$ g mol^{-1}
$\quad A_R(\text{Na}) = 23.0$

$\quad$ Therefore: 0.318 g of NaCl contains $\dfrac{0.318 \times 23.0}{58.46}$ g Na $= 0.125$ g Na

Molar mass (KCl) $= 74.55$ g mol^{-1}
$A_R(K) = 39.10$
500 mg l^{-1}K $\equiv$ 125 mg K in 250 cm^3 of water

Therefore: $0.125 \text{ g K} = \dfrac{y \times 39.10}{74.55}$

which gives: $y = \dfrac{74.55 \times 0.125}{39.10} = 0.2383 \text{ g KCl}$

Issues to Consider for your Practical Report

- What are the potential sources of error in this experiment? How could they be overcome?
- How many significant figures should you use when reporting your results?
- Identify alternative methods for determining sodium and potassium in water? How do they compare to this method?
- What is the source of the unknown river and lake-water samples?
- What are the sodium and potassium concentrations in rivers, streams, lakes and drinking water in your local area? How do your data compare with these values?
- Identify and comment upon the emission lines for the Group I and Group II metals.
- What are the legal limits (if any) of alkali metals in drinking water? Do your values exceed such legal limits?
- What are the potential human health and environmental effects of excess salts in potable and irrigation water? Give examples to illustrate your answers.

Useful References

West, P.W., Folse, P. and Montgomery, D. (1950). 'Application of flame spec-tophotometry to water analysis'. *Analytical Chemistry*, **22**, 667–670.

Greenberg, A.E., Clesceri, L.S. and Eaton, A.D. (Eds), (1992). *Methods for the Examination of Water and Wastewater*. American Public Health Association, Washington, DC, USA. ISBN 0–87553–207– 1.

Experiment 5.3 Determination of Phenols in River Water by UV–Visible Spectrophotometry

Introduction

Phenols are compounds containing a hydroxyl group (OH) attached to an aromatic ring. The simplest phenol, benzenol (C_6H_5OH), is always called phenol, by convention. A mixture of phenol and water is commonly known as carbolic acid since phenol dissociates in water to produce a weak acid as follows:

$$C_6H_5OH(aq) + H_2O(l) \rightleftharpoons C_6H_5O^-(aq) + H_3O^+(aq)$$

Dissociation occurs to only a slight extent; phenol is a very weak acid with $pK_a = 10.0$.[†] Phenol in water is a powerful antiseptic, although it is not convenient (indeed safe) for use as its vapour is toxic and the liquid causes skin burns. Most of the simpler phenolic compounds possess similar chemical characteristics and are toxic to aquatic organisms. They also show similar UV/visible spectral behaviour and therefore it is possible to report a concentration of 'total phenols'. This is achieved by making use of the change in the spectra of phenols when made alkaline. This is due to the formation of the phenate ion ($C_6H_5O^-$) in alkaline solution for example:

$$C_6H_5OH(l) + NaOH(aq) \rightleftharpoons C_6H_5O^-Na^+(aq) + H_2O(l)$$

By finding the wavelength at which this change of intensity of absorption is an optimum, it is possible to use 'difference spectrometry' to analyse for phenols in the presence of other pollutants, e.g. surfactants, that do not show this spectral shift.

Time Required

Approximately three hours, depending upon the number of water samples determined.

Equipment

UV/visible single-beam spectrophotometer
  100 cm^3 beakers
  A selection of pipettes (25.0, 10.0, 5.0, 2.0 and 1.0 cm^3) plus pipette filler
  100 and 1000 cm^3 volumetric flasks
  Petri dish
  Tweezers

Reagents

Aqueous phenol solution ($C_6H_5OH(aq)$) (0.100 g in 1 litre or 100 mg dm^{-3})
  Sodium hydroxide (NaOH) pellets
  Double-distilled, deionized water
  River water samples: (a) must be surfactant-free and (b) must contain a linear, anionic surfactant. Both samples should contain >1 mg dm^{-3} total phenols.

[†] pK_a is the negative logaritham of the acidity (dissociation) constant (K_a), where the latter is a measure of the proton-donating strength of the acid to water. A small acidity constant indicates a weak acid (cf.pH).

Significant Experimental Hazards

- Students should be aware of the hazards associated with the use of all glassware (cuts) and electrical equipment (shocks, burns).

- Phenol is harmful by ingestion and if swallowed may cause burns. It is toxic in contact with skin and eyes; if such contact occurs, you must therefore remove contaminated clothing immediately after contact and wash the affected area for at least 10 minutes with plenty of cold water. Seek medical attention immediately.

- Sodium hydroxide is harmful by ingestion and may cause severe burns to eyes and skin.

Procedure

(1) The phenol solution provided has a concentration of 100 mg dm^{-3} or 10^{-3} M if it is assumed that the average molecular weight of the simpler phenolics is 100. Dilute this solution to 5×10^{-4} M by pipetting 50 cm^3 into a 100 cm^3 volumetric flask and making up to the mark with double-distilled deionized, water.

(2) Transfer 25 cm^3 of this diluted solution to a 100 cm^3 beaker and add a pellet of sodium hydroxide by using a pair of tweezers. This will produce an alkaline solution with negligible change in volume.

(3) By using silica cells, obtain the spectra (270–350 nm) of (a) the phenol solution with respect to water as a blank and (b) the alkaline phenol with respect to water as a blank. The spectra should be obtained by recording the absorbance values for each 5 nm.

(4) From your graph, locate λ_{OPT} i.e. where the difference between the absorbances of (a) and (b) is greatest. It is also possible to calculate the effective molar absorbtivity for the absorbance at λ_{OPT}.

(5) Prepare the following dilutions from the phenol solution provided: 10.0, 5.0, 2.0 and 1.0 mg dm^{-3}. For example, the 10.0 mg dm^{-3} dilution may be performed by pipetting 10.0 cm^3 of the standard phenol solution (100 mg dm^{-3}) into a 100 cm^3 volumetric flask and diluting to the mark with double-distilled, deionized water (see Chapter 1).

(6) Transfer approximately 50 cm^3 of the 1.0 mg dm^{-3} solution into a 100 cm^3 beaker and make alkaline by the addition of one pellet of sodium hydroxide. Use this as the sample in one of a pair of silica cells, with the neutral phenol as the 'blank'. Zero the blank in a single-beam spectrophotometer at the identified optimum wavelength and then determine the absorbance of the 'sample'.

(7) Repeat Step 6 for the other standard solutions and plot a calibration graph of concentration against absorbance.

(8) Repeat Step 6 for your surfactant-free river water sample. Determine the concentration of 'total phenols' in this sample by using your calibration graph (see Chapter 1).

(9) Repeat Step 6 for your river water sample that is contaminated with surfactants. Determine the concentration of 'total phenols' in this sample by using your calibration graph. The surfactant should absorb slightly at the optimum wavelength, but its absorbance does not vary with pH, unlike that of the phenolic compounds. Since the same concentration of surfactant is present in both 'sample' and 'blank', its absorbance is effectively cancelled out.

Issues to Consider for your Practical Report

- What are the potential sources of error in this experiment? How could they be overcome?

- Are there alternative methods for determining phenols in water? If so, how do they compare to this method?

- What is the source of the unknown river water samples?

- What are typical phenol concentrations in rivers, streams, lakes and drinking water in your local area? How do your data compare with these values?

- How are phenols made industrially and what are they used for?

- What are the potential sources of phenols in environmental waters? What quantity of phenolic substances enter environmental waters annually?

- What are the legal limits (if any) of phenols in drinking and river waters? Do your values exceed such legal limits?

- What are the potential human health and environmental effects (if any) of excess phenols in potable and river waters?

Useful References

Ochynski, F.W. (1960). The absorptiometric determination of phenol. *Analyst*, **85**, 278–281.

Greenberg, A.E., Clesceri, L.S. and Eaton, A.D. (Eds), (1992). *Standard Methods for the Examination of Water and Wastewater*. American Public Health Association, Washington, DC USA. ISBN 0–87553–207–1.

Self-Study Exercises

Introduction to Minerals, Metals and Fossil Fuels

5.1   Define the following terms:

    (a)   minerals;

    (b)   coal;

    (c)   igneous rocks;

    (d)   metallurgy;

    (e)   precious metals;

    (f)   fossil fuels.

5.2   Decide whether the following statements are true or false, giving reason(s) for your answers.

    (a)   Sedimentary rocks result from physical and chemical alteration of other rocks.

    (b)   Aluminium can occur in nature as a native metal.

    (c)   The chemical properties of metals are determined by how easily they lose electrons.

    (d)   Sodium, potassium and lithium are known as the alkali metals.

    (e)   All mineral ores contain metals.

    (f)   Crude oil is thick and greenish-brown in appearance.

    (g)   Natural gas is often found together with coal.

Metal Ore Formation

5.3   Define the following terms:

    (a)   hydrothermal processes;

    (b)   magmatic concentration;

    (c)   placer deposits;

    (d)   Evaporites;

    (e)   Residual concentration.

5.4   Decide whether the following statements are true or false, giving reason(s) for your answers.

    (a)   Hydrothermal minerals tend to concentrate on beaches and in river beds.

    (b)   Autotrophic bacteria are dependent on organic material for carbon.

    (c)   Placer deposits are often found in hot, tropical climates.

    (d)   Minerals formed from cooling magma form deposits by crystallizing at different temperatures.

(e) Heterotrophic bacteria produce hydrogen sulfide gas that reacts to form insoluble sulfides.

5.5 Give examples of minerals formed by the following processes:

(a) magmatic concentration;

(b) evaporation of seawater;

(c) mechanical concentration;

(d) microbiological processes;

(e) hydrothermal processes;

(f) ore enrichment.

Metal Extraction Processes

5.6 Define the following terms:

(a) enthalpy;

(b) sublimation;

(c) solvation energy;

(d) Hess's Law;

(e) endothermic;

(f) electrode potential;

(g) activation energy.

5.7 Draw a complete, fully labelled Born–Haber cycle for the formation of solid calcium oxide ($CaO(s)$) from its elements in their standard states by using the following information.

$$Ca(s) + 1/2O_2(g) \longrightarrow Ca^{2+}O^{2-}(s) \qquad \Delta H^{\ominus}_{r,m} = -636 \text{ kJ mol}^{-1}$$
$$Ca(s) \longrightarrow Ca(g) \qquad \Delta H^{\ominus}_{s1,m} = +177 \text{ kJ mol}^{-1}$$
$$1/2O_2(g) \longrightarrow O(g) \qquad \Delta H^{\ominus}_{s2,m} = +248 \text{ kJ mol}^{-1}$$
$$Ca(g) \longrightarrow Ca^+(g) + e^- \qquad \Delta H^{\ominus}_{s3,m} = +590 \text{ mol}^{-1}$$
$$Ca^+(g) \longrightarrow Ca^{2+}(g) + e^- \qquad \Delta H^{\ominus}_{s4,m} = +1100 \text{ mol}^{-1}$$
$$O(g) + 2e^- \longrightarrow O^{2-}(g) \qquad \Delta H^{\ominus}_{s5,m} = +745 \text{ kJ mol}^{-1}$$

(a) Calculate the lattice energy for calcium oxide.

(b) Explain why $\Delta H^{\ominus}_{s4,m} > \Delta H^{\ominus}_{s3,m}$.

5.8 Draw a complete, fully labelled Born–Haber cycle for the formation of solid potassium chloride ($KCl(s)$) from its elements in their standard states by using the following information. In addition calculate the standard enthalpy of formation of solid potassium chloride.

$$K(s) \longrightarrow K(g) \qquad \Delta H^{\ominus}_{s1,m} = +90 \text{ kJ mol}^{-1}$$
$$1/2Cl_2(g) \longrightarrow Cl(g) \qquad \Delta H^{\ominus}_{s2,m} = +122 \text{ kJ mol}^{-1}$$
$$K(g) \longrightarrow K^+(g) + e^- \qquad \Delta H^{\ominus}_{s3,m} = +418 \text{mol}^{-1}$$
$$Cl(g) + e^- \longrightarrow Cl^-(g) \qquad \Delta H^{\ominus}_{s4,m} = -348 \text{ kJ mol}^{-1}$$
$$K^+(g) + Cl^-(g) \longrightarrow KCl(s) \qquad \Delta H^{\ominus}_{r,m} = -718 \text{ kJ mol}^{-1}$$

5.9   Arrange the following metals in order of increasing strength as reducing agents.

(a)   Li, Na, K, Mg;

(b)   Pb, Sn, Au, Ag;

(c)   Cu, Fe, Ca, Ag.

5.10   A mining company finds a large deposit of magnesium carbonate. A mining engineer suggests using the following method for extracting the magnesium from this deposit:

$$MgCO_3(s) \xrightarrow{\text{heat}} MgO(s) + CO_2(g)$$

$$MgO(s) + CO(g) \xrightarrow{\text{heat}} Mg(s) + CO_2(g)$$

Use Figure 5.5 to explain why his method will not work.

5.11   Use Figure 5.5 to help you answer the following questions:

(a)   Explain the change in gradient in the MgO graph at 1400 K.

(b)   Explain the following observations – when silver nitrate is heated, the product is silver, while when zinc nitrate is heated, the product is zinc oxide.

(c)   Name an oxide that you predict will not be reduced by aluminium.

(d)   Explain the utility of coke (carbon) as a reducing agent for metal oxides.

(e)   Give the temperature range over which magnesium oxide can be reduced by coke.

5.12   Determine whether titanium dioxide can be reduced by carbon at 1000 K in the following reactions (at 1000 K; $\Delta G^{\circ}_{CO_2(g)} = -396 \text{ kJ mol}^{-1}$, $\Delta G^{\circ}_{CO(g)} = -200 \text{ kJ mol}^{-1}$, $\Delta G^{\circ}_{TiO_2(s)} = -762 \text{ kJ mol}^{-1}$).

(a) $TiO_2(s) + C(s) \longrightarrow Ti(s) + CO_2(g)$

(b) $TiO_2(s) + 2C(s) \longrightarrow Ti(s) + 2CO(g)$

Extraction Technology for Selected Metals

5.13   Decide whether the following statements are true or false, giving reason(s) for your answers.

(a)   Sodium is extracted from sodium chloride by thermal reduction during the Pidgeon Process.

(b)   Group I metals cannot be extracted by reduction of their oxides or electrolysis of aqueous solutions.

    (c) Manganese may be extracted from seawater.

    (d) The raw materials for the manufacture of iron are coke, limestone and iron pyrites.

    (e) Steel is an alloy of iron with carbon and other elements.

    (f) Nickel is extracted from its oxide by the Kroll process.

    (g) Limestone neutralizes acidic impurities during the manufacture of iron.

    (h) Iron silicate is removed as an impurity during the manufacture of copper.

    (i) Aluminium is extracted by the electrolysis of cryolite.

    (j) Magnesium nodules found on the sea floor are highly enriched in many metals.

5.14 Write complete, balanced equations for the extraction of the following metals from their ores, listing all reagents and physical conditions.

    (a) potassium from potassium chloride;

    (b) magnesium from dolomite;

    (c) aluminium from bauxite;

    (d) titanium from rutile;

    (e) manganese from pyrolusite;

    (f) copper from chalcopyrite;

    (g) nickel from nickel (II) sulfide;

    (h) magnesium from seawater.

5.15 Draw fully labelled, schematic diagrams of the following industrial plants:

    (a) A blast furnace for the extraction of iron from its ores.

    (b) The Hall–Heroult cell for the production of aluminium.

    (c) An electric arc furnace.

5.16 Relate the properties of aluminium, nickel and sodium to their uses.

5.17 What are the advantages and disadvantages of steel as a structural material compared with aluminium?

5.18 Explain in detail why many companies are developing technology for harvesting manganese nodules from the seabed.

Corrosion of Metals

5.19 Give the approximate chemical formula of rust and explain why oxygen and water are both required for iron to rust.

5.20 Explain in detail why rusting is accelerated in polluted areas.

5.21 A chromium-plated steel scooter handlebar is scratched in a collision with a wall. Will the rusting of the iron in the steel be encouraged or retarded by the chromium? Give reason(s) for your answer.

5.22 Suggest two metals that could be used for the cathodic protection of a chromium pipeline. What factors, other than electrochemical properties, need to be considered in practice?

5.23 Explain why magnesium and zinc are often used as sacrificial anodes in the cathodic protection of buried steel pipelines and identify the electrolyte solution.

5.24 Explain why galvanizing, as opposed to tinning, is a more permanent method of protecting iron from corrosion.

5.25 Search for corrosion and corrosion costs on the internet. Look at the range of companies and services provided for corrosion protection.

Fossil Fuels

5.26 Define the following terms:

  (a) kerosene;.

  (b) sapropelic coal;

  (c) unconventional natural gas;

  (d) dispersant;

  (e) anthracite;

  (f) oil boom;

  (g) anaerobic decomposition.

5.27 Explain, in detail, why crude oil is unlikely to have been formed simply by the accumulation of biogenic material.

5.28 Explain how coal becomes richer in carbon as it is formed from peat, using chemical equations to support your argument.

5.29 Identify the typical fractions obtained from crude oil and list their uses.

5.30 Discuss the potential environmental consequences of mining and burning coal.

5.31 Outline the potential health and environmental effects of oil spills at sea.

5.32 List the uses of both petroleum and coal and their by-products.

5.33 Find out about oil shale and tar sands and comment on their future use as fossil fuels.

5.34 Outline the geological environment you think would favour the large-scale accumulation of oil.

Challenging Exercises

5.35 A heap of metal ore, estimated at 300 million tonnes and containing 0.8% by mass of copper, is situated in a remote mountainous region of an African country. Eventually, it becomes economically exploitable as other sources of copper become poorer. Copco Ltd propose to extract the copper in a three-stage process, as follows:

  • dissolving the crushed ore in sulfuric acid to obtain an impure copper (II) sulfate solution from the copper (II) oxide ore;

- solvent extraction of the aqueous copper at pH 2 and 25°C into an organic phase;

- extraction of the organic-phase copper into a new aqueous solution at low pH, followed by electrolysis.

(a) Identify the energy needs of the proposed process.

(b) Explain how Copco Ltd could safely dispose of the hazardous waste that arises.

(c) Explain why three stages are proposed for the extraction of the metal from its ore.

(d) Identify potential markets for the extracted copper.

5.36 Do you think that the European Union countries should stockpile oil against potential future shortages caused by political disruptions? Identify the advantages and disadvantages of such a strategy.

5.37 Minamata is a town on the west coast of Kyushu Island (Japan) where an extreme case of heavy metal poisoning from methyl mercury occurred in the 1950s. The mercury was ingested in the diet of the inhabitants, thus causing severe disablement and poisoning. Write a brief account of the Minamata incident, identifying the source of the mercury pollution and commenting on the subsequent health and environmental effects.

Further Sources of Information

Further Reading

Belousova, A.P., Krainov, S.R. and Ryzhenko, B.N. (1999). Evolution of groundwater chemical composition under human activity in an oilfield. *Environmental Geology*, **38**(1), 34–46.

Bonatti, E. (1978). The origin of metal deposits in the oceanic lithosphere. *Scientific American*, **238**(2), 54–61.

Brimhall, G. (1991). The genesis of ores. *Scientific American*, **264**(5), 48–55.

Franks, F. (1983). *Water*. The Royal Society of Chemistry, Cambridge, UK. ISBN 0-85186-473-2.

Harada, M. (1995). Minamata Disease: Methylmercury poisoning in Japan caused by environmental pollution. *Critical Reviews in Toxicology*, **25**(1), 1–24.

Hodges, C.A. (1995). Mineral resources, environmental issues and land use. *Science*, **268**(5315), 1305–1312.

McBeath, M.K., Rock, P.A., Casey, W.H. and Mandell, G.K. (1998). Gibbs energies of formation of metal carbonate solid solutions, Part 3: The $Ca_xMn_{1-x}CO_3$ system at 298 K and 1 bar. *Geochimica et Cosmochimica Acta*, **62**(16), 2799–2808.

Meyer, C. (1985). Ore metals through geologic history. *Science*, **227**(4693), 1421–1428.

Nriagu, J., (1990). Global metal pollution, poisoning the biosphere. *Environment*, **32**(7), 7–11.

Nriagu, J., (1996). A history of global metal pollution. *Science*, **272**(5259), 223–224.

Philippi, G.T., (1974). The influence of marine and terrestrial source material on the composition of petroleum. *Geochimica et Cosmochimica Acta*, **38**, 947–966.

Press, F. and Siever, R. (1986). *Earth* 4th Edn. W.H. Freeman and Company, New York. ISBN 0-7167–1776-X.

Preston, A. (1973). Heavy metals in British waters. *Nature*, **242**, 95–97.

Useful reference books

Carmichael, R.S. (1989). *CRC Practical Handbook of Physical Properties of Rock and Minerals. CRC Press Inc.*, Boca Raton, FL, USA. ISBN 0–8493-3703–8.

Cox, P.A. (1995). *The Elements on Earth.* Oxford University Press, Oxford, UK. ISBN 0–19–855903–8.

Smith, J.E. (Ed.), (1968). *Torrey Canyon: Pollution and Marine Life.* Cambridge University Press, Cambridge. UK ISBN 0-5210-7144-50.

Tylecote, R.F. (1976). *A History of Metallurgy.* The Metals Society, London. ISBN 0–904357–06–6.

Useful websites

Coal

Kentucky Coal Council
http://www.coaleducation.org/ (good links site)

Petroleum

US Department of Energy
http://www.eia.doc.gov/energy/petrol.html

Oil spill research
http://www.lib.lsu.edu/osradp/osradp.html

6 Natural Cycles

6.1 Introduction

The total amount of matter on Planet Earth is fixed; inputs of matter (e.g. from meteors) and outputs of matter (such as satellites) are negligible. Consequently, the Earth is a closed system and the chemicals essential for life must continuously cycle through the biosphere. These cyclical movements of chemicals are the *biogeochemical* cycles (bio for living; geo for water, rocks, soil, etc.; chemical for the reactions involved). There are three types of natural cycle:

- the hydrological (or water) cycle;
- the gaseous cycles;
- the sedimentary cycles.

The energy for these cycles predominantly comes from the Sun, although some is supplemented from the internal energy of the Earth. The cycles are interlinked and dependent upon one another.

Chemical species that are actively taken up by organisms and used to maintain their functions (life, growth and reproduction) are termed *nutrients*. The chemical elements vital for life are known as *essential elements* or *bioelements*. About 30 bioelements have been identified, although they are not equally important and may be subdivided into the *macro-nutrient* and the *micro-nutrient* elements.

The macro-nutrient elements are needed in greater quantities than other essential elements. These are oxygen, carbon, hydrogen, nitrogen, calcium, phosphorus, sulfur, potassium and magnesium. All living organisms are made up of a combination of some, or all, of these nine elements. Micro-nutrient elements are present in living organisms in smaller (trace) quantities. The most important micro-nutrients are vanadium, chromium, molybdenum, manganese, iron, cobalt, nickel, copper, zinc, tin, boron, silicon, arsenic, selenium, fluorine, chlorine and iodine. Elements such as cadmium, germanium, barium and bromine have been identified as potential micro-nutrients. The functions of the essential elements and associated chemical species are summarized in Table 6.1.

Table 6.1 Functions of some essential elements and associated chemical species

Element	Important species or substance	General importance or function
Macronutrients		
Calcium	Ca^{2+}	Constituent of bone, shells and enamel. Key to the formation of cell walls and cell wall development. Activates enzymes during muscle contraction
Chlorine	Cl^-	Important in anion/cation and osmotic balance. Constituent of hydrochloric acid in stomach juice. Involved in carbon dioxide transport in blood
Magnesium	Mg^{2+}	Part of the structure of chlorophyll. Constituent of bone and tooth. Cofactor for many enzymes
Nitrogen	NO_3^-, NH_4^+	Vital for synthesis of proteins, nucleic acids, etc.
Phosphorus	PO_4^{3-}, $H_2PO_4^-$	Vital for synthesis of ATP[a], nucleic acids and some proteins. Constituent of bone and tooth enamel
Potassium	K^+	Important in anion/cation and osmotic balance, conduction of nervous impulses and maintenance of electrical potentials across membranes. Cofactor in photosynthesis and respiration
Sodium	Na^+	As for potassium
Sulfur	SO_4^{2-}	Vital for synthesis of proteins and other organic compounds
Micronutrients		
Cobalt	Vitamin B_{12}	Red blood cell development
Copper	Haemocyanin	Oxygen carrier in some invertebrates
	Plastocyanin	Electron carrier in photosynthesis
	Cytochrome oxidase	Electron carrier in respiration
	Tyrosinase	Vital for synthesis of melanin
Fluorine	Calcium fluoride	Constituent of tooth enamel and bone
Iodine	Thyroxine	Hormone that controls metabolic rate
Iron	Haem group in haemoglobin and myoglobin	Responsible for carrying oxygen in the body
	Cytochromes	Electron carriers in photosynthesis and respiration
	Porphorins	Vital for the synthesis of chlorophyll
Manganese	Phosphatases	Vital for bone development
Molybdenum	Nitrate reductase	Reduction of nitrate to nitrite during synthesis of amino acids in plants
	Nitrogenase	Vital for nitrogen fixation
Zinc	Alcohol dehydrogenase	Anaerobic respiration in plants
	Carbonic anhydrase	Transport of carbon dioxide in blood of vertebrates

[a]ATP, adenosine triphosphate.

In this present chapter, we will consider the bioelements involved in the gaseous and sedimentary cycles; the hydrological cycle is discussed in more detail in Chapter 7. The abundance, physico-chemical properties, cycling and uses of each element are discussed together with any environmental implications that may arise as a result of human interventions.

6.2 Iron

6.2.1 Abundance and properties

Iron (Fe) is a hard, silver-grey metallic element. It is a *transition metal*–a metal with two electrons in its outer energy shell and an incomplete penultimate shell of similar energy. Transition metals possess certain characteristic features, including high boiling and melting points, high densities, variable oxidation states, an ability to catalyse specific reactions and notable malleability and ductility. Iron is a moderately reactive element, stable in dry air at ordinary temperatures, although it rusts rapidly in moist air (see Chapter 5). The physical and chemical properties of iron are summarized in Table 6.2.

Iron is the second most abundant metal in the crust of the Earth and the fourth most abundant of all elements. As we saw above in Chapter 4, the Earth's core is predominantly composed of liquid iron and the mantle contains large quantities of iron-containing silicates. The crust contains much lower concentrations of iron; the most important minerals are haematite (Fe_2O_3), magnetite (Fe_3O_4), siderite ($FeCO_3$) and iron pyrites (FeS_2). It is clear from these observations that the *oxidation numbers* of iron change from 0 in the core to +2 in the mantle to a mixture of +2 and +3 in the crust. (The term oxidation number is explained in Box 6.1.) The changes in iron's oxidation number are of key importance to understanding its mobility in the Earth's crust.

Table 6.2 Physical and chemical properties of iron

Property	Value/Example
Atomic number	26
Atomic weight (g)	55.8470
Density at 298 K (g cm^{-3})	7.86
Naturally occurring isotopes	^{54}Fe (5.84%); ^{56}Fe (91.75%); ^{57}Fe (2.12%); ^{58}Fe (0.28%)
Radioactive isotopes	^{52}Fe, ^{53}Fe, ^{55}Fe, ^{59}Fe, ^{60}Fe, ^{61}Fe, ^{62}Fe
Oxidation numbers	
0	Free iron (Fe)
+2	Iron (II) oxide (FeO); iron (II) hydroxide ($Fe(OH)_2$); iron (II) chloride ($FeCl_2$); iron (II) sulfate ($FeSO_4$); iron pyrites (FeS_2)
+3	Iron (III) oxide (Fe_2O_3); iron (III) hydroxide ($Fe(OH)_3$); iron (III) chloride ($FeCl_3$); iron (III) sulfate ($Fe_2(SO_4)_3$)
Mixed	Iron(II), di-iron (III) oxide, $Fe^{2+}O^{2-}.Fe_2^{3+}O_3^{2-}$ or Fe_3O_4

Box 6.1 Oxidation Numbers

The term *redox* is used by chemists as an abbreviation for the processes of *red*uction and *ox*idation. Reduction and oxidation normally occur simultaneously. In modern chemistry, redox reactions include all electron transfer processes, including combustion, respiration and rusting. Many redox processes–usually those involving ions–involve a complete transfer of electrons from one substance to another. However some redox reactions (e.g. the combustion of carbon, $C + O_2 \longrightarrow CO_2$) do not involve a complete transfer of electrons between substances. In order to decide whether redox has occurred in reactions involving transfer or re-sharing of electrons, we use *oxidation numbers*.

The oxidation number (or oxidation state) of an element is equal to its combining power with oxygen, i.e. its relative state of oxidation or reduction. Elements in both ionic and covalent substances can have oxidation numbers. The rules for assigning oxidation numbers are:

(1) The oxidation number of atoms in uncombined elements is zero, e.g. the oxidation numbers of both Cu and $N_2 = 0$.

(2) The oxidation number of oxygen in all its compounds (except peroxides O_2^- and some oxygen halides) is -2.

(3) The oxidation number of hydrogen in all its compounds (except metal hydrides) is $+1$.

(4) The algebraic sum of the oxidation numbers of all the atoms in a species is equal to its total charge, e.g. the total charge of potassium chloride (KCl) is zero since K^+ has an oxidation number of $+1$ and Cl^- has an oxidation number of -1.

(5) In any substance, the more electronegative atom has the negative oxidation number.

When an element has more than one oxidation number, the numerical value is written as a Roman numeral after the element's name, e.g. iron (II), Fe^{2+}.

Oxidation numbers are used to explain the chemical processes occurring during redox reactions. Atoms are reduced when their oxidation number decreases and oxidized when their oxidation number increases. Consider the reaction:

$$2K \quad + \quad Cl_2 \quad 2KCl$$

Oxidation numbers $\qquad 0 \qquad\quad 0 \quad +1 -1$

Potassium has been oxidized as its oxidation number has increased from 0 to $+1$, whereas chlorine has been reduced because its oxidation number has decreased from 0 to -1.

Millions of years ago, before the advent of oxygenic photosynthesis, the oceans were rich in dissolved iron (II). However, when oxygen first began to appear in the biosphere, iron (II) spontaneously reacted with it, forming iron (III) as follows:

$$Fe^{2+}(aq) + 1/4O_2(g) + H^+(aq) \longrightarrow Fe^{3+} + 1/2H_2O(l)$$

The iron (III) ions subsequently reacted with hydroxide ions, producing an insoluble, reddish coloured precipitate of iron (III) hydroxide:

$$Fe^{3+} + 3OH^-(aq) \longrightarrow Fe(OH)_3(s) \tag{6.1}$$

These two reactions have proved very useful to scientists since the presence of a series of formations in which brown bands of iron (III) hydroxide occur in the Earth's crust indicates the time period when oxygen first appeared on the planet.

Dissolved iron is still found in trace quantities in the hydrosphere and insoluble iron compounds are present in streams and oceans as suspended solids. Iron is also found together with manganese in manganese nodules at the bottom of the oceans. These nodules consist of iron oxide (Fe_2O_3), manganese oxide (MnO_2), small amounts of clays, $CaCO_3$, SiO_2 and organic matter. Trace quantities of Ni, Cu and Co may also be present in the nodules (see Chapter 5). In the atmosphere, iron is found in suspended particles of dust. Typical concentrations of iron-containing substances in environmental (and biological) systems are given in Table 6.3, although you should be aware that environmental concentrations can vary widely from place to place.

6.2.2   Cycling of iron

A schematic diagram of the iron cycle is shown in Figure 6.1. Dust particles containing iron may be blown by the wind to and from the crust and also may be deposited on the crust and in the oceans by rainout. The cycling of iron between the land and the oceans occurs mainly through transfer of suspended solids because of the low solubility of most iron compounds.

In the lithosphere, compounds of iron occur in two important oxidation states, Fe^{2+} (iron(II)) and Fe^{3+} (iron(III)). In order to understand iron's properties and behaviour, including mobility, in the Earth's crust, we need to be aware of the environmental circumstances that influence its oxidation state. Iron (II) compounds are generally more soluble but less stable than compounds of iron (III), although their relative stabilities are heavily dependent upon environmental conditions.

Table 6.3 Typical concentrations of iron-containing substances in environmental and biological systems

Location	Concentration
Earth's crust (by weight)	5%
Sea water (by weight)	
Surface	$0.01–0.1~\mu g~dm^{-3}$
Deep	$0.1–0.4~\mu g~dm^{-3}$
Stream water (by weight)	$670~\mu g~dm^{-3}$
Drinking water (by weight)	$0.01–10.0~mg~dm^{-3}$
Human body (by weight)	
Average	$60~mg~dm^{-3}$
Blood	$450~mg~dm^{-3}$
Atmosphere	
In airborne particulate matter (by mass per unit volume)	$0.5~\mu g~m^{-3}$

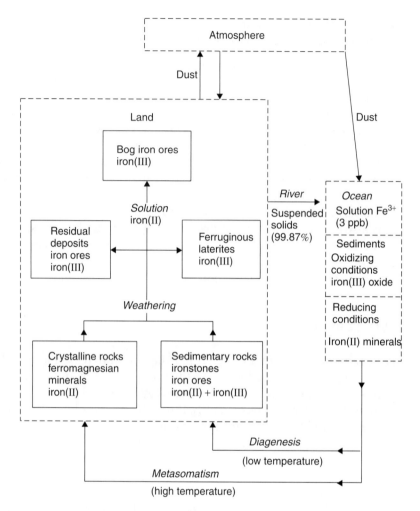

Figure 6.1 Schematic representation of the iron cycle

Small changes in pH (proton activity) or redox potential (electron activity) can cause Fe^{2+} to be oxidized to Fe^{3+} or Fe^{3+} to be reduced to Fe^{2+}, with resultant effects on solubility (and thus mobility).

Both free iron (Fe^0 or iron(0)) and iron (II) are generally unstable in the oxygen-rich atmosphere of the Earth and may be oxidized to iron (III):

$$4Fe + 3O_2 \longrightarrow 4Fe^{3+} + 6O^{2-}$$

$$4Fe^{2+} + O_2 \longrightarrow 4Fe^{3+} + 2O^{2-}$$

The oxidation of Fe^{2+} to Fe^{3+} occurs fairly slowly, but at pH 7–8, the reaction is driven forward by the removal of Fe^{3+} by precipitation. In moist soil, Fe^{3+} may be precipitated as iron (III) hydroxide, $Fe(OH)_3$, which can accumulate in the subsoil of waterlogged or alkaline soils (see Equation 6.1). The removal of (alkaline) hydroxide ions by precipitation with Fe^{3+} effectively results in soil acidification. Iron (III) hydroxide dissolves in acids, but on standing (e.g. in waterlogged soil), it changes to a less soluble form:

$$2Fe(OH)_3 \longrightarrow Fe_2O_3.3H_2O \longrightarrow Fe_2O_3 + 3H_2O$$

Iron (III) compounds do not dissolve in water as long as there is oxygen present. In anaerobic conditions, insoluble iron (III) is converted to soluble iron (II), which may be leached away. Groundwater contains appreciable amounts of iron (II) because a limited amount of dissolved oxygen and a high level of carbon dioxide are present. However, in river waters that are well aerated, iron is always present as insoluble (suspended) iron (III). Consequently, the presence of iron in drinking water is unlikely to cause a threat to human health at typical concentrations in domestic water supplies. In water treatment plants, insoluble iron (III) compounds are removed by initial filtration of suspended material. Aeration is then used to oxidize (soluble) Fe^{2+} to (insoluble) Fe^{3+}, which can then be removed by filtration. The European Union (EU) Drinking Water Directive sets the maximum admissible concentration (MAC) of iron in potable water at 0.2 mg dm^{-3}, although the desirable level is only 0.05 mg dm^{-3}.

The Fe^{2+}/Fe^{3+} system plays an important part in many other environmental reactions, as illustrated by Examples 6.1 and 6.2 below. However, in natural systems there are many dissolved species other than iron that can influence the behaviour of iron compounds by reacting with them, and hence the Fe^{2+}/Fe^{3+} system is very complex.

Example 6.1

The cycling of sulfur (see Section 6.7.2):

$$SO_4{}^{2-} + 9H^+ + 8e^- \longrightarrow HS^- + 4H_2O$$
$$8Fe^{2+} - 8e^- \qquad\qquad \longrightarrow 8Fe^{3+}$$

Overall $\quad SO_4{}^{2-} + 8Fe^{2+} + 9H^+ \longrightarrow HS^- + 8Fe^{3+} + 4H_2O$

Example 6.2

The weathering of iron silicate:

$$Fe_2SiO_4(s) + 5/2O_2(g) + 5H_2O(l) \longrightarrow 2Fe(OH)_3(s) + H_4SiO_4(aq)$$

In the atmosphere, transition metal ions such as Fe^{2+} are important catalysts for the production of free radicals. Transition metal ions may enter raindrops in trace concentrations because airborne dust particles can act as nuclei upon which water vapour may condense. Fe^{2+} is very reactive with $O_2(g)$ because it has four unpaired d electrons which are readily donated, forming a dioxygen adduct:

$$Fe^{2+} + O_2 \rightleftharpoons (FeO_2)^{2+} \tag{6.2}$$

This adduct can dissociate to produce a superoxide ion, O_2^- (see equation 6.3), that can either react with a proton to form the hydroperoxyl radical, HO_2^-, or two protons to form hydrogen peroxide, H_2O_2 (see equation 6.4). Hydrogen peroxide can react further to generate a hydroxyl radical, HO^- (see equation 6.5):

$$(FeO_2)^{2+} \rightleftharpoons Fe^{3+} + O_2^- \tag{6.3}$$

$$2O_2^- + 2H^+ \rightleftharpoons O_2 + H_2O_2 \tag{6.4}$$

$$Fe^{2+} + H_2O_2 + H^+ \rightleftharpoons Fe^{3+} + H_2O + HO^- \tag{6.5}$$

Under typical atmospheric conditions, the Fe^{3+} ions (from Equation 6.5) form $Fe^{3+}OH^-$ complexes that regenerate the Fe^{2+} ion upon absorption of ultraviolet radiation (h_v) from the sun:

$$Fe^{3+}OH^- + h_v \rightleftharpoons Fe^{2+} + HO^-$$

By these and other similar reaction mechanisms, Fe^{2+} can induce atmospheric photocatalytic cycles that generate reactive free radicals in the presence of sunlight and $O_2(g)$.

6.2.3   Iron in biological and environmental systems

Iron has many important functions in biological systems. In vertebrates, one of the functions of iron in red blood cells is to act as an oxygen carrier. Red blood cells carry oxygen from the respiratory organ (the lungs) to the tissues. Each red blood cell is filled with the red pigment haemoglobin (see Figure 6.2), a complex protein containing four iron haem groups. Oxygen diffuses into the red blood cell across its thin membrane and combines with haemoglobin (Hb) to form oxyhaemoglobin (HbO_8). A molecule of oxygen gas can combine with each of the four iron-containing haem groups in the haemoglobin molecule. The reaction is similar to that described in Equation 6.2 in that the Fe^{2+} ion is not oxidized by the attachment of the oxygen molecule and remains in the divalent state throughout the process:

$$Hb + 4O_2(g) \underset{\text{tissues}}{\overset{\text{lungs}}{\rightleftharpoons}} HbO_8$$

Haemoglobin has a high affinity for oxygen molecules, which are rapidly attached to it in the lungs. However, the cells of the body contain myoglobin, which has a higher affinity for oxygen than haemoglobin. The myoglobin therefore accepts the oxygen from the haemoglobin and stores it in the tissues until it is required by the muscles. You can see from Figures 6.2 and 6.3 that myoglobin is chemically similar to haemoglobin.

Unfortunately, haemoglobin combines even more readily with carbon monoxide (CO) than with oxygen, forming carboxyhaemoglobin (COHb):

$$HbO_2 + CO(g) \underset{\text{tissues}}{\overset{\text{lungs}}{\rightleftharpoons}} HbCO + O_2(g)$$

Carboxyhaemoglobin is much more stable than oxyhaemoglobin, and therefore the transport of oxygen from the lungs to the tissues will be inhibited in atmospheres

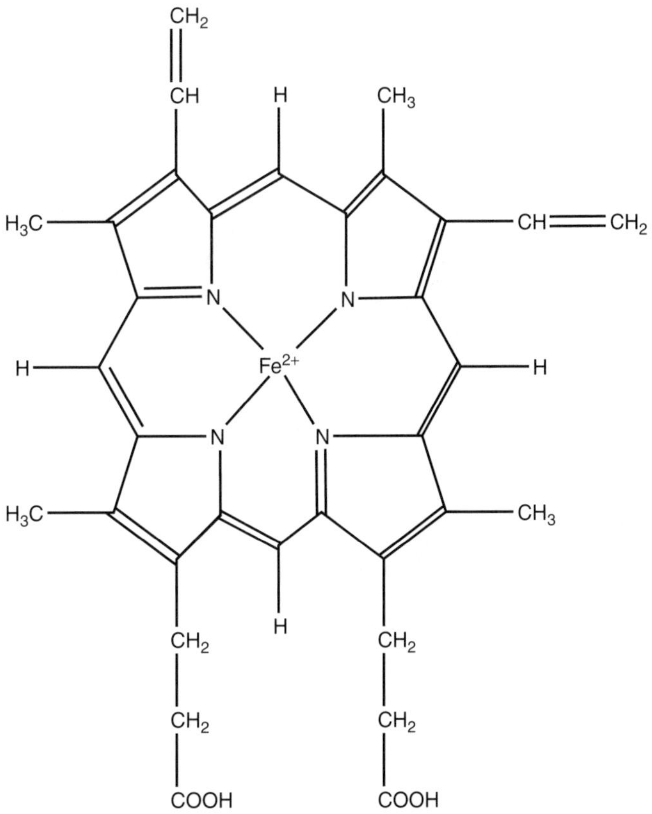

Figure 6.2 Structure of haemoglobin

with high CO concentrations. This is the reason for the high toxicity of CO. Cigarette smoking will usually lead to elevated levels of COHb in the blood of smokers. Recovery from mild CO poisoning may be effected by providing the person with pure oxygen to breathe, which will move the equilibrium of Equation 6.6 from right to left.

Iron is an essential micro-nutrient in animals. Most of the iron in the body is continuously recycled, although iron deficiency can arise from poor diets or excessive bleeding. Animals have only a limited ability to excrete metabolic iron, and maintain the correct iron balance by regulating the rate of dietary absorption. Iron deficiency reduces the level of haemoglobin in red blood cells, causing anaemia. Any shortage of haemoglobin in the blood decreases oxygen transport to the tissues, and hence energy generation in the cells. The main symptoms of anaemia are fatigue, palpitations, breathlessness and poor resistance to infection.

Iron compounds are also essential micro-nutrients for plants. When soil is iron deficient or its pH is too high, iron availability decreases. Iron deficiency in plants becomes noticeable when their leaves become light in colour or turn yellow. The problem can be overcome by the use of organic compounds to form a chelate (a ring compound where the metal atom becomes part of the ring) with iron. The chelate ring holds the iron tightly (forms a stable complex with it) but releases it at the root of the plant, thus making it bio-available.

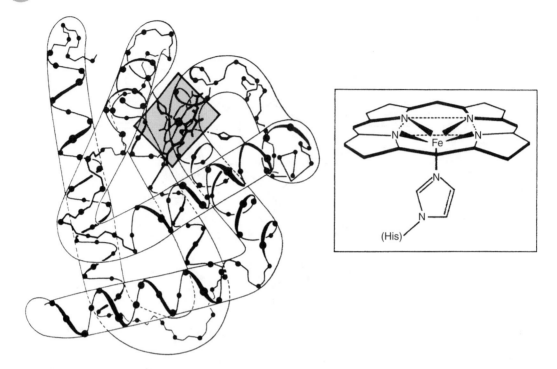

Figure 6.3 Structure of myoglobin. Reproduced from Dickerson (1964) by permission of Academic Press, Orlando, Florida

Iron is an essential component of the catalyst involved in the formation of chlorophyll, the green pigment present in most plants. Chlorophyll is responsible for the absorption of light energy during photosynthesis, and is similar in structure to haemoglobin, except with magnesium rather than iron as the reactive metal centre of the molecule.

6.2.4 Uses of iron

The extraction of iron from its ores, and its usefulness in making pig, wrought and cast iron, as well as carbon and alloy steels, has been discussed in detail in Section 5.5.2. Iron and steel are widely used as structural materials in buildings, bridges, pipes, railways, roads and vehicles, as well as in furniture, household items, etc.

6.3 Carbon

6.3.1 Abundance and properties

Carbon is a non-metal that exists naturally in three pure forms—diamond, graphite and buckminsterfullerene. Diamond is the hardest substance known and is found as small, usually colourless, transparent crystals. In contrast, crystalline graphite is black, shiny, opaque and very soft. Buckminsterfullerene (C_{60}), known only since

1985, consists of five- and six-membered rings arranged in a sphere. Well-known varieties of carbon, such as coke, soot and charcoal, are composed of minute crystals of graphite and are said to be microcrystalline. Pure carbon is very stable–it is unaffected by air, water, acids and alkalis at ordinary temperatures.

A carbon atom has four electrons in its outer energy level and may thus form four covalent bonds by sharing with four electrons from other atoms. Carbon atoms readily form multiple covalent bonds with each other and with other atoms, thus resulting in long chains and rings, e.g. in alkanes, alkenes, alkynes and aromatic compounds. The study of the chemistry of carbon compounds is called *organic chemistry*. The physical and chemical properties of carbon are summarized in Table 6.4.

Carbon compounds in the atmosphere include carbon dioxide (CO_2), methane (CH_4), carbon monoxide (CO) and particulate elemental carbon (PEC) (in dust). On land, carbon compounds include biotic and abiotic organic matter, carbonaceous rocks (chalk and limestone) and hydrocarbons in fossil fuels (petroleum, coal and natural gas). In the oceans, carbon compounds include dissolved CO_2, the carbonate and bicarbonate (HCO_3^-) ions and a variety of organic substances. The vast majority of carbon is present in the form of inorganic rocks. The carbon content of the atmosphere, land biota, soil humus, fossil fuels and marine biota represent less than 1 per cent of total carbon. Typical concentrations of carbon-containing substances in environmental and biological systems are presented in Table 6.5.

6.3.2 Cycling of carbon

The carbon cycle, illustrated in Figure 6.4, includes four main reservoirs of stored carbon:

- carbon dioxide in the atmosphere;
- dissolved carbon dioxide in the oceans and other water bodies;
- organic compounds in living or recently dead organisms (plants and animals);
- calcium carbonate in limestone and carbon in buried organic matter (e.g. humus, peat, coal, petroleum and natural gas).

Table 6.4 Physical and chemical properties of carbon

Property	Value/Example
Atomic number	6
Atomic weight (g)	12.0107
Density at 298 K ($g\,cm^{-3}$)	
Diamond	3.53
Graphite	2.25
Naturally occurring isotopes	^{12}C (98.93%), ^{13}C (1.07%)
Radioactive isotopes	$^{9}C, ^{10}C, ^{11}C, ^{14}C, ^{15}C, ^{16}C, ^{17}C$
Oxidation numbers	
-4	Methane (CH_4)
0	Graphite (C)
-2	Carbon monoxide (CO)
$+4$	Carbon tetrafluoride (CF_4); carbon tetrachloride (CCl_4); carbon dioxide (CO_2); carbonate (CO_3^{2-})

Table 6.5 Typical concentrations of carbon-containing substances in environmental and biological systems

Location	Concentration
Earth's crust (by weight)	200 ppm
Sea water (by weight)	
Inorganic	26–30 mg dm^{-3}
Organic	1–2 mg dm^{-3}
Drinking water (by weight)	1.0–1000 mg dm^{-3} (as HCO_3^-)
	0.01–10.0 mg dm^{-3} (as CO_3^{2-})
Human body (by weight)	230 000 mg dm^{-3}
Atmosphere	
Gases (by volume)	
CO	0.04–0.08 ppm (remote); 3–15 ppm (urban)
CO_2	370 ppm
CH_4	1.3–1.7 ppm
Non-methane hydrocarbons	1–5 ppm
In airborne particulate matter (by mass per unit volume)	
Elemental carbon	3 μg m^{-3}
Organic carbon	5 μg m^{-3}

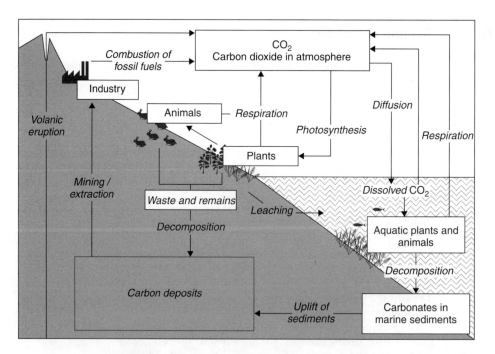

Figure 6.4 Schematic diagram of the carbon cycle

Hundreds of billions of tons of carbon (as $CO_2(g)$) is absorbed from or emitted to the atmosphere annually through natural processes. The carbon cycle is largely based on the conversion of CO_2 into organic carbon compounds in living organisms via photosynthesis, the release of CO_2 during aerobic respiration, the decomposition of plants and animals, oceanic absorption and release of CO_2, and the uplift of minerals from sediments.

Although these annual carbon flows are large, they are small in comparison to the various carbon reservoirs. For example, the atmosphere contains about 765×10^9 tons of carbon (as $CO_2(g)$) and about $1\,020 \times 10^9$ tons are dissolved in the surface layers of the world's oceans. Terrestrial carbon stocks are more difficult to approximate. It is estimated that almost $1\,680 \times 10^9$ tons of carbon are present in ground litter and soils, with terrestrial organisms, primarily vegetation, accounting for 610×10^9 tons of carbon, mostly as cellulose in the stems and branches of trees. However, the largest carbon reservoirs are the deep oceans and fossil fuel deposits, which account for approximately 38 000 and $5\,000 \times 10^9$ tons of carbon, respectively.

Both aquatic and terrestrial producers (self-feeding organisms or autotrophs) make the nutrients they need to survive from substances in their environment. On land, green plants are the producers, while in the oceans the producers are microscopic unicellular algae called phytoplankton that float and drift with ocean currents. Producers continuously extract carbon (in the form of carbon dioxide) from the atmosphere and use it to synthesize carbohydrates and sugars through the process of *photosynthesis*, which is summarized in the following equation:

$$6CO_2(g) + 6H_2O(l) \xrightarrow{\text{energy from sunlight } (h_v)} C_6H_{12}O_6(s) + 6O_2(g)$$

The sugar (glucose) made by this process can be converted into more complicated organic compounds (carbohydrates) such as cellulose and starch. All other organisms in ecosystems (consumers) obtain the nutrients they require by feeding on the tissues of producers or other consumers. For example, animals consume plants and use the carbohydrates and sugars in their metabolism. Organisms release energy from the organic compounds they synthesize, as and when it is required, by a process known as *aerobic respiration*:

$$C_6H_{12}O_6(s) + 6O_2(g) \longrightarrow 6CO_2(g) + 6H_2O(l) + \text{energy}$$

Aerobic respiration is, in effect, the reverse of photosynthesis. This linkage between photosynthesis in photoautotrophs and aerobic respiration in both producers and consumers is a significant part of the global carbon cycle.

When plants and animals die, the organic compounds that make up their tissues combine with $O_2(g)$ during decay to form $CO_2(g)$, which may be released back into the biosphere. However, not all of the organic compounds are oxidized. A small fraction is transported, re-deposited as sediment and trapped where it can form deposits of coal and petroleum (see Chapter 5).

Carbon dioxide from the atmosphere also dissolves in oceans and other water bodies. This enables the creation of a dynamic equilibrium that involves the bicarbonate (HCO_3^-) and carbonate ions (CO_3^{2-}). The overall reactions for this process are shown below:

$$CO_2(g) \rightleftharpoons CO_2(aq)$$

$$H_2O(l) + CO_2(aq) + CO_3^{2-}(aq) \rightleftharpoons 2HCO_3^-(aq) \qquad (6.7)$$

In seawater that is slightly acidic (e.g. due to input of industrial effluent), carbonic acid ($H_2CO_3(aq)$) may be formed, followed by a series of equilibrium reactions that again involves the bicarbonate and carbonate ions:

$$H_2O(l) + CO_2(aq) \rightleftharpoons H_2CO_3(aq)$$

$$H_2CO_3(aq) \rightleftharpoons H^+(aq) + HCO_3^-(aq)$$

$$HCO_3^-(aq) \rightleftharpoons H^+(aq) + CO_3^{2-}(aq)$$

However, most oceans have a slightly alkaline pH (8.0–8.3) due to the presence of hydroxide ions produced from the following reactions:

$$HCO_3^-(aq) \rightleftharpoons OH^-(aq) + CO_2(aq)$$

$$H_2O(l) + CO_3^{2-}(aq) \rightleftharpoons HCO_3^-(aq) + OH^-(aq)$$

Many aquatic animals use dissolved carbon dioxide to make shells from calcium carbonate ($CaCO_3(s)$). The calcium carbonate is precipitated from seawater when $[Ca^{2+}][CO_3^{2-}] > K_{sp}(CaCO_3)$ (see Section 7.8 below) – you can see from Equation 6.7 that the solubility of the carbonate ion will be affected by the relative concentrations of $CO_2(aq)$, $CO_3^{2-}(aq)$ and $HCO_3^-(aq)$. The shells of dead organisms (e.g. phytoplankton or coral reefs) accumulate on the sea floor and can form limestone which is part of the sedimentary cycle.

Carbon dioxide, like water vapour, methane and nitrous oxide, is a trace gas in the Earth's atmosphere. Nevertheless, these gases, known as *greenhouse gases*, have a substantial effect on Earth's heat balance. The greenhouse gases allow infrared, visible and some ultraviolet radiation from the sun to pass through the troposphere to the surface of the Earth. Much of this energy is absorbed by the Earth's surface, before being re-emitted, at longer wavelengths (i.e. at lower energy), back into the troposphere. Some of this re-emitted energy escapes into space and some is absorbed by the greenhouse gases, thus warming the air. This trapping of heat energy in the troposphere is called the *greenhouse effect*, and without it, the Earth would be too cold to support life. So-called global warming is due to the *enhanced greenhouse effect*. This is the extra absorption of re-emitted heat due to anthropogenic emissions of greenhouse gases, including chlorofluorocarbons (CFCs).

During the growing season, CO_2 decreases in the atmosphere of temperate latitudes due to the increased sunlight and temperatures that help plants to increase their rate of carbon uptake and growth. During the winter dormant period, more CO_2 enters the atmosphere than is removed by plants, and the concentration rises because plant respiration and the decay of dying plants and animals occurs faster than photosynthesis. Because the landmass in the Northern Hemisphere is greater than in the Southern Hemisphere, the global concentration of CO_2 tends to follow the seasonal growth of terrestrial vegetation in the Northern Hemisphere.

6.3.3 Anthropogenic effects on the carbon cycle

There is no doubt that human activities have unbalanced the global carbon cycle. In preindustrial times, the atmospheric concentration of CO_2 was about 280 ppm; indeed; it had been relatively steady at this value for millennia. However, as the Earth's population has increased, human activities have significantly disturbed the global carbon cycle in two ways. First, the increased burning of fossil fuels and wood for energy has released huge amounts of CO_2 into the atmosphere. Secondly, the large-scale removal of vegetation without replacement has left less to remove CO_2 via photosynthesis. Activities that remove vegetation include deforestation and brush clearing (for permanent pasture, cropland or human settlements), the development of infrastructure (e.g. dams and roads), accidental and intentional forest burning, and unsustainable logging.

These activities have added more $CO_2(g)$ to the atmosphere than plants and oceans can remove, and by the year 2 000, the atmospheric CO_2 concentration had increased to 370 ppm. During the 1990s, the average concentration of CO_2 increased by about 1.5 ppm per year. It is estimated that the Earth's atmosphere now contains over 200×10^9 tons more carbon than it did two centuries ago. If we continue to exploit the world's fossil fuel reserves at the same rate, the atmospheric CO_2 concentration will reach 1 100–1 200 ppm in 400–800 years time, i.e. about three times today's values. Computer models estimate that this would raise the Earth's surface temperature by anything from 3–10°C. Such computer models contain large uncertainties–they are unable to accurately account for absorption in the oceans and biotic uptake, and they cannot account for future human interventions. Additional studies have suggested that over the last few decades the combined warming effect of other gases, such as CH_4, N_2O and CFCs, should have been comparable to that of CO_2. These data and observations indicate that future global surface temperatures will be higher than anything the Earth has experienced for over a million years.

6.3.4 Uses of carbon

The different forms of carbon and its compounds have a huge assortment of uses. Diamond's unique physical properties determine its uses as a gemstone and as a cutting agent in drills. Graphite is used to make brushes in electric motors, inert electrodes in electrochemical processes, "lead" in pencils and furnace linings. It is also used as a lubricant and as a moderator in nuclear reactors.

Other forms of carbon find a variety of uses, depending upon their source and properties. Wood charcoal is used to absorb gases in industrial and chemical processes because it is very porous. Carbon black is added to rubber tyres in order to increase its strength and is also used in the manufacture of black shoe polish, ink and carbon paper. Coke, produced by the destructive distillation of coal, is used (together with carbon monoxide) during the extraction of metals from their ores (see Chapter 5). Carbon dioxide is produced in the manufacture of quicklime and in fermentation processes, and is used in fire extinguishers and as dry ice. Hydrocarbons, produced by the fractional distillation of crude oil and the cracking of petroleum naphtha, are mainly used as fuel for power stations and vehicles. However, they are also used to make a range of chemicals and in products such as cosmetics, detergents, dyes, fertilizers, food additives, pharmaceuticals, pesticides, polymers (including paint and plastics), solvents and textiles.

In a separate context, the radioactive isotope, carbon-14, is used to date archaeological, anthropological and geological samples by using a technique known as *carbon dating*. When a plant or an animal dies, it stops absorbing carbon-14 and consequently the amount present gradually decreases due to radioactive decay. The half-life of carbon-14 is approximately 5 600 years, and so the level of radioactivity drops to half its original value over this time. The degree to which the radioactivity has fallen is used to estimate the age (or at least the time since death) of the plant or animal.

6.4 Nitrogen

6.4.1 Abundance and properties

Nitrogen (N) is a non-metal that readily combines with itself to form a colourless, odourless and tasteless gas (dinitrogen gas, $N_2(g)$) that is only slightly soluble in water. Dinitrogen gas is extremely resistant to chemical attack under typical atmospheric conditions (we say it is inert or unreactive) due to the very strong covalent bond that exists between two nitrogen atoms. In air, $N_2(g)$ does not burn or support combustion and it is unaffected by water, acids and alkalis. The physical and chemical properties of nitrogen are summarised in Table 6.6.

Nitrogen is a vital component of the global ecosystem and is present in a wide variety of organic and inorganic substances. It is the major chemical element in the atmosphere, where it exists in several environmentally important forms. In the form of dinitrogen gas, it makes up 78 per cent by volume and 76 per cent by weight of the Earth's atmosphere. It also exists in the atmosphere as oxidized gases (e.g. nitrous oxide, nitric oxide and nitrogen dioxide), reduced gases (e.g. ammonia) and in aerosols (e.g. nitrates, nitrites, nitric acid, ammonium sulfate and nitrate). However, these nitrogenous species contribute less than 0.001 per cent of the atmosphere (by volume).

In natural waters, nitrogen occurs in small concentrations in organic and inorganic forms. The most important inorganic forms of nitrogen are the ammonium ion (NH_4^+), nitrate (NO_3^-) and nitrite (NO_2^-) ions. These ions play a vital role in the nitrogen cycle, but as we shall see, excess concentration in rivers and streams can lead to damaging environmental consequences. Nitrogen is incorporated into all

Table 6.6   Physical and chemical properties of nitrogen

Property	Value/Example
Atomic number	7
Atomic weight (g)	14.0067
Density at 77 K (g cm^{-3})	0.81
Naturally occurring isotopes	^{14}N (99.63%), ^{15}N (0.37%)
Radioactive isotopes	^{12}N, ^{13}N, ^{16}N, ^{17}N, ^{18}N, ^{19}N, ^{20}N
Oxidation numbers	
−3	Ammonia (NH_3)
0	Dinitrogen gas (N_2)
+1	Nitrous oxide (N_2O)
+2	Nitrogen (II) fluoride (N_2F_4); nitric oxide (NO)
+3	Nitrogen (III) chloride NCl_3; nitrite (NO_2^-)
+4	Dinitrogen tetroxide N_2O_4
+5	Nitrate (NO_3^-)

plants and animal tissues as amino acids and proteins and in humans it is excreted in urea $((NH_2)_2CO)$.

Nitrogen is the 31st most abundant element in the Earth's crust and apart from nitrate deposits, it is rarely found combined in mineral ores. Typical concentrations of nitrogen-containing substances in environmental and biological systems are presented in Table 6.7.

6.4.2 Cycling of nitrogen

Nitrogen, in its gaseous form of $N_2(g)$, is the major chemical element in the atmosphere and consequently this constitutes the major nitrogen reservoir, as can be seen in Figure 6.5.

Some of the dinitrogen gas is converted in soils and waters to ammonia (NH_3), ammonium (NH_4^+), or many other nitrogen compounds. The process is known as *nitrogen fixation*, and, in the absence of industrial fertilizers, is the primary source of nitrogen to all living things. Specialized nitrogen-fixing bacteria and algae mediate biological nitrogen fixation. On the land, these bacteria often live on nodules on the roots of legumes where they use energy from plants to do their work; such examples include *Rhizobium, Actinomycetes, Azospirillum* and *Azotobacter*. In freshwater and, possibly, in marine systems, cyanobacteria (blue–green algae) fix nitrogen.

The actual chemical processes involved in the fixation of nitrogen are extremely complicated and are not fully understood. They involve the breakdown of a very stable substance (dinitrogen gas) into less stable substances and require the natural cycles of many elements, including carbon, oxygen, sulfur, phosphorus and iron, to

Table 6.7 Typical concentrations of nitrogenous substances in environmental and biological systems

Location	Concentration
Earth's crust (by weight)	20 ppm
Sea water (by weight)	
Dissolved N_2	15 mg dm^{-3}
Soluble compounds	0.7 mg dm^{-3}
Drinking water (by weight)	
Nitrate	0.01–10.0 mg dm^{-3}
Nitrite (NO_2)	0.1 mg dm^{-3a}
Ammonia (NH_4^+)	0.5 mg dm^{-3a}
Human body (by weight)	26 000 mg dm^{-3}
Atmosphere (by volume)	
N_2	78.09%
N_2O	310 ppb
NO	0–6 ppb (background); ppb (urban)
NO_2	0.2–5 ppb (background); 10–50 ppb (urban)
In airborne particulate matter (by mass per unit volume)	
Nitrate ion (NO_3^-)	2–10 µg m^{-3}
Ammonium ion (NH_4^+)	2–6 µg m^{-3}

[a]Maximum value.

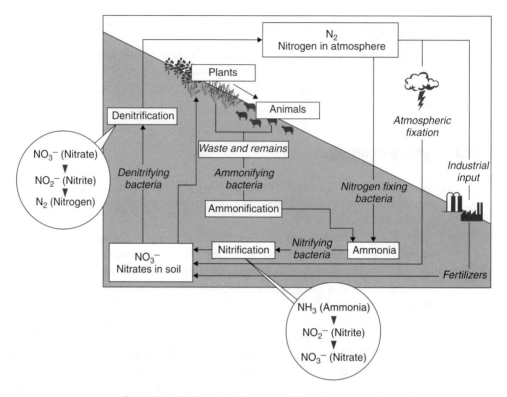

Figure 6.5 Schematic diagram of the nitrogen cycle

come together in order to create the necessary thermodynamic conditions. However, we can summarize the overall process of nitrogen fixation by using the following equation:

$$3CH_2O + 3H_2O + 2N_2 + 4H^+ \longrightarrow 3CO_2 + 4NH_4^+ \tag{6.8}$$

Once nitrogen has been fixed in the soil or aquatic system, it can follow two different pathways. It can be oxidized for energy in a process called *microbial nitrification* or assimilated by an organism into its biomass in a process called *ammonia assimilation*. Plants convert ammonium ions into nitrate ions that they can use to manufacture amino acids and subsequently, proteins.

$$4NH_4^+ + 6O_2 \longrightarrow 4NO_2^- + 8H^+ + 4H_2O$$

$$4NO_2^- + 2O_2 \longrightarrow 4NO_3^-$$

When plants and animals die and decay, the complex organic molecules are broken down into simpler molecules and ions by a process called *ammonification*. For example, urea may be broken down to ammonia and carbon dioxide:

$$(NH_2)_2CO + H_2O \longrightarrow 2NH_3 + CO_2$$

Nitrogen fixed as proteins in the bodies of living organisms eventually returns via the nitrogen cycle to its original form of nitrogen gas in the air. The process of *denitrification* starts when plants containing the fixed nitrogen are either eaten or die. Fixed nitrogen products in dead plants, animal bodies and animal excreta encounter denitrifying bacteria that use nitrate to replace dioxygen as their source of respiratory energy:

$$5CH_2O + 4NO_3^- + 4H^+ \longrightarrow 2N_2 + 5CO_2 + 7H_2O$$

Generally, N_2 is the end product of denitrification, but nitrous oxide (N_2O) is also produced in much smaller quantities (up to 10 per cent).

The chemical reactions in the nitrogen cycle are summarized in the sequence below:

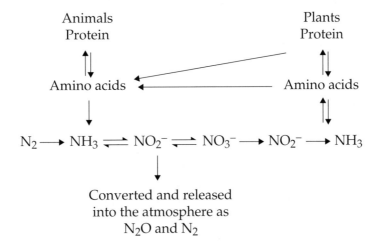

6.4.3   Nitrogen and the environment

The disruption of the nitrogen cycle by human activities plays an important role in a wide range of environmental problems, ranging from the production of tropospheric smog to the perturbation of stratospheric ozone and the contamination of groundwater.

Natural sources of nitrogen oxides are all around us – including lightning, volcanic eruptions and bacterial action in the soil – and annual natural global emissions of NO_x far outweigh anthropogenic NO_x emissions. N_2O is the most abundant of the nitrogen oxides to be naturally emitted, being formed by natural microbiological processes in the soil, but it is not normally considered a pollutant, although it does have an effect on stratospheric ozone (O_3) and is a potent greenhouse gas. The use of nitrogen-based fertilizers may be leading to increased levels of atmospheric N_2O.

Anthropogenic sources of NO_x are dominated by the combustion of fossil fuels, mainly by vehicles and power stations. Compounds such as NO, NO_2, N_2O, nitrous (HNO_2) and possibly nitric (HNO_3) acids are emitted into the atmosphere from road traffic. These oxides are formed in combustion processes as nitrogen in the air and the fuel combines with oxygen at temperatures of $1800°C$ or higher. Nitric oxide is the major contributor to oxidized nitrogen pollution, but at ambient temperatures NO is oxidized to the more toxic secondary pollutant NO_2, a process which is accelerated by the presence of reactive hydrocarbons and ozone (see Section 6.5.2).

Nitrogen dioxide can be decomposed by ultraviolet light to NO, thus producing an equilibrium situation. It has been estimated that NO contributes 90–95 per cent by volume of the total NO_x emissions, although this figure varies considerably from one source to another.

Emissions of NO_x from anthropogenic sources have increased rapidly since the Second World War. In the United States, NO_x emissions rose rapidly from 1940–1970 and then steadied due to the implementation of emission controls. In the UK, NO_x emissions rose steadily from 1970 to 1989, although some peaks and troughs occurred due to events such as the Organization of Petroleum Exporting Countries (OPEC) oil crisis in 1974 and particularly cold winters. However, UK emissions have declined since 1990 because of:

- reduced emissions from motor vehicles due to increases in road fuel tax, increased use of diesel vehicles, improved engine efficiencies and the fitting of catalytic converters to new petrol-engined cars;

- reduced emissions from power stations as a result of a move from coal to gas generation of electricity.

The ubiquity of nitrogen oxides in urban areas has led to considerable research into their health and environmental effects. Nitrogen dioxide is regarded as a pulmonary irritant, and acute exposures to NO_2 have been shown to cause breathing problems. These respiratory symptoms can cause irritation and eventually lead to diseases such as oedema or emphysema. Research has indicated that NO_2 from car exhausts damages the sensitive skin that lines the nose, increasing susceptibility to hay-fever, asthma and eczema. Nitrogen dioxide may cause plant damage, although this effect is more pronounced when NO_2 and SO_2 occur simultaneously. Exposure to NO_2 has also been shown to have detrimental effects on a range of materials, including textile dyes, and it also contributes to pollutant haze.

Nitrogen oxides play important roles in atmospheric chemistry (as in the participation of NO_2 in photochemical smog formation–see Section 6.5.2) and have a major role in acid deposition (see Section 7.10.2). Nitrous oxide is a greenhouse gas like carbon dioxide and water vapour that can trap heat near the earth's surface. It also destroys stratospheric ozone. Eventually, nitrous oxide in the stratosphere is broken down by ultraviolet light into NO_2 and NO, which can catalytically reduce ozone.

The nitrogen cycle has also been disrupted by inputs of nitrogenous materials to freshwaters, mainly from agricultural run-off (nitrogenous fertilizers and animal excreta), but also from sewage effluent (urea, proteins, amino acids and ammonia) and industrial effluent (nitric acid, and ammonia). Many countries have a particular problem with high concentrations of nitrates in their fresh waters. For example, in the UK, nitrate concentrations rose significantly from 1930 to the mid-1970s due to changing agricultural practices. In addition, the UK use of nitrogenous fertilizers increased by about 50 per cent from 1970 to the early 1990s, thus causing nitrate problems in a number of specific catchment areas. A high nitrate concentration in freshwaters contributes to environmental problems such as eutrophication and health problems such as methaemoglobinaemia and the formation of carcinogenic nitrosamines in the human digestive system. Methaemoglobinaemia is most common in young children and is also known as "blue-baby syndrome". This name arises from the apparent colour of the child due to the reduction in the ability of its haemoglobin to carry oxygen because of the presence of excess nitrogenous material.

6.5 Oxygen

6.5.1 Abundance and properties

Oxygen (O) is a non-metallic element that readily combines with itself to form a colourless, odourless and tasteless gas, dioxygen gas, $O_2(g)$, (see Example 4.3). Oxygen is very reactive and consequently single atoms of oxygen have a very short atmospheric lifetime. Its reactivity may be attributed to its small atomic radius, high electronegativity and ability to form stable double or multiple bonds with itself and other elements. Although oxygen does not react with water, dilute acids or alkalis, it combines with all elements except the noble gases and fluorine. Oxygen is a very powerful oxidizing agent; combustion of elements or compounds in air or oxygen is one of the most important chemical reactions. Respiration may be regarded as a slow form of combustion. The physical and chemical properties of oxygen are summarized in Table 6.8.

Oxygen is an unique component of the global ecosystem in that it is the only element present in high concentrations in every environmental compartment in a wide variety of organic and inorganic substances. Oxygen is the most abundant element in the lithosphere and is the most widely distributed. It is present in combined form as oxides, mainly water, silica, alumina, iron and titanium oxides, silicates, aluminosilicates, carbonates, sulfates and oxo-salts. In many of these species (e.g. SiO_4^{4-}, AlO_4^{5-}, CO_3^{2-}), the oxyanions do not decompose, and may be transported unaltered between the crust, atmosphere and hydrosphere.

Dioxygen gas is only slightly soluble in water and its solubility varies with temperature (see Table 6.14 below). Nevertheless, its presence in water is vital to underwater plant and animal life. In the atmosphere, it exists in several environmentally important forms. In the form of dioxygen gas ($O_2(g)$), it makes

Table 6.8 Physical and chemical properties of oxygen

Property	Value/Example
Atomic number	8
Atomic weight (g)	15.9994
Density at 90 K (g cm^{-3})	1.14
Naturally occurring isotopes	^{16}O (99.757%), ^{17}O (0.038%), ^{18}O (0.205%)
Radioactive isotopes	^{14}O, ^{15}O, ^{19}O, ^{20}O, ^{21}O, ^{22}O
Oxidation numbers	
−2	Oxide (O^{2-}); water (H_2O); carbon dioxide (CO_2); sulfate (SO_4^{2-}); perchlorate (ClO_4^-)
−1	Peroxide (O_2^{2-}), e.g. hydrogen peroxide (H_2O_2)
0	Dioxygen gas (O_2)
+2	Oxygen (II) fluoride (OF_2)

up 21% by volume and 23% by weight of the Earth's atmosphere. In the form of tropospheric ozone ($O_3(g)$), it acts as the Earth's "global sunscreen", filtering out harmful ultraviolet rays from the Sun. It also exists in the atmosphere as:

- oxides of carbon (carbon monoxide and carbon dioxide);
- oxides of nitrogen (e.g. nitrous oxide, nitric oxide and nitrogen dioxide);
- oxides of sulfur (e.g. sulfur dioxide);
- acids (e.g. nitric acid and sulfuric acid);
- atmospheric particles (e.g. nitrates and sulfates);
- free radicals (e.g. the hydroxyl and superoxide radicals).

6.5.2 Cycling of oxygen

The ubiquity and reactivity of oxygen on Earth means that the oxygen cycle is extremely complicated. However, since oxygen is mainly present in combined form in the Earth's crust, it can be regarded as being chemically inert. The role of oxygen-containing compounds in the crust and hydrosphere are considered in detail elsewhere in this book, namely:

- the formation, weathering and reactions of silicates, aluminates and aluminosilicates (Chapter 4);
- the formation and accumulation of fossil fuels (Chapter 5);
- the extraction and reaction of metal oxides (Chapter 5);
- corrosion chemistry and protection (Chapter 5);
- the reactions of water (Chapter 7).

In this present chapter, therefore, we will consider only the atmospheric chemistry of oxygen. At present, the rates of formation and utilization of atmospheric dioxygen

Table 6.9 Typical concentrations of oxygen-containing substances in environmental and biological systems

Location	Concentration
Earth's crust (by weight)	46.6%
Sea water (by weight)	
Dissolved O_2	6 mg dm^{-3} (15°C)
Total	85.4%
Stream water (by weight)	–[a]
Human body (by weight)	610 000 mg dm^{-3}
Atmosphere (by volume)	
O_2	20.94%
O_3	0.005–0.04 ppm (remote)

[a]See Table 6.14 below.

are in dynamic equilibrium–if they were not, the concentration of atmospheric $O_2(g)$ would constantly change. As you will see from the processes outlined below, the carbon and oxygen cycles are interlinked and dependent upon each other. Life on Earth probably depends upon the ability of these two cycles to continuously counterbalance each other.

Respiration

Respiration is a biochemical process in which organic compounds (food) are progressively broken down (oxidized) to release energy in the form of adenosine triphosphate (ATP) (see Section 6.6.3). Most organisms use oxygen for this process, but some bacteria use nitrate or sulfate as the oxidant.

$$C_6H_{12}O_6(s) + 6O_2(g) \longrightarrow 6CO_2(g) + 6H_2O(l) + energy \qquad (6.9)$$

Respiration occurs in all organisms, and in every living cell, since it is the only way a cell can obtain usable energy. Although Equation 6.9 gives an overall summary of the process, it is misleading in that it suggests that respiration occurs in one chemical step, when in fact it is a much more complicated process (see Section 6.6.3).

Photosynthesis

Photosynthesis is the process by which green plants trap light energy and use it to drive a series of chemical reactions that lead to the formation of carbohydrates. The overall reaction for photosynthesis may be summarized as follows:

$$6CO_2(g) + 6H_2O(l) \xrightarrow{h_v} C_6H_{12}O_6(s) + 6O_2(g) \qquad (6.10)$$

The glucose made by this process can be converted into more complicated carbohydrates such as cellulose and starch. Carbohydrates make up 60–90% all solid plant material and provide a source of food (energy) for animals. You can see from Equations 6.9 and 6.10 that respiration is effectively the reverse of photosynthesis. However, photosynthesis is a more complicated process than this reaction mechanism might suggest–it may, in fact, be divided into "light" and "dark" reactions. During the light reaction, sunlight is used to split water into oxygen, hydrogen ions (photons) and electrons:

Light reaction $\qquad 12H_2O \xrightarrow{h_v} 6O_2 + 24H^+ + 24e^-$

This reaction depends upon the ability of chlorophyll to trap the energy from sunlight. Sunlight is not required for the dark reaction, in which the photons and electrons are used to convert carbon dioxide into carbohydrates:

Dark reaction $\qquad 6CO_2 + 24H^+ + 24e^- \longrightarrow C_6H_{12}O_6 + 6H_2O$

Combustion

Combustion may be defined as the rapid combination of a substance with oxygen, accompanied by the evolution of heat and usually light. Although some combustion reactions occur spontaneously, most require an energy barrier to be overcome before the reaction can start. Examples of combustion reactions common in the environment and in industry are given below.

Dihydrogen sulfide $\qquad 2H_2S(g) + 3O_2(g) \longrightarrow 2SO_2(g) + 2H_2O(l)$

Methane $\qquad CH_4(g) + 2O_2(g) \longrightarrow CO_2(g) + 2H_2O(g)$

Butane (lighter fluid) $\qquad 2C_4H_{10}(g) + 13O_2(g) \longrightarrow 8CO_2(g) + 10H_2O(g)$

Octane (gasoline) $\qquad 2C_8H_{18}(l) + 25O_2(g) \longrightarrow 16CO_2(g) + 18H_2O(l)$

Tristearin (fat) $\qquad 2C_{57}H_{110}O_6(s) + 163O_2(g) \longrightarrow 114CO_2(g) + 110H_2O(l)$

Hydrazine (rocket fuel) $\quad N_2H_4(aq) + O_2(g) \longrightarrow N_2(g) + 2H_2O(l)$

Magnesium $\qquad 2Mg(s) + O_2(g) \longrightarrow 2MgO(s)$

White phosphorus $\qquad P_4(s, white) + 3O_2(g) \longrightarrow P_4O_6(s)$

Ammonia

(Haber process) $\qquad 4NH_3 + 5O_2(g) \xrightarrow{\text{850°C, 5 atm, Pt/Rh catalyst}} 4NO(g) + 6H_2O(g)$

Incomplete combustion occurs when a substance burns in a limited supply of oxygen, resulting in the formation of carbon monoxide:

Carbon (steel manufacture) $\quad 2C(s) + O_2(g) \longrightarrow 2CO(g)$

Octane (gasoline) $\qquad 2C_8H_{18}(l) + 17O_2(g) \longrightarrow 16CO(g) + 18H_2O(l)$

Tropospheric reactions

The troposphere is the inner layer of the atmosphere and extends for about 17 km above sea level at the equator and 8 km over the poles. It contains about 75% of the mass of the earth's air. The chemical pathways for the formation and removal of gaseous pollutants in the troposphere are controlled largely by *atoms* and *free radicals*, despite the fact that they are present at low concentrations.

The term "free radical" refers to any atom or group of atoms that has an odd number of electrons and is capable of independent existence. Some free radicals can carry an electrical charge that will influence their properties. The high reactivity of free radicals is responsible for their low concentrations in the troposphere and also for their importance as intermediates in tropospheric reactions.

Free radicals are denoted by a 'dot' next to the appropriate chemical symbol – the dot indicating the presence of an unpaired electron in an outer orbit around the nucleus. Examples of atmospheric free radicals include:

$O^{\bullet}$ oxygen radical
$^{\bullet}OH$ hydroxyl radical
$^{\bullet}O_2^{-}$ superoxide radical
$^{\bullet}F_2^{-}$ fluoride radical

Of these, the most important free radical in the troposphere is the hydroxyl radical, which is formed continuously by a series of photochemical reactions. There are two main pathways for such reactions.

- Ozone (O_3) is a high-energy form of oxygen that mainly exists in the stratosphere. There are very few O_3 molecules in the troposphere and they rarely encounter UV photons (from UV light that has passed through the stratosphere). However, when they do, ozone is photolysed:

$$O_3 \xrightarrow{h_v} O^{\bullet} + O_2 \qquad \text{(given that } \lambda < 320 \text{ nm)}$$

The $O^{\bullet}$ is rapidly quenched by O_2 and N_2 in the atmosphere. With O_2, it reforms O_3, which is then photolysed, and the cycle is repeated:

$$O^{\bullet} + O_2 \longrightarrow O_3$$

The $O^{\bullet}$ atoms have sufficient energy to react with water, producing hydroxyl radicals:

$$O^{\bullet} + H_2O \longrightarrow 2\,^{\bullet}OH$$

- The hydroxyl radical is also formed by the photolysis of carbonyl compounds in the presence of NO:

$$HCHO \xrightarrow{h_v} CHO^{\bullet} + H^{\bullet}$$

$$H^{\bullet} + O_2 \longrightarrow HO_2^{\bullet} \quad \longleftarrow \quad \text{this is the hydroperoxyl radical}$$

and then

$$HO_2^{\bullet} + NO \longrightarrow NO_2 + {}^{\bullet}OH$$

There are very few hydroxyl radicals in the troposphere. However, they are so reactive that they constitute the main pathways for the removal of most oxidizable molecules in this region of the atmosphere, including methane (CH_4), carbon monoxide (CO) and sulfur dioxide (SO_2).

Although tropospheric O_3 is rare, its reactions are important, particularly in urban areas. Tropospheric ozone is formed by the reaction of nitrogen dioxide with sunlight, producing oxygen atoms that react subsequently with oxygen molecules in the presence of a third body (M). The nitrogen dioxide is formed by the oxidation of nitric oxide, which is largely emitted from combustion sources:

$$NO + O_3 \longrightarrow NO_2 + O_2$$
$$NO_2 + h_v \longrightarrow NO + O$$
$$O + O_2 + M \longrightarrow O_3 + M$$

Although these processes cannot produce more ozone than is originally present (a so-called "stationary state"), the following reactions occur in the presence of a reactive hydrocarbon (RH):

$$RH + OH + O_2 \longrightarrow RO_2 + H_2O$$
$$RO_2 + NO + O_2 \longrightarrow Carbonyl + 2NO_2 + HO_2$$
$$HO_2 + NO \longrightarrow NO_2 + OH$$

Overall $RH + 2NO + 2O_2 \longrightarrow Carbonyl + 2NO_2 + H_2O$

The overall equation for these reactions shows that NO_2 is produced without consuming ozone. This disturbs the stationary state, which is restored by the decomposition of some of the NO_2 to NO and O_3, so resulting in net ozone formation. Since there are many reactive hydrocarbons in the atmosphere, many different reactions can occur with nitrogen oxides and free radicals. In the right circumstances, a "chemical soup" can form in the atmosphere–*photochemical smog*. The main requirements for the formation of photochemical smog are:

- strong sunlight;
- stable meteorological conditions;
- the presence of NO_x;
- the presence of hydrocarbons, particularly unsaturated ones.

Stratospheric reactions

The stratosphere is the atmosphere's second layer and extends from about 17–48 km above the earth's surface. Although the stratosphere contains less matter than the troposphere, its composition is similar, with two key exceptions:

- its volume of water vapour is about 1 000 times less;
- its volume of ozone is about 1 000 times greater.

Ozone in the stratosphere absorbs ultraviolet radiation from the sun, preventing UV-C radiation–which is lethal to man–reaching the earth's surface and reduces the amount of harmful UV-B passing through the stratosphere. Atmospheric conditions in the stratosphere are very stable and, hence, foreign material introduced into it may persist for a very long time. Any environmental problems in this region are therefore, compounded by this characteristic, with the main such problem being ozone depletion.

Even though it is found only in very small quantities, ozone is the most important compound in the stratosphere. If all stratospheric ozone could be brought down to sea level and spread evenly around the globe, the pressure of the atmosphere above it would squeeze it into a layer of 3 mm thickness. Its importance, though, is due to the fact that it absorbs all radiation between wavelengths of 240–290 nm–radiation that would otherwise reach the earth and be harmful to plants and animals. The absorbed radiation also warms the stratosphere and provides energy for circulation.

As a consequence, concern about pollution in the stratosphere centers on damage to the ozone layer, and ozone depletion is rightly considered as one of the greatest threats to the environment. The first clear sign of damage to the stratospheric ozone layer–over the Antarctic–was reported by Joe Farman of the British Antarctic Survey in 1982. There is now evidence of stratospheric ozone depletion occurring everywhere except at the equatorial regions. Ozone "holes", where O_3 concentrations are down to around 200 μg m^{-3}, occur most frequently in the Antarctic region, and most often in spring when rising temperatures create the right conditions for the release of certain reactive gases and their subsequent chemical reactions.

The chemistry of ozone destruction is complex, although it is possible to simplify it in order to obtain an overview. If we could envisage an "oxygen-only" stratosphere, ozone generation and consumption would have to be due only to the following reactions:

(a) Ozone formation

$$O_2(g) + h_v \longrightarrow 2O(g)^{\bullet} \quad \text{(given } \lambda = 242 \text{ nm)}$$

$$O_2(g) + O(g)^{\bullet} + M \longrightarrow O_3(g) + M \quad \text{(where M is any third body)}$$

(b) Ozone destruction

$$O_3(g) + h_v \longrightarrow O_2(g) + O(g)^{\bullet}$$

$$O_3(g) + O(g) \longrightarrow 2O_2(g)$$

The free oxygen atoms produced in reaction (a) are very reactive, and some will react with other oxygen molecules to make ozone. This can only happen if some other molecule (M) is present to take up the kinetic energy produced in the reaction–usually the other molecule will be nitrogen, but it could be almost anything. As well as producing ozone, this pair of reactions provides energy to molecule M, thus making it move faster. When molecules of gas move faster, the gas becomes hotter. Therefore, UV radiation is absorbed, so converting oxygen into ozone and warming the atmosphere.

Under the conditions of reactions (a) and (b), there would be a dynamic balance between formation and consumption–a steady-state situation. However, there are many other substances present in the atmosphere that will react with ozone. For example, ozone can be destroyed in catalytic chain reactions in which it is converted to dioxygen by a chain carrier, $X^{\bullet}$, as follows:

General reactions $\quad O_3 + X^{\bullet} \longrightarrow XO + O_2$

$$XO + O^{\bullet} \longrightarrow X^{\bullet} + O_2$$

Overall reaction $\quad O_3 + O^{\bullet} \longrightarrow 2O_2$

Research has shown that the main substances that can act as the chain carrier X in the stratosphere are nitrogen oxides, halogen compounds and the hydroxyl radical.

6.5.3 Uses of oxygen

Oxygen is produced industrially by the fractional distillation of liquefied air. It is principally used for the purification of iron in the steel industry (see Section 5.5.5.2) and in welding and cutting. It is also used in the chemical industry for the production of many chemicals, including nitric acid, methanol and ethylene oxide, in rockets for the combustion of fuel, and to aid respiration in hospitals and aircraft.

6.6 Phosphorus

6.6.1 Abundance and properties

Phosphorus (P) is a highly reactive, non-metallic element that exhibits *allotropy*; that is, it exists in several physically different but chemically identical forms. The two most common forms are white phosphorus (a yellowish, translucent, soft, waxy solid) and red phosphorus (a non-volatile, non-toxic, reddish powder). White phosphorus exists as P_4 molecules with a tetrahedral structure (see Figure 6.6), whereas red phosphorus has a giant three-dimensional macromolecular structure. The third allotrope of phosphorus, black phosphorus, is an iron-grey crystalline solid that resembles graphite in appearance, properties and structure.

Like nitrogen, phosphorus is essentially covalent in all its chemistry. However, unlike nitrogen compounds, phosphorus only forms compounds in one oxidation state that is stable (+5). The reactivity of phosphorus is dependent upon its physical structure. For example, white phosphorus is relatively unstable compared to red phosphorus because the tetrahedral P–P bonds are under a certain amount of strain, thus making it more reactive. Hence, white phosphorus can spontaneously

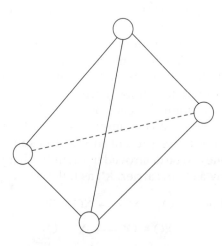

Figure 6.6   Structure of white phosphorus

combust in air while red phosphorus is slowly oxidized to orthophosphoric acid. The physical and chemical properties of phosphorus are summarized in Table 6.10.

Phosphorus is an essential element in living organisms and a main constituent of bones, nerve and brain tissue, teeth, etc. In the environment, phosphorus mainly occurs as phosphate minerals of the *apatite* family, $Ca_9(PO_4)_6.CaX_2$, where $X = F$, Cl or OH, or as inorganic phosphates of aluminium, calcium and iron. The actual composition of phosphate minerals is very complex due to a high degree of isomorphous substitution between Ca^{2+}, Fe^{2+}, Fe^{3+} and Al^{3+}. Phosphates occur throughout the natural environment–in phosphate rocks, in suspended solids in rivers and oceans and (to a lesser extent) as dust particles in the air. Phosphate minerals are essential for plant growth and are also rich in the other trace elements required for plant growth. It is impossible to conceive of any substitute for phosphorus as a plant nutrient. Typical concentrations of phosphorus-containing substances in environmental and biological systems are presented in Table 6.11.

Table 6.10 Physical and chemical properties of phosphorus

Property	Value/Example
Atomic number	15
Atomic weight (g)	30.973 76
Density at 298 K (g cm^{-3})	
Red phosphorus	2.20
White phosphorus	1.82
Black phosphorus	2.70
Naturally occurring isotopes	^{31}P (100%)
Radioactive isotopes	^{29}P, ^{30}P, ^{32}P, ^{33}P
Oxidation numbers	
0	Red phosphorus (P_4)
+2	Phosphorus (II) chloride (P_2Cl_4); phosphorus (II) bromide (P_2Br_4)
+3	Phosphorus (III) fluoride (PF_3); phosphorus (III) hydride (PH_3); phosphorus (III) oxide (P_4O_6)
+5	Phosphate (PO_4^{3-}); phosphorus (V) oxide (P_4O_{10}); phosphorus (V) iodide (PI_5)

Table 6.11 Typical concentrations of phosphorus-containing substances in environmental and biological systems

Location	Concentration
Earth's crust (by weight)	1 050 ppm
Sea water (by weight)	
Total	70–90 µg dm^{-3}
Drinking water (by weight)	0.0001–0.1 mg dm^{-3}
Human body (by weight)	11 000 mg dm^{-3}
Atmosphere	
In airborne particulate matter (by mass per unit volume)	<1 µg m^{-3}

6.6.2 Cycling of phosphorus

Phosphorus moves through the hydrosphere, Earth's crust and living organisms in the phosphorus cycle, as shown in Figure 6.7. Apart from dust transfer, there is relatively little circulation of phosphorus between the atmosphere and other environmental compartments. This is because the mobility of naturally occurring phosphorus compounds is low, due to their low solubilities and volatilities. Consequently, the geochemical cycling of phosphorus between the land and the oceans occurs mainly through transfer of suspended solids from rocks and sediments to living organisms and back again.

Soil and aquatic microbial processes are important in the cycling of phosphorus, as it is the most common limiting nutrient in water, particularly for the growth of photosynthetic algae. Phosphorus is a *biolimiting element*, which means that its concentration limits biological growth. In the oceans, the variation in concentration of phosphorus is characteristic of biolimiting elements (see Figure 6.8); phosphate concentrations are low at the surface, where photosynthesis is at a maximum, and increase with increasing depth, due to decreasing biological uptake and photosynthesis.

The low solubility of most inorganic phosphates means that phosphorus is not always available as a nutrient because, like nitrogen, it must be present in a simple, inorganic form before plants can take it up. The main phosphate species present in soils are the orthophosphate (PO_4^{3-}), hydrogenphosphate (HPO_4^{2-}) and dihydrogenphosphate ($H_2PO_4^{-}$) ions. The orthophosphate ion is relatively insoluble because it is bulky and triply charged, thus making it strongly attracted

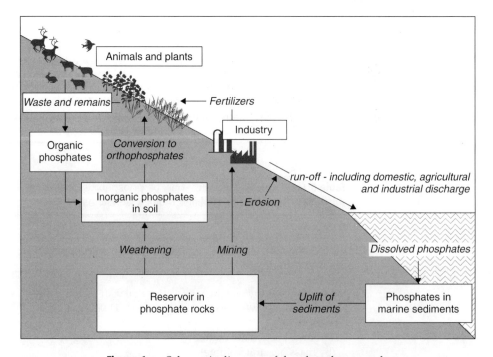

Figure 6.7 Schematic diagram of the phosphorus cycle

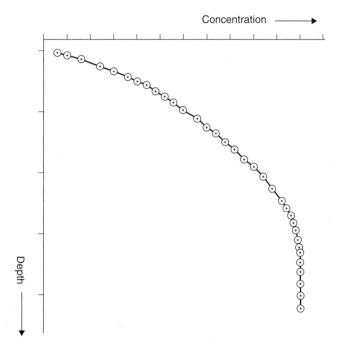

Figure 6.8 Variation in concentration with depth for biolimiting elements in oceans

to cations and difficult for water to remove from solids. The hydrogen phosphates are more soluble than orthophosphate because they have lower charges.

Orthophosphate is most available to plants at nearly neutral pH values. In acidic soils, the PO_4^{3-} ion reacts with aluminium (III) and iron (III) to form insoluble compounds of aluminium phosphate ($AlPO_4$) and iron phosphate ($FePO_4$). In alkaline soils, the PO_4^{3-} ion may react with calcium carbonate to form insoluble calcium hydroxyapatite ($Ca_5(PO_4)_3(OH)_4$). As a result of these reactions, many soils are deficient in bio-available phosphate. The more soluble hydrogen phosphates are commonly used as nutrients in fertilizers, although soluble phosphorus rarely migrates far from a fertilizer particle. Phosphate fertilizers may be obtained by treating phosphate rocks with concentrated sulfuric acid to produce superphosphate:

$$Ca_3(PO_4)_2(s) + 2H_2SO_4(aq) + 4H_2O(l) \longrightarrow \underbrace{Ca(H_2PO_4)_2(s) + 2CaSO_4.2H_2O(s)}_{\text{superphosphate}}$$

Triple superphosphate, a concentrated form of $Ca(H_2PO_4)_2$, is obtained by treating rock phosphate with phosphoric acid (H_3PO_4).

Phosphate ions can undergo condensation reactions and form di-, tri- and polyphosphates, e.g. $P_2O_7^{4-}$, $P_3O_{10}^{5-}$. Polyphosphates are more soluble than simple orthophosphates and hence are used as *builders* in detergents. Detergent molecules are prevented from removing dirt and grease from clothing, dishes, etc. by the presence of cations such as Ca^{2+} and Mg^{2+} in hard water. Builders interact with these divalent ions and prevent them from precipitating detergent molecules, thus allowing them to do their job effectively. Sodium tripolyphosphate (STP), $Na_5P_3O_{10}$, is an efficient commercial builder since it is relatively cheap and decomposes in the environment into sodium phosphate, a naturally occurring mineral and plant nutrient.

However, the widespread use of phosphates as builders in detergents has led to the build-up of phosphates in natural water bodies and the eutrophication of many lakes (see Section 6.6.3 below). It is possible to produce phosphate-free detergents, but more detergent is needed for the wash. Attempts have been made to replace phosphates by other substances (such as silicates, borates, polycarboxylates and zeolites), but no adequate replacement has been found—alternatives are either not as effective or have some environmental impact.

6.6.3 Phosphorus in biological and environmental systems

We have already seen that phosphorus is a macro-nutrient for plants and animals and is frequently a biolimiting element. This is because phosphorus is a key element in many biochemical processes. In order to stay alive, organisms must release energy in a controlled and usable form. This is achieved in living cells by the decomposition (hydrolysis) of adenosine triphosphate (ATP). The latter compound is a nucleotide—a complex organic molecule (adenosine) to which three phosphate groups are attached (see Figure 6.9). ATP is found in all cells and is believed to be the universal supplier of energy in biological systems. In the presence of an appropriate enzyme (ATPase), ATP is hydrolysed to adenosine diphosphate (ADP) and phosphoric acid, releasing 34 kJ mol^{-1} of energy:

$$\text{ATP} + \text{H}_2\text{O} \xrightarrow{\text{ATPase}} \text{ADP} + \text{H}_3\text{PO}_4 + 34 \text{ kJ mol}^{-1}$$

Although some of this energy is lost as heat, a proportion is used directly for biological activities such as muscle contraction, nerve transmission and synthesis of materials. ATP is continuously synthesized from ADP by using the energy released

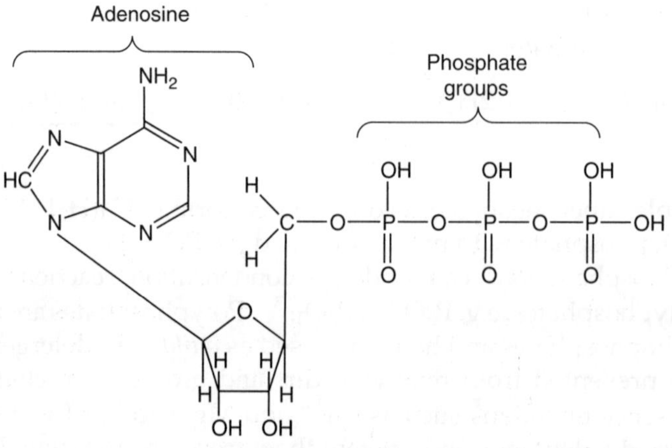

Figure 6.9 The chemical structure of adenosine triphosphate (ATP)

during respiration:

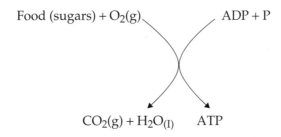

$$\text{Food (sugars)} + O_2(g) \qquad ADP + P$$

$$CO_2(g) + H_2O_{(l)} \qquad ATP$$

Phosphates are also an essential component of deoxyribonucleic acid (DNA), the complex giant molecule that contains all the chemical codes required to build, control and maintain a living organism.

In the environment, the concentration of phosphorus compounds in unpolluted waters is generally low (see Table 6.11). However, additional phosphorus can get into the water system from various sources, including:

- soil erosion caused by deforestation and agriculture;
- sewage treatment plant effluent;
- detergents and water softeners which use polyphosphates;
- animal manure (water run-off from farmland);
- water treatment;
- organophosphate (OP) pesticides, etc.

In aquatic systems, phosphorus is frequently the biolimiting element to plant growth. The presence of excess soluble phosphates promotes the rapid growth of algae in lakes (algal blooms), thus leading to reduced light transmission and consequently reduced photosynthesis; less oxygen is therefore produced. When the algae die, their decay consumes the available oxygen and the conditions deteriorate further. This process is known as *eutrophication*–excessive enrichment of streams, lakes and shallow sea areas by nutrients leading to the rapid growth of algae and bacteria and the consequent reduction of dissolved oxygen in the water. Eutrophication due to the presence of excess nutrients has led to severe environmental problems in many watercourses, including Lake Erie (Canada/USA), Lough Neagh (Ireland) and the Norfolk Broads (UK). However, as algae are part of the aquatic food chain and are consumed by higher organisms, a certain level of phosphate is necessary or fish, plants and other aquatic life would starve. What is needed is a self-balancing equilibrium between nutrients, algae, plants and other organisms.

The UK has suffered serious environmental problems from elevated concentrations of phosphates. In 1989, the deaths of sheep and dogs at Rutland Water and the hospitalization of two soldiers after canoe exercises at Rudyard Lake was attributed to the toxic effects of algal blooms. In 1994, it was reported that giant stinging nettles were growing all over England on canal and riverbanks. Nettles have become the most widespread flowering plants in the last 40 years. Previously, stinging nettles were confined to field edges and patches of ground rich in phosphates (like graveyards where decaying bones provide a constant supply of phosphates).

Phosphates have become a serious problem in many UK rivers designated as sites of special scientific interest (SSSI). Phosphates can be removed by tertiary treatment in sewage treatment plants, using lime, aluminium and iron compounds to precipitate the respective metal phosphates, but this is a costly process. However, Sweden has solved most of its eutrophication problems by installing phosphate removal equipment at sewage treatment plants.

Organophosphate chemicals (OPs), such as Parathion and Malathion, have been used for many years as non-persistent insecticides. Although they are designed to break down into harmless, water-soluble products in the environment, some OPs are highly toxic and have caused severe injury and death to agricultural workers.

6.6.4 Uses of phosphorus

Phosphorus is often burned to form phosphorus (V) oxide, which in turn is used to manufacture phosphoric acid. This acid may be used to rustproof steel (see Chapter 5) and for chemical and electrolytic polishing of alloys. Dilute phosphoric acid is used as an acidity regulator in foodstuffs (e.g. jellies and soft drinks) and in phosphate fertilizers, although these are predominantly prepared from phosphate rock.

Calcium phosphate is used in baking powder and in dietary supplements, while sodium tripolyphosphate is used in washing powders and water softeners. Red phosphorus is used to coat the side of matchboxes as it ignites a striking match when the frictional force is sufficiently high.

Toxic organophosphorus compounds are used as insecticides and chemical warfare agents ("nerve gases").

6.7 Sulfur

6.7.1 Abundance and properties

Sulfur (S) is a brittle, non-metallic element that exists in three different allotropic forms (two crystalline and one amorphous). Rhombic sulfur is a yellow solid that forms octahedral crystals and is stable at ordinary temperatures. The other crystalline allotrope of sulfur, monoclinic sulfur, is unstable at ordinary temperatures and slowly reverts to the rhombic form. These two allotropes are composed of ring-shaped S_8 molecules, although the rings are arranged in a different manner in each form. Sulfur melts to a pale yellow, mobile liquid upon heating, which turns black and viscous at about 200°C. Other forms of sulfur are also known, including flowers of sulfur, roll sulfur, plastic sulfur and colloidal sulfur. Pure sulfur is insoluble in water.

Like oxygen, sulfur is a Group VI element and is fairly reactive. It has six electrons in its outer energy level and may attain a stable octet by:

- gaining two electrons into its outer energy level, forming a bivalent anion (S^{2-});
- sharing electrons to form two covalent bonds, as in $H_2S(g)$, or one double bond, as in $SO_2(g)$;
- forming univalent hydrosulfide ions, HS^-.

However, sulfur can also form compounds with oxidation states other than -2, as shown in Table 6.12. In general, sulfur species may be divided into two major categories:

- Oxidized forms (where O_2 is available), mainly in the atmosphere, soils and surface waters. The most oxidized form of sulfur is the sulfate ion, SO_4^{2-}, and other species include sulfur dioxide (SO_2), gypsum ($CaSO_4.2H_2O$) and sulfuric acid (H_2SO_4).

- Reduced forms (where O_2 is low or absent), mainly in rocks, sediments and the deep sea. Examples of reduced species include sulfide minerals and the gases, dihydrogen sulfide (H_2S) and dimethyl sulfide ($(CH_3)_2S$).

Sulfur (S) is found in a variety of forms in the environment and consequently the sulfur cycle is complex. It is found as deposits of elemental sulfur in several countries, including the USA, Mexico, Japan and Italy. Sulfur is found in sulfate ores such as gypsum and in sulfide ores such as zinc blende (ZnS), galena (PbS), cinnabar (HgS) and iron pyrites (FeS_2). Elements in the Periodic Table found naturally as sulfides are geochemically classified as *Chalcophiles*. Sulfur is present in the oceans as sulfate (SO_4^{2-}), which comprises 7.8% (by weight) of seawater (the third most abundant species), dihydrogen sulfide (H_2S) and dimethylsulfide ($(CH_3)_2S$).

In living organisms, sulfur is an essential constituent of proteins and certain amino acids, namely cysteine and methionine. Sulfur is often regarded as an "organic" element in biology and can form self-linkages to produce long $-S-S-S-S-$chains. The S$-$S bonds are particularly important in proteins, forming intermolecular and intramolecular bonds.

In the atmosphere, sulfur exists as:

- $H_2S(g)$ from volcanoes and from biological decay of organic matter;

- $(CH_3)_2S(g)$ from marine phytoplankton (algae that float near the surface of the sea);

Table 6.12 Physical and chemical properties of sulfur

Property	Value/Example
Atomic number	16
Atomic weight (g)	32.0666
Density at 298 K (g cm^{-3})	
Rhombic sulfur	2.07
Monoclinic sulfur	1.96
Naturally occurring isotopes	^{32}S (94.93%), ^{33}S (0.76%), ^{34}S (4.29%), ^{36}S (0.02%)
Radioactive isotopes	^{30}S, ^{31}S, ^{35}S, ^{37}S, ^{38}S, ^{39}S, ^{40}S
Oxidation states	
-2	Sulfide, S^{2-} e.g. dihydrogen sulfide (H_2S), Galena (PbS), dimethylsulfide ($(CH_3)_2S$)
-1	Disulfide, S_2^{2-}, e.g. iron (II) pyrites (FeS_2)
0	Free sulfur, S_8
$+4$	Sulfur dioxide (SO_2), sulfite (SO_3^{2-}); bisulfite (HSO_3^-)
$+6$	Sulfuric acid (H_2SO_4); sulfur trioxide (SO_3); sulfate (SO_4^{2-}), e.g. gypsum ($CaSO_4.2H_2O$)

- volatile compounds such as carbon disulfide ($CS_2(g)$), carbonylsulfide ($OCS(g)$), methylmercaptan ($CH_3SH(g)$) and dimethydisulfide ($CH_3SCH_3(g)$) released from plants and microbial organisms;

- sulfate aerosol ($SO_4^{2-}(aq)$), from marine and anthropogenic sources;

- sulfur dioxide ($SO_2(g)$), from natural and anthropogenic sources.

Sulfur dioxide is one of the most widely monitored of air pollutants, with national sampling networks in many parts of the world. Typical concentrations of $SO_2(g)$ are given in Table 6.13, together with values for other sulfur-containing substances.

6.7.2 Cycling of sulfur

Sulfur moves through the atmosphere, hydrosphere, Earth's crust and living organisms in the sulfur cycle, as shown in Figure 6.10. Globally, most sulfur is present in the lithosphere and in seawater, with only minor amounts in the biosphere and trace quantities in the atmosphere. In contrast to the extremely slow sedimentary-cycle processes (erosion, sedimentation and uplift of sulfur-containing rocks), the lifetime of most sulfur compounds in the air is relatively short (days). The cycling of sulfur greatly increased after the Industrial Revolution and again after the Second World War, due to increased consumption of fuel, increased metal extraction and the production of fertilizers. In natural processes, sulfur compounds tend to be reduced to sulfides, whereas in anthropogenic processes they get oxidized to sulfites and sulfates.

As with the nitrogen cycle, microorganisms play an important role in the biogeo-chemical cycling of sulfur. Indeed, there are certain similarities between the cycling of these two elements. The most important biogenically produced sulfur compound in the oceans is dimethylsulfide, generated by the decomposition of dimethylsul-fonopropionate (DMSP), an organic compound produced by marine phytoplankton for osmoregulation during their seasonal bloom. (Phytoplankton produce DMSP to protect themselves from the negative effects of high salinity and freezing.) In terrestrial environments, dihydrogen sulfide is the main biogenically produced

Table 6.13 Typical concentrations of sulfur-containing substances in environmental and biological systems

Location	Concentration
Earth's crust (by weight)	260 ppm
Sea water (by weight)	905 mg dm^{-3}
Total	
Stream water (by weight)	1–100 mg dm^{-3}
Drinking water (by weight)	1.0–1000 mg dm^{-3} (as SO_4^{2-})
Human body (by weight)	2000 mg dm^{-3}
SO_2	1–3 ppb (background); 3–20 ppb (urban)
In airborne particulate matter (by mass per unit volume)	
Sulfate ion (SO_4^{2-})	5–10 µg m^{-3}

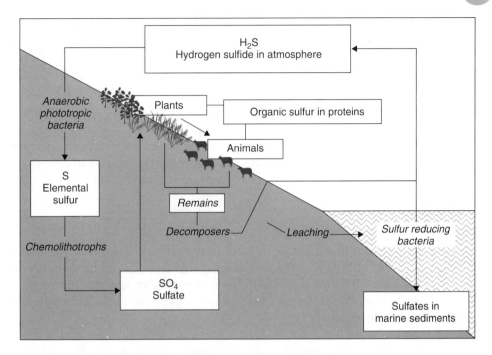

Figure 6.10 Schematic diagram of the sulfur cycle

sulfur compound. Both gases are oxidized to sulfur dioxide, and eventually, to sulfate:

$$H_2S \longrightarrow (S) \xrightarrow{O_2} SO_2 \xrightarrow{O_2} SO_3 \longrightarrow SO_4^{2-} \tag{6.11}$$

The sulfur cycle can be coupled to the carbon cycle to make some reactions, which require an input of energy, feasible. For example, energy is released during the oxidation of a carbohydrate ($C_x(H_2O)_y$) by a microorganism, but is required for the reduction of sulfate:

$$SO_4^{2-} + 8e^- + 9H^+ \longrightarrow HS^- + 4H_2O \qquad +171.2 \text{ kJ mol}^{-1}$$

$$CH_2O + H_2O \longrightarrow CO_2 + 4H^+ + 4e^- \qquad -188.2 \text{ kJ mol}^{-1}$$

Coupling the above two reactions provides an output of energy (-205.0 kJ mol^{-1}):

$$SO_4^{2-} + 2CH_2O + H^+ \longrightarrow HS^- + 2H_2O + 2CO_2 \tag{6.12}$$

It can be seen that in the reduction of sulfate to hydrogen sulfide there is a free energy change (an energy release) of -205.0 kJ mol^{-1} (an eight-electron process), although not all of this energy is available to the microorganism. H_2S is typically formed under anaerobic conditions in swamps and marine sediments at great depth. The H_2S formed may escape as a gas or it may react with metal ions in the sediments to form insoluble sulfides:

$$2Fe(OH)_3(s) + 3H_2S(g) \longrightarrow 2FeS(s) + S(s) + 6H_2O(l)$$

and then

$$FeS(s) + S(s) \longrightarrow FeS_2(s) \tag{6.13}$$

The characteristic black colour of sediments is due to the presence of these iron sulfides and partially decomposed organic matter. If these sediments are exposed to the atmosphere, perhaps because of mining or geological uplift followed by weathering, oxidation of the S_2^{2-} (oxidation number $= -1$) to SO_4^{2-} (oxidation number $= +6$) will produce sulfuric acid in the presence of water and microorganisms:

$$2FeS_2(s) + 2H_2O(l) + 7O_2(g) \longrightarrow 2FeSO_4(aq) + 2H_2SO_4(aq)$$

The Fe^{2+} ion (from $FeSO_4(aq)$) can further react with oxygen and water to form a brown precipitate of iron (III) hydroxide, $Fe(OH)_3$, which can seriously contaminate ground and surface waters near to mine workings. The overall reaction for this process:

$$FeS_2(s) + 7/2H_2O(l) + 15/4O_2(g) \longrightarrow Fe(OH)_3(s) + 2H_2SO_4(aq)$$

demonstrates that a single mole of iron pyrite produces one mole of $Fe(OH)_3$ and two moles of H_2SO_4. Streams receiving this seepage can become very acidified and contaminated by toxic concentrations of metals. In acid environments, many metals may be leached away from soils and sediments and bioaccumulated by plants and aquatic organisms.

Soil acts as a source and sink for various S species and a medium for the organic–inorganic sulfur cycle. In humid regions, soil sulfur is predominantly in an organic form that is concentrated in the topsoil. This organically bound sulfur is gradually oxidized to sulfate that is rapidly taken up by plants or leached to the subsoil. Inorganic sulfur occurs almost exclusively as sulfate in the subsoil, although in arid regions, gypsum occurs in topsoil and often forms a surface coating. Many of the world's soils are sulfur-deficient, and fertilizers such as ammonium sulfate $((NH_4)_2SO_4)$ are used to provide soil sulfur, especially to agricultural land. Further research into the soil-sulfur cycle is vital as the Earth's population increases and high-yielding crops remove ever-greater amounts of soil nutrients.

In summary, the major steps in the sulfur cycle include:

- reduction of SO_4^{2-} into HS^- groups in proteins (Equation 6.12);
- release of HS^- to form H_2S during excretion, decomposition, etc.;
- oxidation of H_2S to form sulfur (S) and SO_4^{2-} (Equation 6.11);
- reduction of SO_4^{2-} by anaerobic respiration of sulfate-reducing bacteria (mainly in marine environments);
- anaerobic oxidation of H_2S and S by phototrophic bacteria (Equation 6.13).

6.7.3 Sulfur and the environment

The sulfur cycle has been heavily perturbed by human activity. Globally, emissions of sulfur compounds into the atmosphere by natural sources approximately equal those derived from anthropogenic sources. Natural sources of SO_2 include volcanoes and wind-blown aerosols from the sea surface. Man-made sources of SO_2 are dominated by emissions resulting from the combustion of fossil fuels for heating and energy production. All fuels used by humans (oil, coal, natural gas, peat, wood and other organic matter) contain sulfur, which upon burning will form $SO_2(g)$:

$$S(s) + O_2(g) \longrightarrow SO_2(g)$$

The other important anthropogenic source of SO_2 is the processing of sulfur-bearing ores. For example, the basic scheme for obtaining copper from its ore is:

$$CuFeS_2(s) + 5/2O_2(g) \longrightarrow Cu(s) + FeO(s) + 2SO_2(g)$$

Similar processing occurs for the principal ores of lead, zinc and nickel, resulting in the release of $SO_2(g)$. The SO_2 may be used to manufacture sulfuric acid or may escape into the atmosphere, returning eventually to ground with precipitation or sometimes as a gas. Small quantities of SO_2 are further oxidized to SO_3, which immediately dissolves in the existing water vapour to form a sulfuric acid aerosol. Sulfuric acid is eventually formed in the atmosphere from the emitted SO_2. The aqueous phase chemistry of SO_2 is discussed in detail in Section 7.10.1 below.

Concentrations of sulfate aerosols have also significantly increased due to industrial activities. Aerosols are very small, solid or liquid particles, ranging from a few molecules to 20 μm in radius, which are suspended in the atmosphere. Sulfate aerosols reflect sunlight back into space, thereby counteracting the greenhouse effect and cooling the Earth.

Internationally, emissions of SO_2 are decreasing from a peak in the early 1970s. For example, SO_2 emissions in the US rose steeply from 1940–1970, largely as a result of increases in coal-fired electricity production without any thought as to the consequent emissions. Since 1970, however, SO_2 emissions have reduced and stabilized due to the introduction of control technology, energy efficiency strategies and fuel modifications. In the longer term, the prospects are that international emissions of SO_2 in developed countries will decline still further as a result of:

- increased regulatory control of SO_2 emissions;

- the progressive introduction of improved emission control technology in power stations;

- increased usage of low-sulfur fuel;

- increased use of sulfur-free fuels, such as most natural gas, in place of coal and fuel oil (which typically contain 0.5–4.0% S);

- lower industrial demand;

- energy conservation measures that reduce demand for electricity;

- fiscal incentives, such as emissions taxes.

The link between ambient concentrations of pollutants and atmospheric emissions is not clear-cut. In the UK, for example, there was a decrease in SO_2 emissions of 25% from 1976 to 1986. However, the corresponding levels of SO_2 concentrations declined from 70 to 38 μg m^{-3}, a drop of 46%. The discrepancy between these figures is due to a shift from low-level SO_2 sources (for example, domestic chimneys that pollute the urban environment) to high-level sources (for example, power stations with tall stacks) which disperse pollutants over a large area. This shift to high-level sources effectively transferred the SO_2 problem from a local to an international and global scale.

The atmospheric dispersion of SO_2 from anthropogenic sources depends upon the climatic conditions, as well as on engineering and technical details (e.g. chimney height). The venting of SO_2 into the atmosphere has historically caused considerable human health and environmental problems, most notably the "Great London Smog of 1952" and acid precipitation. Consequently, the health and environmental effects of atmospheric SO_2 have been widely studied.

Particulate-associated SO_2 can aggravate existing respiratory diseases, such as bronchitis and emphysema, and contribute to the development of others. There is some evidence to suggest that exposure to SO_2 alone can cause slight effects on respiratory function at 2.03 mg m^{-3}. Some synergistic effects on pulmonary function from joint exposures to SO_2 and hydrogen peroxide, and to SO_2 and O_3, have been also reported.

Direct action of gaseous SO_2 on vegetation can cause plant damage, while other effects include damage to paints, dyes, textiles, leather and photographic materials, and discolouration of paper. Sulfuric acid mists can also cause plant injury and contribute to visibility reduction. Sulfur oxides play a major role in the phenomenon of acid precipitation that has caused extensive environmental damage in Europe (this topic is discussed in detail later in Section 7.10.2).

The ability of soil to cope with acid precipitation depends primarily upon the underlying rock–carbonates can neutralize acid better than quartz or granite. The soil's cation exchange capacity also plays an important part (see Section 4.11); metal released from the soil (through ion exchange) can cause environmental and health problems.

Many governments and international authorities have introduced legislation over the years in an attempt to abate the SO_2 problem-examples from the UK go back over 150 years! Legislation usually requires an industrial process to change or the addition of control technology to the emission source. Reduction of SO_2 emissions may be achieved by a variety of strategies, including reducing the demand for products that involve the generation of these pollutants in their production.

Sulfur may be directly removed from liquids or gases through a series of steps:

(1)   A catalytic process is used to convert the S contained in the liquid or gas to H_2S;

(2)   The H_2S is then removed by scrubbing with an alkaline solution;

(3)   The removed H_2S is converted to elemental S, for example through the following process:

$$H_2S + 1/2O_2 \longrightarrow S + H_2O$$

(4)   The residual elemental sulfur may now be converted into a useful product–such as sulfuric acid–or placed in a landfill.

The process of treating and removing sulfur dioxide from gases differs according to its concentration. In the case of *rich waste gases*, such as those arising from the smelting of metal sulfide ores where the concentration of SO_2 generally falls in the range 2–12%, economic treatment is possible through the production of sulfuric acid by the following reactions:

$$SO_2 + \tfrac{1}{2}O_2 \xrightarrow{\text{vanadium catalyst}} SO_3$$

and then:

$$SO_3 + H_2O \longrightarrow H_2SO_4$$

However, the major sources of SO_2, except near smelters, are the stacks of large coal- or oil-burning power stations. Here, the SO_2 concentration of the exhaust gas is typically 0.1%, which is too low for profitable recovery as sulfuric acid. For such *lean waste gases*, the approach is to reduce the level of SO_2 emissions. There are two main techniques for achieving this. *Pre-combustion techniques* usually involve either some modification of furnace design and/or fuel treatment. *Post-combustion techniques* are the most common means of removing SO_2 from the waste gases of coal and high-sulfur oil-burning power stations. The basic approach is to burn the fuel and then treat the plant's exhaust gas – most successfully with limestone or lime in a wet scrubber. This process converts the SO_2 to $CaSO_4$, which can be purified for use as plasterboard, or disposed of in landfill. The technique, known generally as *Flue-Gas Desulfurisation* (FGD), typically removes over 90% of SO_2 from flue gases.

6.7.4 Uses of sulfur

Sulfur is an important industrial element and has a large number of uses. It is chemically combined with rubber using a process called vulcanization in order to make rubber harder, stronger and more elastic. Sulfur is used to manufacture a number of industrial gases and chemicals, including sulfur dioxide, hydrogen sulfide, carbon disulfide and sulfuric acid. The latter is the most heavily produced industrial chemical in the world, and is used for the production of fertilizers, detergents, dyestuffs and petrochemicals. Sulfur dioxide is used in the manufacture of sulfuric acid, as bleach for flour and textiles, as a preservative for dried fruit, wine and soft drinks, and for the removal of excess chlorine following disinfection of drinking water.

Other sulfur-containing compounds are used as industrial and laboratory chemicals and in the manufacture of paper, gunpowder, fireworks, medicines, skin ointments, fungicides and insecticides. Sulfates are used in the manufacture of explosives, matches and medicines. Sodium thiosulfate is used as a fixing agent in the photographic process, and is a widely used laboratory reagent.

6.8 Summary

After carefully studying this chapter, you should have a clear understanding of the abundances, physico-chemical properties, cycling and uses of those bioelements involved in natural cycles. In addition, we have outlined a few of the environmental implications arising from anthropogenic activities and detailed the environmental chemistry of several important terrestrial and atmospheric processes. Your wider understanding of environmental chemistry and your practical skills will be further developed by carrying out Experiments 6.1–6.4.

Experiment 6.1 Determination of Ammonia by Nessler's Method

Introduction

Ammonia in water bodies results from the decomposition of nitrogenous organic and inorganic matter in soil and in water. It is also discharged into water by industrial processes (e.g. paper production) and sewage effluent. When ammonia is dissolved in water, the following equilibrium is set up:

$$NH_3(g) + H_2O(l) \rightleftharpoons NH_4^+(aq) + OH^-(aq)$$

At high pH values, the equilibrium is shifted to the left, increasing the concentration of free ammonia, which is more toxic than the NH_4^+ ion to aquatic life. Unpolluted waters contain small amounts of ammonia, typically less than $0.1 \, \text{mg} \, l^{-1}$ as nitrogen. Higher concentrations indicate organic pollution from domestic sewage, fertilizer run-off, industrial waste, etc.

Nessler's Method is based on converting ammonia into a reddish-brown coloured complex compound using Nessler's Reagent. This is made up of potassium iodide and mercuric chloride–the approximate composition of the ammonia complex is $Hg_2NH_2I_3$. The intensity of the colour, which is proportional to the ammonia concentration, is measured by comparison with permanent glass colour standards, called *Lovibond standards*, by using a simple optical device called a *Nesslerizer*.

Time Required

Two hours (assuming all reagents are made up in advance).

Equipment

Nesslerizer
Nesslerizer glasses
Standard Lovibond Nesslerizer discs, NAA, NAB, NAC, NAD

   Disc NAA covers the range 0.001–0.01 mg of ammonia
   Disc NAB covers the range 0.01–0.026 mg of ammonia
   Disc NAC covers the range 0.028–0.06 mg of ammonia
   Disc NAD covers the range 0.06–0.1 mg of ammonia

Reagents

To make Nessler's Reagent–dissolve 35.00 ± 0.01 g of potassium iodide (KI) and 12.50 ± 0.01 g of mercuric chloride ($HgCl_2$) in 800 cm^3 of double-distilled deionized water. Add a cold, saturated solution of mercuric chloride until, after repeated shaking, a slight red precipitate remains. Add 120.00 ± 0.01 g of sodium hydroxide (NaOH), shake until dissolved, and finally add a little more of the saturated mercuric chloride solution and sufficient double-distilled deionized water to produce one litre in a volumetric flask. Shake occasionally during several days and then allow to stand. Use the clean supernatant liquid for the experiment.

Several suitable water samples (e.g. tap, lake or river waters, urban and agricultural run-off, wastewaters, etc.) which must not be acidic.

Significant Experimental Hazards

- Students should be aware of the hazards associated with the use of all glassware (cuts) and electrical equipment (shock, burns).

- Nessler's Reagent is toxic by ingestion and may burn eyes and skin.

- Mercuric chloride is extremely poisonous. It is toxic by ingestion, inhalation and skin absorption and may burn eyes and skin.

Procedure

(1) Fill one of the Nesslerizer glasses to the 50 cm^3 mark with the unknown solution under examination and place it in the left-hand compartment of the Nesslerizer.

(2) Fill a second Nesslerizer glass to the 50 cm^3 mark with the same unknown solution; add 2 cm^3 of Nessler's Reagent, mix and place in the right-hand compartment of the Nesslerizer.

(3) Place the standard Lovibond Nesslerizer disc covering the lowest concentration range into the Nesslerizer viewer.

(4) Switch on the standard white light in the Nesslerizer column and compare the colour produced in the test solution with the colours produced in the standard disc, rotating the disc until a colour disc is replaced. If a colour match is not obtained, replace the Nesslerizer disc with the disc covering the next concentration range.

With small quantities of ammonia (1–5 µg) the colour develops slowly and in order to obtain accurate results the solution should stand for 15 minutes before matching the colour. In other cases, 5 minutes is normally sufficient for the colour to develop.

Issues to Consider for your Practical Report

- What are the potential sources of error in this analytical determination? How could they be overcome?

- What are the limitations of this analytical method?

- Are there alternative methods for determining ammonia? If so, how do they compare to this method?

- What is the source of the unknown river and lake-water samples?

- What are typical ammonia values in natural and wastewaters? How do your data compare with these values?

- How can ammonia be removed from domestic drinking water?

- What are the legal limits (if any) of ammonia in drinking water? Do your values exceed such legal limits?

- What are the potential human health and environmental effects (if any) of excess ammonia in potable water?

Useful References

Greenberg, A.E., Clesceri, L.S. and Eaton, A.D. (Eds), (1992). *Standard Methods for the Examination of Water and Wastewater*. American Public Health Association, Washington, DC, USA. ISBN 0-87553-207-1.

Experiment 6.2 Determination of the Phosphate Levels in Clean and Polluted Waters

Introduction

Phosphorus is an essential element in all living organisms. It is an essential constituent of the energy-transferring molecules, adenosine triphosphate (ATP), adenosine diphosphate (ADP) and adenosine monophosphate (AMP), and of the genetic and information-carrying molecules, deoxyribonucleic acid (DNA) and ribonucleic acid (RNA). It is found in bones, teeth, nerve and brain tissue and is an essential element for plants. Phosphorus is often present as the orthophosphate ion, PO_4^{3-}, in the environment.

Ammonium molybdate and potassium antimonyl tartrate in an acid medium combine with dilute solutions of orthophosphates to form an antimony–phosphomolybdate complex. By addition of ascorbic acid, the complex is reduced and an intense blue colour, which is proportional to the orthophosphate concentration, is developed. Samples are examined colorimetrically by using a spectrophotometer at a wavelength of 880 nm in order to determine their concentrations.

Time Required

Two–three hours, depending upon the number of water samples determined (assuming all reagents are made up in advance).

Equipment

$100 \, cm^3$ volumetric flasks
5, 10, 25 and $50 \, cm^3$ pipettes and pipette filler
$100 \, cm^3$ conical flasks
Dropping pipette
Spectrophotometer (able to operate at 880 nm)
Cuvettes

Reagents

Standard phosphate solution containing $2.50 \, \mu g$ phosphate per $1.00 \, cm^3$ (dissolve 0.2195 g of anhydrous KH_2PO_4 in one litre of double-distilled deionized water–this gives $50 \, \mu g \, cm^{-3}$ phosphate; diluting 20 times gives $2.50 \, \mu g \, cm^{-3}$).

Reagents needed to make the combined reagent:

- potassium antimonyl tartrate $(K(SbO)C_4H_4O_6)$ solution (4.3888 g dissolved in $200 \, cm^3$ double-distilled, deionized water);

- ammonium molybdate $((NH_4)_6Mo_7O_{24}.4H_2O)$ solution (20 g dissolved in $500 \, cm^3$ double-distilled, deionised water);

- a solution of ascorbic acid $(C_6H_8O_6)$ (1.76 g dissolved in $100 \, cm^3$ double-distilled, deionized water);

- 5N sulfuric acid ($70 \, cm^3$ of concentrated H_2SO_4 made up to $500 \, cm^3$ with double-distilled, deionized water)

To make up $100 \, cm^3$ of combined reagent, mix the above reagents in the following order:

(1) $50 \, cm^3$ of 2.5 M sulfuric acid;

(2) $5 \, cm^3$ of potassium antimonyl tartrate solution;

(3) $15 \, cm^3$ of ammonium molybdate solution;

(4) $30 \, cm^3$ of ascorbic acid solution.

The mixture will give a yellowish colouration—any phosphate contamination will give a blue colouration. The reagent will remain stable for a week if stored in a refrigerator.

Phenolphthalein indicator
Suitable water sample(s) or dilute phosphate solutions

Significant Experimental Hazards

- Students should be aware of the hazards associated with the use of all glassware (cuts), Bunsen burners (burns, fire), tongs to move hot objects (burns, dropping hazards) and furnaces.

- Phosphate solutions may irritate the eyes.

- Combined reagent may be harmful by ingestion, irritating to eyes and skin and may cause burns and dermatitis.

- Sulfuric acid is harmful by ingestion and may cause severe burns to eyes and skin.

- Phenolphthalein indicator may be harmful if ingested in quantity and may irritate eyes and skin.

Procedure

(1) Prepare, in turn, a series of calibration standards by pipetting the following amounts of standard phosphate solution (5, 10, 25 and 50 cm^3) into clean 100 cm^3 volumetric flasks and diluting to the mark with distilled water. Stopper the flask and mix the contents thoroughly. The concentrations of the prepared dilutions will be successively: 0.125, 0.250, 0.625 and 1.250 mg l^{-1}.

(2) Pipette 50 cm^3 of the 0.125 mg l^{-1} standard solution into a clean conical flask and add one drop of phenolphthalein indicator. If a red colour develops, add 5N sulfuric acid one drop at a time until the solution turns clear. After each addition of acid, swirl the flasks thoroughly to completely mix the contents. Repeat this procedure for each of the other three prepared standard dilutions, for a reagent blank (by measuring out 50 cm^3 of distilled water) and for 50 cm^3 of each of the unknown samples.

(3) To each of the conical flasks, add 8 cm^3 of the combined reagent (measured accurately). Completely mix the contents of the flask and allow it to stand for 10 minutes (but no longer than 30 minutes) to enable a blue colour to fully develop.

(4) Prepare a spectrophotometer for operation at a wavelength of 880 nm and fill the sample cell with the reagent blank solution. With this solution in position in the sample holder of the spectrophotometer,

set the absorbance reading to zero. Without changing the zero setting, introduce each of the calibration standards into the sample cell and record the absorbance values. Plot a calibration curve of absorbance (*y*-axis) against concentration (*x*-axis).

(5) Record the absorbance of each sample, employing an appropriate dilution ratio if necessary. Use the derived calibration curve to calculate the phosphate concentration in each of the unknown samples as mg dm^{-3}.

Issues to Consider for your Practical Report

- What are the potential sources of error in this analytical determination? How could they be overcome?

- How many significant figures should you use when reporting your results?

- What other methods could be used to quantitatively determine phosphate in natural waters?

- What is the source of the unknown water samples?

- What are typical concentrations of phosphates in UK rivers, streams, lakes and drinking water? How do your data compare with these values?

- What are the legal limits (if any) of phosphates in the environment? Do your values exceed such legal limits?

- What are the main sources of additional phosphates in natural waters?

- What are the potential human health and environmental effects of excess phosphates in the environment?

Useful References

Greenberg, A.E., Clesceri, L.S. and Eaton, A.D. (Eds), (1992). *Standard Methods for the Examination of Water and Wastewater*. American Public Health Association, Washington, DC, USA. ISBN 0-87553-207-1.

ISO Compendia on the Environment: Water Quality. Volume 2, *Chemical Methods* (1992). 'Water quality–Determination of phosphate', ISO6878/1-1986(E), 149–158. International Organization for Standardization (ISO), Geneva, Switzerland. ISBN 92-67-10193-5.

Methods for the examination of waters and associated materials, (1981). *Phosphorus-Waters, Effluents and Sewages*. HMSO, London. ISBN 0-11-751-582-5.

Experiment 6.3 Biochemical Oxygen Demand using Winkler's Method

Introduction

Biochemical oxygen demand (BOD) is a measure of the oxygen required to oxidize the bioavailable organic matter in a water sample to stable inorganic compounds and gives an approximate index of organic pollution. Biochemical oxidation is brought about by microorganisms that utilize organic matter (often polluting effluent) as a source of carbon, and in doing so consume dissolved oxygen. The consumption of oxygen in water by the degradation of organic matter can be represented by the following equation:

$$(CH_2O)(s) + O_2(g) \longrightarrow CO_2(g) + H_2O(l)$$

The process of biochemical oxidation is an inherent, natural process of self-purification carried out by aquatic systems when an effluent discharge (e.g. domestic sewage) enters a water body. The activity of microorganisms results in the rapid oxidation of readily degradable compounds present in organic wastes until they are all depleted.

The BOD test is commonly carried out by measuring the dissolved oxygen concentration in water samples before and after incubation in the dark at 20°C for five days. Preliminary dilution and aeration of the sample may be necessary to ensure that not all of the oxygen is used during incubation. Oxygen has a limited solubility in water (about 9 mg dm^{-3} at 20°C), and thus samples absorbing more than 6 mg dm^{-3} of oxygen must be diluted to ensure that dissolved oxygen is present at the end of the test. In order to make the test quantitative, samples must be protected from air during storage to prevent re-aeration as the dissolved oxygen diminishes.

The BOD value recorded after five days is not the ultimate BOD value as biochemical oxidation continues after this time period. However, an incubation period of five days is recommended in order to minimize interference from the actions of nitrifying organisms and because a reasonably large percentage of the total BOD is consumed in five days (70–80% for domestic sewage). The length of the BOD test can depend on the amount of organic pollution thought to be present and can be denoted by, for example, BOD$_{3.5..7...20}$, where the subscripts represent the length of the incubation period in days. The degradation of organic matter is considered to be complete at 20 days, although theoretically an infinite time can be required for complete decomposition. For convenience, many teaching experiments utilize an incubation period of seven days.

BOD is an important parameter used for the determination of water quality and the organic loading of domestic and industrial wastes. In water treatment and pollution control processes, it provides an indirect means of evaluating the amount of organic matter that needs to be oxidized by microbial action before the treated water is discharged into streams and rivers. This quantity of biodegradable matter cannot be measured directly. BOD measurements are also used to analyse the purification capacity of streams and rivers and to assist regulating authorities to check compliance with water quality regulations. A natural water body is generally regarded as seriously polluted if its BOD exceeds 5 mg dm^{-3}.

The Winkler Method is a means of measuring the dissolved oxygen in a water sample by using a redox titration. In this method, a white precipitate of manganese (II) hydroxide is produced by the action of sodium hydroxide on manganese (II) sulfate and the dissolved oxygen converts this into an orange-brown precipitate of manganese (IV) hydroxide:

$$Mn^{2+}(aq) + 2OH^-(aq) \rightleftharpoons Mn(OH)_2(s) \tag{6.14}$$
$$\text{manganese (II) hydroxide}$$

$$Mn(OH)_2(s) + 1/2O_2(aq) \rightleftharpoons MnO(OH)_2(s) \tag{6.15}$$
$$\text{manganese (IV) hydroxide}$$

When potassium iodide is present in the solution, the manganese (IV) hydroxide liberates iodine upon addition of an acid (H_2SO_4) in an amount equivalent to the quantity of dissolved oxygen originally present:

$$MnO(OH)_2(s) + 2I^-(aq) + 4H^+(aq) \rightleftharpoons Mn^{2+}(aq) + I_2(g) + 3H_2O(l) \tag{6.16}$$

Measurement of the amount of iodine allows the concentration of dissolved oxygen to be determined. Sodium thiosulfate converts the iodine back into iodide ions and this reaction can be monitored volumetrically:

$$I_2(g) + 2S_2O_3^{2-}(aq) \rightleftharpoons S_4O_6^{2-}(aq) + 2I^-(aq) \tag{6.17}$$

Time Required

Two sessions of two hours duration, separated by a period of either five or seven days, depending upon whether you decide to determine BOD$_5$ or BOD$_7$ and upon the number of samples determined (assuming all reagents are made up in advance).

Equipment

Incubation bottles of known capacity (250–300 cm^3) with ground glass stoppers (bottles should be thoroughly cleaned before use)
Incubator thermostatically controlled at 20°C ± 1°C from which all light should be excluded to prevent formation of dissolved oxygen by algae in the sample
250 cm^3 conical flasks
2.0 cm^3 pipettes and pipette filler
50 cm^3 burette
10.0 cm^3 measuring cylinder
Boss, clamp, retort stand and white tile

Reagents

Phosphate buffer solution. (Weigh the following reagents into a one litre volumetric flask: 8.50 ± 0.01 g potassium dihydrogen phosphate (KH_2PO_4), 21.750 ± 0.001 g dipotassium hydrogen phosphate (K_2HPO_4), 33.40 ± 0.01 g disodium hydrogen phosphate ($Na_2HPO_4.7H_2O$), and 1.70 ± 0.01 g ammonium chloride (NH_4Cl). Dissolve in double-distilled, deionized water to give one litre of solution.)

Magnesium sulfate solution. (Dissolve 22.50 ± 0.01 g $MgSO_4.7H_2O$ in double-distilled, deionized water to give one litre of solution.)

Calcium chloride solution. (Dissolve 27.50 ± 0.01 g anhydrous $CaCl_2$ in double-distilled, deionized water to give one litre of solution.)

Ferric chloride solution. (Dissolve 0.250 ± 0.001 g $FeCl_3.6H_2O$ in double-distilled, deionized water to give one litre of solution.)

Dilution water. (Double-distilled, deionized water saturated with air and kept at 20°C in cotton plugged bottles. When required for use, add 1.0 cm^3 of each of the following solutions, phosphate buffer, magnesium sulfate solution, calcium chloride solution and ferric chloride solution, to each litre of distilled water.)

0.0125 M sodium thiosulfate solution ($Na_2S_2O_3$)

50% manganese sulfate solution

Alkaline iodide–azide solution. (Prepared from 500.0 ± 0.1 g sodium hydroxide (NaOH) and 140 ± 0.1 g sodium iodide (NaI) in one litre of distilled water with 300 cm^3 of 2.5% sodium azide (NaN_3) solution added.)

Starch indicator

Concentrated sulfuric acid (H_2SO_4)

Suitable water samples (e.g. samples of settled sewage and final effluent from a local water treatment plant)

Significant Experimental Hazards

- Students should be aware of the hazards associated with the use of all glassware (cuts), electrical equipment (shock, burns) and fume cupboards.

- Sewage and final effluent samples contain harmful bacteria and may contain toxic organic substances and metals (harmful if ingested)

- Sulfuric acid is harmful by ingestion and may cause severe burns to eyes and skin.

- Sodium thiosulfate may be harmful if ingested in quantity and may irritate eyes and skin.

- Alkaline iodide–azide solution causes severe burns and may be harmful if ingested.

- Manganese sulfate solution is be harmful by ingestion and may irritate eyes and skin.

Procedure

(1) Ensure that the initial dissolved oxygen content of the water sample is near to the saturation level and aerate if necessary, avoiding supersaturation.

(2) Depending on the potential BOD value of the sample it will need to be diluted. The final effluent from a water treatment plant typically requires a fivefold dilution, while untreated settled sewage often needs to be diluted fifty times. If you are not sure what dilution factor to use, then seek advice. To complete the dilution, add the required volume of undiluted sample to each of two incubation bottles and make up the volume with dilution water. Ensure that the bottles are completely filled and insert the stoppers, making sure that no air is trapped in the bottle.

(3) Holding the stopper tightly, invert the incubation bottles several times to ensure thorough mixing.

(4) You will now need to determine the dissolved oxygen content of one of the bottles by using Winkler's Method. The other bottle should be incubated in the dark in a thermostatically controlled incubator at $20 \pm 1°C$. Ensure that you have fully labelled the incubation bottle with your name, sample type and number, and the date.

(5) In order to determine the initial dissolved oxygen content of your sample, you first need to record the temperature of the sample with an accurate thermometer.

(6) Place the incubation bottle in a fume cupboard on a layer of disposable absorbent paper. Using a pipette for each addition, add 2 cm^3 of manganese sulfate solution and 2 cm^3 of alkaline iodide–azide solution. In each case, ensure that the tip of the pipette is well below the surface of the liquid in the bottle. Incline the bottle and carefully replace the stopper so as to avoid the inclusion of air bubbles. Some of the water sample will spill out of the bottle as you replace the stopper; ensure that your hands do not come into direct contact with the liquid and dry the outside of the bottle with a disposable paper towel.

(7) Thoroughly mix the contents of the bottle by inverting the bottle at least 20 times and then allow the precipitate to settle. Repeat the mixing and allow the precipitate to settle again to leave a completely clear supernatant liquid.

(8) In a fume cupboard, add 2 cm^3 of concentrated sulfuric acid very slowly and carefully down the neck of the bottle. (Take all sensible precautions when performing this procedure, including lowering the sash on the fume cupboard and wearing protective glasses and gloves.)

(9) Replace the stopper in the bottle and mix its contents thoroughly by inversion until all of the precipitate has dissolved. If the precipitate

does not dissolve instantaneously, allow the bottle to stand for a few minutes and repeat the mixing procedure. In either case, mix the contents of the bottle again immediately before measurement. (Some of the liquid will spill out of the bottle as you replace the stopper; follow the precautions outlined in Step 6.)

(10) Accurately transfer 200 cm³ of the solution to a 250 cm³ conical flask and immediately titrate the iodine present with 0.0125 M sodium thiosulfate solution. When the brown colour, due to iodine, has changed to a pale straw colour, add starch indicator to produce a blue colouration. The titration is then continued until the blue colour just disappears.

(11) After the desired incubation period (generally five or seven days), repeat Steps 5–10 in order to determine the dissolved oxygen content of the incubated sample. The difference between the initial dissolved content and that after the incubation period represents the BOD value of the sample.

NB–The above procedure is for the determination of one sample. It is desirable to perform the determination in duplicate if possible.

Calculation

From Equations 6.14–6.17, it can be seen that 2 moles of thiosulfate are equivalent to 0.5 mole of dissolved oxygen. The concentration of the thiosulfate solution is 0.0125 M and therefore 1 cm³ of this solution contains 0.0125×10^{-3} moles. Thus:

1 cm³ of 0.0125 M thiosulfate solution $\equiv 1/4(0.0125 \times 10^{-3})$ moles dissolved oxygen
1 cm³ of 0.0125 M thiosulfate solution $\equiv 1/4(0.0125 \times 10^{-3}) \times 32$ g dissolved oxygen
1 cm³ of 0.0125 M thiosulfate solution $\equiv 0.1$ mg dissolved oxygen

From the above relationship, the concentration of dissolved oxygen (in mg dm^{-3}) in the original solution can be calculated by using Table 6.14.

The BOD value may then be calculated by using the following equation:

$$BOD(\text{mg dm}^{-3}) = F(D_0 - D_1)$$

where D_0 is the initial dissolved oxygen concentration (mg dm^{-3}), D_1 is the dissolved oxygen concentration after the incubation period (mg dm^{-3}), and F is the dilution factor (e.g. if five times dilution, $F = 5$)

Issues to Consider for your Practical Report

- What are the potential sources of error in this analytical determination? How could they be overcome?

Table 6.14 Dissolved oxygen (mg dm^{-3}) in fresh water as a function of temperature

Temp (°C)	0	0.1	0.2	0.3	0.4	0.5	0.6	0.7	0.8	0.9
5	12.80	12.77	12.74	12.70	12.67	12.64	12.61	12.58	12.54	12.51
6	12.48	12.45	12.42	12.39	12.36	12.33	12.29	12.26	12.23	12.20
7	12.17	12.14	12.11	12.08	12.05	12.02	11.99	11.96	11.93	11.90
8	11.87	11.84	11.81	11.79	11.76	11.73	11.70	11.67	11.65	11.62
9	11.59	11.56	11.54	11.51	11.49	11.46	11.43	11.41	11.38	11.36
10	11.33	11.31	11.28	11.26	11.23	11.21	11.18	11.16	11.13	11.11
11	11.08	11.06	11.03	11.01	10.98	10.96	10.93	10.91	10.89	10.85
12	10.83	10.81	10.78	10.76	10.74	10.72	10.69	10.67	10.65	10.62
13	10.60	10.58	10.55	10.53	10.51	10.49	10.46	10.44	10.42	10.39
14	10.37	10.35	10.33	10.30	10.28	10.26	10.24	10.22	10.19	10.17
15	10.15	10.13	10.11	10.09	10.07	10.05	10.03	10.01	9.99	9.97
16	9.95	9.93	9.91	9.89	9.87	9.84	9.82	9.80	9.78	9.76
17	9.74	9.72	9.70	9.68	9.66	9.64	9.62	9.60	9.58	9.56
18	9.54	9.52	9.50	9.48	9.46	9.44	9.43	9.41	9.39	9.37
19	9.35	9.33	9.31	9.30	9.28	9.26	9.24	9.22	9.21	9.19
20	9.17	9.15	9.13	9.12	9.10	9.08	9.06	9.04	9.03	9.01
21	8.99	8.97	8.96	8.94	8.93	8.91	8.89	8.88	8.86	8.85
22	8.83	8.82	8.80	8.79	8.77	8.76	8.74	8.73	8.71	8.70
23	8.68	8.67	8.65	8.64	8.62	8.61	8.59	8.58	8.56	8.55
24	8.53	8.52	8.50	8.49	8.47	8.46	8.44	8.43	8.41	8.40
25	8.38	8.36	8.35	8.33	8.32	8.30	8.28	8.27	8.25	8.24

- Are there alternative methods and indicators available for measuring the organic matter content of water samples? If so, how do they compare to this method?

- Why is an incubation temperature of 20°C used in this determination?

- Why is the bottle incubated in the dark during the incubation period?

- What is the source of the water samples? What are typical BOD values for these water samples? How do your data compare with these values?

- How can organic matter be removed from water used for domestic and industrial purposes?

- What are the legal limits (if any) of BOD in natural and drinking waters? Do your values exceed such legal limits?

- What are the potential human health and environmental effects (if any) of excess organic matter in potable and natural waters?

Useful References

Greenberg, A.E., Clesceri, L.S. and Eaton, A.D., (Eds), (1992). *Standard Methods for the Examination of Water and Wastewater*. American Public Health Association, Washington, DC, USA. ISBN 0-87553-207-1.

ISO Compendia on the Environment: Water Quality. Volume 2, *Chemical Methods* (1992). 'Water quality–Determination of BOD', ISO5813-1983(E), 36–40.

International Organization for Standardization (ISO), Geneva, Switzerland. ISBN 92-67-10193-5.

Montgomery, H.A.C. and Cockburn, A. (1964). 'Errors in sampling for dissolved oxygen (DO)'. *Analyst*, **89**, 679–681.

Montgomery, H.A.C., Thorn, N.S. and Cockburn, A. (1964). 'Determination of dissolved oxygen by the Winkler Method and the solubility of oxygen in pure water and seawater'. *Journal of Applied Chemistry*, **14**, 280–296.

Hart, I.C. (1967). 'Nomograms to calculate dissolved oxygen contents and exchange (mass-transfer) coefficients'. *Water Research*, **1**, 391–395.

Experiment 6.4 Determination of Chemical Oxygen Demand in Water Samples

Introduction

The chemical oxygen demand (COD) of domestic and industrial wastewater is a measure of the amount of oxygen required to oxidize (decrease) the organic matter content of a sample that is susceptible to oxidation by a strong oxidant (such as potassium dichromate). Most organic compounds will oxidize under the influence of oxidizing agents in acid conditions–oxidation of most organic compounds is 95–100% of the theoretical value. BOD values are often lower than COD values as some materials (such as cellulose) will react with the dichromate present in COD tests, but not the oxygen present under biological conditions.

COD and BOD values are directly comparable, and are often used side-by-side to allow for differentiation between biologically oxidizable matter and biologically inert matter. If the COD value of a water sample is much larger than the BOD value, it suggests that the sample contains large amounts of organic substances that are not easily biodegradable. COD tests have the advantage of being relatively quick to perform, with results available in about three hours, although they are subject to some interferences and errors.

The COD determination is based on the principle that organic material in water can be oxidized by potassium dichromate in acidic solution in the presence of a silver catalyst. The dichromate reflux method is preferred over procedures using other oxidants because of its superior oxidizing ability, applicability to a wide range of samples and ease of manipulation. Under these experimental conditions, some of the dichromate is reduced by the organic matter in the water sample and the remainder is titrated with ferrous ammonium sulfate using a ferroin (1,10-phenanthroline) indicator to indicate the end point of the reaction. The ferroin forms an intense red colour with ferrous (Fe^{2+}) ions but no colour with ferric (Fe^{3+}) ions. During the reaction, most of the organic matter in the sample is oxidized to carbon dioxide and water, while the dichromate is reduced to trivalent chromium.

$$Cr_2O_7{}^{2-} + 6Fe^{2+} + 14H^+ \longrightarrow 2Cr^{3+} + 6Fe^{3+} + 7H_2O$$

The more organic matter present in the sample, then the more dichromate is reduced. The same test procedure is used for a blank sample of double-distilled, deionized water so that any errors arising from the presence of extraneous organic matter in the reagents can be accounted for.

Time Required

Three hours, depending upon the number of water samples determined (assuming all reagents are made up in advance).

Equipment

1, 5 and 10 cm^3 pipettes and pipette filler
20 cm^3 measuring cylinder
50 cm^3 round-bottomed flasks
Cork rings (for safely holding round-bottomed flasks)
Glass beads (or anti-bump granules)
Large water-cooled reflux condenser
Heating equipment—Bunsen burners, tripod stands and metal gauzes, or appropriate heating mantles
Dropping pipettes
50 cm^3 burette
100 cm^3 conical flasks
Bosses, clamps, retort stands and white tile

Reagents

All reagents should be *AnalaR* grade
0.025 M ferrous ammonium sulfate (FAS) ((NH$_4$)$_2$SO$_4$.FeSO$_4$.6H$_2$O)
0.020 83 M potassium dichromate (K$_2$Cr$_2$O$_7$)
20% mercuric sulfate (HgSO$_4$) in 10 vol per cent sulfuric acid (20.00 $\pm$ 0.01 g HgSO$_4$ dissolved in 10 cm^3 of concentrated H$_2$SO$_4$ and made up to 100.0 cm^3 with double-distilled, deionized water)
1% silver sulfate (Ag$_2$SO$_4$) in concentrated sulfuric acid (10.00 $\pm$ 0.01 g of Ag$_2$SO$_4$ dissolved in one litre of concentrated sulfuric acid)
Ferroin indicator (1,10-phenanthroline ferrous complex)
1% sulfuric acid (in a wash bottle)
Suitable water samples (e.g. samples of settled sewage and final effluent from a local water treatment plant)

Significant Experimental Hazards

- Students should be aware of the hazards associated with the use of all glassware (cuts), Bunsen burners (burns and fire), tongs to move hot objects (burns and dropping hazards) and fume cupboards. This method requires the handling and boiling of a strong solution of sulfuric acid and potassium dichromate and so great care is necessary.

- Sewage and final effluent samples contain harmful bacteria and may contain toxic organic substances and metals (Harmful if ingested).

- Sulfuric acid is harmful by ingestion and may cause severe burns to eyes and skin.

- Ferroin indicator may be harmful if ingested in quantity and may irritate eyes and skin. It may also stain clothing and skin.

- Ferrous ammonium sulfate may be harmful if ingested in quantity and may irritate eyes and skin.

- Potassium dichromate is harmful by ingestion, inhalation and skin contact. There is also danger of combustion with organic matter.

- Mercuric sulfate is toxic by ingestion and skin contact. It may also irritate and burn eyes and skin.

- Silver sulfate is toxic by ingestion and may irritate and burn eyes and skin.

Procedure

(1) Set up two reflux condensers, side-by-side, on retort stands above Bunsen burners or heating mantles.

(2) Place two labelled round-bottomed flasks on cork rings. Pipette 10.0 cm^3 of your water sample into one flask and 10.0 cm^3 of double-distilled, deionized water into the other.

(3) Into each flask, pipette 5.0 cm^3 of 0.020 83 M potassium dichromate solution, followed by 1.0 cm^3 of the mercuric sulfate reagent. Swirl each flask carefully to ensure thorough mixing.

(4) Measure out 16.0 cm^3 of 1% silver sulfate in concentrated sulfuric acid in a measuring cylinder and add in small portions to the sample in the flask. Swirl the flask carefully after each addition. If the flask gets very hot (if its sides cannot be touched easily with fingers) cool the flask carefully in ice or by holding it (with tongs) under a tap of running cold water. (Make sure that the water does not splash into the flask.) Repeat this procedure for the flask containing the double-distilled, deionized water.

(5) Add a few glass beads to each flask in order to minimise "bumping" (the sudden movement of a hot solution out of its container).

(6) Attach each flask to its condenser and secure on (or in) the heating equipment.

(7) Reflux each flask for 1–2 hours. The water-cooling must be switched on at all times and the top of the condenser must be open to prevent build-up of pressure. (If steam escapes from the top of the condenser you must stop the determination and fit a larger condenser.)

(8) After refluxing, allow the flasks to cool for 10 minutes and then cool to room temperature in ice or under running water.

(9) Wash the inner tube of each condenser with 1% sulfuric acid, collecting the washings in the appropriate round-bottomed flask. Swirl the flasks to ensure thorough mixing.

(10) Add five drops of ferroin indicator to each flask and titrate the solutions with 0.025 M FAS until the solution just becomes pink. During the titrations, the colour of the solution will change from green to blue to pink. Record the titration values for double-distilled, deionized water and your water sample as T_1 and T_2 cm^3, respectively.

Calculation

From the equation for the titration, we can see that 1 mole of $Cr_2O_7^{2-}$ reacts with 6 moles of Fe^{2+}.

$$Cr_2O_7^{2-} + 6Fe^{2+} + 14H^+ \longrightarrow 2Cr^{3+} + 6Fe^{3+} + 7H_2O$$

1 mole   6 moles

During the refluxing process, the organic carbon will be oxidized to carbon dioxide (CO_2); this will require the release of four electrons (i.e. $C^0 \longrightarrow C^{4+}$). However, six electrons are required to reduce the dichromate (in which Cr has an oxidation state of +6) to $2Cr^{3+}$. Consequently:

$$3/2 \text{ moles } O_2 \equiv 1 \text{ mole } Cr_2O_7^{2-}$$

and hence:

$$3/2 \text{ moles } O_2 \equiv 1 \text{ mole } Cr_2O_7^{2-} \equiv 6 \text{ moles } Fe^{2+}$$

$$1 \text{ mole } O_2 \equiv 4 \text{ moles } Fe^{2+}$$

$$1/4 \text{ mole } O_2 \equiv 1 \text{ mole } Fe^{2+}$$

Using this relationship, we can show that 1 cm^3 of 0.025 M FAS is equivalent to 0.2 mg of O_2. Since 1 cm^3 of 0.025 M FAS contains 0.025×10^{-3} moles:

1 cm^3 of 0.025 M FAS $\equiv 1/4(0.025 \times 10^{-3})$ moles of O_2
1 cm^3 of 0.025 M FAS $\equiv 1/4(0.025 \times 10^{-3}) \times 32$ g of O_2
1 cm^3 of 0.025 M FAS $\equiv 0.2$ mg of O_2
Volume of FAS required by 10.0 cm^3 of water sample $= (T_1 - T_2)$ cm^3

Thus, the COD value may be calculated by using the following equation:

$$COD(mg\ dm^{-3}) = (T_1 - T_2) \times 0.2 \times 100$$

where T_1 is the titration value for double-distilled, deionized water (cm^{-3}), and T_2 is the titration value for the water sample (cm^{-3}).

Issues to Consider for your Practical Report

- What are the potential sources of error and interferences in this analytical determination? How could they be overcome?
- Identify the advantages and disadvantages of this COD test.
- Are there alternative methods for determining the COD of a wastewater sample? If so, how do they compare to this method?
- What is the source of the water sample(s)?
- What are typical COD values for domestic and industrial wastewater? How do your data compare with these values?

Useful References

Greenberg, A.E., Clesceri, L.S. and Eaton, A.D. (Eds), (1992). *Standard Methods for the Examination of Water and Wastewater*. American Public Health Association, Washington, DC, USA. ISBN 0-87553-207-1.

ISO Compendia on the Environment: Water Quality. Volume 2, *Chemical Methods* (1992). 'Water quality–Determination of COD', ISO6060-1989(E), 73–76. International Organization for Standardization (ISO), Geneva, Switzerland. ISBN 92-67-10193-5.

Burns, E.R. and Marshall, C. (1965). 'Correction for chloride interference in the chemical oxygen demand test'. *Journal of the Water Pollution Control Federation*, **37**, 1716–1721.

Dobbs, R.A. and Williams, R.T. (1963). 'Elimination of chloride interference in the chemical oxygen demand test'. *Analytical Chemistry*, **35**, 1064–1067.

Department of the Environment (1986). *Chemical Oxygen Demand (Dichromate Value) of Polluted and Waste Waters*, (2nd Edn) HMSO, London, 48.

Self-Study Exercises

Introduction

6.1   Define the following terms.

(a)   macro-nutrient;

(b)   essential element;

(c) transition element;

(d) micro-nutrient;

(e) oxidation number;

(f) biogeochemical cycle.

Iron

6.2 Outline the main physical properties of iron.

6.3 Explain why a brown band of iron (III) hydroxide in the Earth's crust marks the first appearance of oxygen on the planet.

6.4 Decide whether the following statements are true or false, giving reason(s) for your answers.

(a) In the lithosphere, compounds of iron occur mainly in two oxidation states, Fe^0 and Fe^{2+}.

(b) Iron (III) compounds do not dissolve in water in anaerobic conditions.

(c) Iron is the fourth most abundant metal in the crust of the Earth and the second most abundant of all elements.

(d) In water treatment plants, iron (III) compounds are removed by filtration.

(e) Haemoglobin has a higher affinity for oxygen molecules than myoglobin.

(f) The oxidation of Fe^{2+} to Fe^{3+} is speeded up in neutral and slightly basic conditions.

(g) Fe^{2+} is an important catalyst for the production of free radicals in the atmosphere.

(h) The Earth's core is predominantly composed of liquid iron.

6.5 Explain the main difference(s) between the structures of haemoglobin and myoglobin.

6.6 Identify some of the important functions of iron in biological systems.

6.7 The extraction of iron from its oxide became widespread in about 1200 BC. Identify some modern uses.

Carbon

6.8 Decide whether the following statements are true or false, giving reason(s) for your answers.

(a) Pure carbon is very stable.

(b) Carbon atoms readily form multiple ionic bonds with each other.

(c) The vast majority of carbon is present in the form of inorganic rocks.

(d) In the oceans, carbon only exists in an inorganic form.

(e) Carbon exists naturally in three pure forms.

(f) Buckminsterfullerene consists of rings of graphite atoms.

6.9 Sketch the carbon cycle, indicating the magnitude of the carbon flows from one reservoir to another.

6.10 Identify the main sources of the following atmospheric carbon compounds:

(a) methane;

(b) carbon dioxide;

(c) carbon monoxide;

(d) benzene;

(e) propane;

(f) particulate elemental carbon;

(g) butane.

6.11 Explain the cycling of carbon dioxide during photosynthesis and aerobic respiration, using equations as appropriate.

6.12 Write an equation that shows the dissolution of carbon dioxide in seawater.

6.13 Show how inorganic ions of carbon contribute to the slightly alkaline nature of seawater.

6.14 Outline the difference between the greenhouse effect and the enhanced greenhouse effect.

6.15 Explain why carbon dioxide concentrations in the atmosphere show seasonal variations.

6.16 What evidence do we have that human activities have unbalanced the global carbon cycle?

Nitrogen

6.17 Give the oxidation number of nitrogen in each of the following species:

(a) NO;

(b) NO_2;

(c) NO_2^-;

(d) NO_3^-;

(e) NH_3;

(f) NH_4^+.

6.18 Identify the main species of nitrogen in the following:

(a) the atmosphere;

(b) the hydrosphere;

(c) sewage effluent;

(d) the Earth's crust;

(e) industrial fertilizers.

6.19 Explain, using equations as appropriate, the processes of nitrogen fixation and denitrification.

6.20 Outline how motor vehicles have contributed to the anthropogenic disruption of the nitrogen cycle.

6.21 Identify the main industrial and commercial uses of nitrogen and its compounds.

Oxygen

6.22 Outline the main physical properties of oxygen.

6.23 Using the data contained in Table 6.14 to aid you, sketch the variation of oxygen solubility in water with temperature.

6.24 Decide whether the following statements are true or false, giving reason(s) for your answers.

(a) Oxygen is a unique component of the global ecosystem.

(b) Carbon monoxide is formed when a carbonaceous substance burns in a limited supply of oxygen.

(c) The most important free radical in the troposphere is the hydroxyl radical.

(d) Oxygen does not react with dilute acids or alkalis.

(e) In the Earth's crust, oxygen may be regarded as chemically inert.

(f) Dioxygen gas acts as the Earth's global sunscreen.

6.25 Write balanced chemical reaction(s) to describe the following processes:

(a) aerobic respiration;

(b) the dark reaction in photosynthesis;

(c) the combustion of petroleum;

(d) the rusting of iron;

(e) the photolysis of ozone;

(f) the stratospheric destruction of ozone.

6.26 Explain how "summer ozone" is formed in the troposphere in the presence of a reactive hydrocarbon.

6.27 List the main requirements for the formation of photochemical smog in urban areas.

6.28 Outline the evidence supporting the claim that the stratospheric ozone layer is depleting.

Phosphorus

6.29 Where is phosphorus typically found on Earth?

6.30 Give the oxidation number of phosphorus in each of the following species:

(a) $H_2PO_4^-$;

(b) HPO_4^{2-};

(c) PO_4^{3-};

(d) $P_2O_7^{4-}$;

(e) $P_3O_{10}^{5-}$.

6.31 Decide whether the following statements are true or false, giving reason(s) for your answers.

(a) Phosphorus only forms compounds in one oxidation state that is stable (+3).

(b) Phosphate fertilizers may be obtained by treating phosphate rocks with concentrated sulfuric acid to produce superphosphate.

(c) Effective alternatives to phosphate detergents are available.

(d) Phosphorus exists in several physically different but chemically identical forms.

(e) The orthophosphate ion is relatively soluble because it is bulky and triply charged.

(f) Polyphosphates are less soluble than simple orthophosphates.

(g) The mobility of naturally occurring phosphorus compounds is low.

6.32 Identify some of the important functions of phosphorus in biological systems.

6.33 What do the acronyms ATP, ADP and DNA stand for?

6.34 Many organophosphorus compounds are toxic and used as insecticides or nerve gases. Find and draw the chemical structures of the insecticides Parathion and Aldicarb, and the nerve gas, Sarin.

6.35 Outline the environmental effects of eutrophication.

Sulfur

6.36 Give the oxidation number of sulfur in each of the following species:

(a) SO_2;

(b) SO_3^{2-};

(c) SO_4^{2-};

(d) $S_2O_3^{2-}$;

(e) PbS.

6.37 Summarize the physical properties of sulfur.

6.38 Identify the main species of sulfur in the following:

(a) the atmosphere;

(b) seawater;

(c) the Earth's crust;

(d) industrial fertilizers.

6.39 Explain how the sulfur and the carbon cycles combine in order to allow the biogeochemical cycling of sulfur.

6.40 Write balanced chemical reaction(s) to describe the following processes:

(a) the combustion of sulfur;

(b) the formation of hydrogen sulfide in anaerobic swamps;

(c) the weathering of sulfide rocks.

6.41 Draw the chemical structure of the amino acids, (a) cysteine and (b) methionine.

6.42 Outline the human health and environmental effects of sulfate aerosol.

6.43 Describe the international strategies that have been employed to reduce emissions of sulfur dioxide from industrial sources.

6.44 List some modern uses of sulfur and its compounds.

Challenging Exercises

Iron

6.45 Recently, scientists have suggested that adding powdered iron to certain areas of the ocean would help to absorb carbon dioxide from the atmosphere. Suggest the reason(s) behind this hypothesis.

6.46 Figure 6.8 shows variation in concentration with depth for biolimiting elements in oceans. Sketch the variation in concentration with depth for non-biolimiting elements in oceans.

Carbon

6.47 Apart from the release of CO_2 from deforestation, identify other human activities in tropical forest zones that produce greenhouse gases.

6.48 Make a detailed list of the potential future effects of global warming.

6.49 Locate the data for CO emissions in the UK for the period 1970 to the present. On graph paper, sketch the time-series plot and write a brief explanation of the trends it shows in CO emissions. (The main source of such data is at the following web site–www.aeat.co.uk/netcen/airqual/.)

Nitrogen

6.50 Identify nitrogen-containing substances that could potentially be used as indicators of human sewage.

6.51 Discuss the methods that may be employed to control the emissions of nitrogen oxides from motor vehicles.

6.52 Sketch the variations in NO and NO_2 concentrations you would expect in an urban area over a 24 hour weekday period. (Plot time on the x-axis and pollutant concentration on the y-axis.)

Oxygen

6.53 Describe the development of international legislation to protect the stratospheric ozone layer.

6.54 Outline the human health effects of exposure to high concentrations of ozone.

6.55 Make a detailed list of the potential future effects of stratospheric ozone depletion.

6.56 Locate the data for O_3 concentrations at the London Bloomsbury monitoring station over the last seven days. On graph paper, sketch the hourly average concentrations and comment on any trends that you see. (The main source of such data is at the following web site–www.aeat.co.uk/netcen/airqual/.)

Phosphorus

6.57 Describe in detail the relationship between the phosphate ion and eutrophication.

6.58 Outline the differences between orthophosphate, polyphosphates and organic phosphorus, using examples as appropriate.

Sulfur

6.59 Explain, using equations as appropriate, how the mining of coal and metal sulfide ores can lead to severe pollution problems from acid and toxic metals.

6.60 Describe in detail how sulfur compounds can cause the corrosion of underground sewers.

6.61 Locate the data for SO_2 emissions in the UK for the period 1970 to the present. On graph paper, sketch the time-series plot and write a brief explanation of the trends it shows in SO_2 emissions. (The main source of such data is at the following web site–www.aeat.co.uk/netcen/airqual/.)

Further Sources of Information

Further Reading

Archer, J. (1994). "Policies to reduce nitrogen loss to water from agriculture in the United Kingdom". *Marine Pollution Bulletin*, **29**(6–12), 444–449.

Cox, P.A. (1995). *The Elements on Earth*. Oxford University Press, New York. ISBN 0-19-855903-8.

D'Elia, C.F. (1987). "Too much of a good thing: nutrient enrichment of the Chesapeake Bay". *Environment*, **29**(2), 6–11, 30–33.

Goulding, K.W.T, Bailey, N.J., Bradbury, N.J., Hargreaves, P., Howe, M., Murphy, D.V., Poulton, P.R. and Willison, T.W. (1998). "Nitrogen deposition and its contribution to nitrogen cycling and associated soil problems". *New Phytologist*, **139**(1), pp49–58.

Heathwaite, A.L., Johnes, P.J. and Peters, N.E. (1996). "Trends in nutrients". *Hydrological Processes*, **10**(2), 263–293.

Houghton, J.T., Meira Filho, L.G., Callander, B.A., Harris, N., Kattenberg, A. and Maskell, K. (Eds), (1996). *Climate Change 1995: The Science of Climate Change*. Report of the Intergovernmental Panel on Climate Change. Cambridge University Press, Cambridge, England. ISBN 0-5215-6436-0.

Kinzig, A.P. and Socolow, R.H. (1994). ''Human impacts on the nitrogen cycle''. *Physics Today*, November 1994, 24–31.

O'Dowd, C.D., Smith, M.H., Consterdine, I.E. and Lowe, J.A. (1997). 'Marine aerosol, sea-salt, and the marine sulfur cycle: a short review'. *Atmospheric Environment*, **31**(1), 73–80.

O'Neill, P., (1993). *Environmental Chemistry* (2nd Edn). Chapman & Hall, London. ISBN 0-412-48490-0.

Pearson, J. and Stewart, G.R. (1993). ''The deposition of atmospheric ammonia and its effects on plants''. *New Phytologist*, **125**(2), 283–305.

Young, K., Morse, G.K., Scrimshaw, M.D., Kinniburgh, J.H., MacLeod, C.L. and Lester, J.N. (1999). ''The relation between phosphorus and eutrophication in the Thames Catchment, UK''. *Science of the Total Environment*, **228**(2–3), 157–183.

7 Water – The Lifeblood of the Earth

The wondrous nature of water, its eccentric physical properties, the manner in which it renders the earth fit as a habitat for life, and its involvement in life processes at all levels have inspired poets, painters, composers and philosophers through the ages.

Felix Franks, *Water* , Royal Society of Chemistry, 1983.

7.1 Introduction

Water is unquestionably the lifeblood of the Earth – no animal or plant life would exist without it. Water's unique physical and chemical properties and its ubiquitous presence make it essential for life, transport, engineering, leisure and recreation, and yet it is sometimes frighteningly dangerous – attributes present in no other substance. In this chapter, we will study the properties of water in an attempt to explain its vitality, as well as our potential for contaminating the Earth's most valuable resource.

Water is an oxide of hydrogen (H_2O) and is the only chemical compound that occurs naturally in all three physical states – solid (snow, hail and ice), liquid (rain, lakes and rivers) and vapour (steam). Pure liquid water is colourless, odourless, tasteless and virtually incompressible. Water is the most important inorganic liquid that exists naturally on Earth and it covers about 70% of the Earth's surface. Almost 97% of the Earth's water is found in the Earth's oceans and just over 3% is found as freshwater (Table 7.1). Liquid water can be found as surface water, groundwater and rainwater (see Section 7.2).

When water freezes it forms solid ice; indeed, geologists consider ice to be a type of rock. Ice, like liquid water, is colourless, forming regular hexagonal crystals. Permanent ice sheets cover vast areas of the Earth's surface – the Antarctic ice sheet covers an area twice the size of Australia! Approximately 1.7% of the Earth's water is permanently frozen in glaciers (masses of ice that flow under their own weight on land) and at the polar ice caps. Currently, glaciers cover almost one-tenth of the Earth's land surfaces (over 15 million km^2). The longest glacier in the world is the Lambert Glacier in the Australian Antarctic Territory – up to 64 km wide and

Table 7.1 Water reserves found on Earth (adapted from Gleick, P.H. (Ed.), (1993). *Water in Crisis: A Guide to the World's Fresh Water Resources.* Oxford University Press, Oxford)

Reserve	Distribution area $(10^3 \ km^2)$	Volume $(10^3 \ km^3)$	% of total water[a]	% of fresh water[a]
World ocean	361 300	1 338 000	96.5	–
Groundwater	134 800	23 400	1.7	–
Fresh water	–	10 530	0.76	30.1
Soil moisture	–	16.5	0.001	0.05
Glaciers and permanent snow cover	16 227	24 064	1.74	68.7
Ground ice/permafrost	21 000	300	0.022	0.86
Water reserves in lakes	2 059	176	0.013	–
Fresh	1 236	91	0.007	0.26
Saline	823	85	0.006	–
Swamp water	2 683	11.5	0.0008	0.03
River flows	148 800	2	0.0002	0.006
Biological water	510 000	1	0.0001	0.003
Atmospheric water	510 000	13	0.001	0.04
Total water reserves	510 000	1 385 984	100	–
Total fresh water reserves	148 800	35 029	2.53	100

[a]Percentage of global reserves.

over 400 km long. The freezing point of ice is used as a standard for measuring temperature (0° on the Celsius scale).

When water boils, it forms water vapour. The boiling point of water (under normal pressure) is 100°C/212°F (the lower the pressure, then the lower the boiling point and vice versa). Water vapour (steam) is an invisible gas formed by vaporizing water – any visible cloud is due to minute suspended water particles. Hot water and steam are intermittently ejected from hot springs and geysers in countries such as Iceland, New Zealand and the USA. The energy derived from hot springs and geysers – geothermal energy – has been successfully used in places such as Reykjavik (Iceland) and Rotorua (New Zealand) for many decades. In the USA, the geothermal industry currently has an operating capacity of 2 300 megawatts and annually generates 17 billion kw-hours of energy.

The ubiquitous nature of water and its vital role in maintaining life can be demonstrated by considering the human body. Water makes up 60–70% of the human body – about 40 litres – of which 25 litres are inside the cells and 15 litres are outside in tissue fluid and blood plasma. Humans lose about 1.5 litres of water per day through breathing, perspiration and faeces. An additional amount is lost in urine that keeps the balance between input and output. In general, a person can live for over a month without food but only a week without water. This statement may seem surprising given the high water content of many foods (Table 7.2), but there is a simple explanation. Felix Franks (in his book, *Water*) has estimated that 0.3 kg of water are synthesized daily in the body by the oxidation of food and that the heart must pump 7000 l of blood around the vascular system to generate enough energy for this oxidation reaction. This sort of calculation illustrates clearly humans' need

Table 7.2 Typical water content of selected foods

Food	Percentage water (by mass)
Lettuce (raw)	96
Tomatoes (raw)	95
Mushrooms (raw)	92
Milk (whole)	87
Oranges (raw)	86
Apples (raw)	84
White fish	82
Potatoes (raw)	78
Eggs (boiled)	75
Beef (raw)	64
Chicken (roasted)	60
Cream cheese	54
Pizza	48
Whole wheat bread	38
Jam	29
Butter	16
Potato crisps	2

for clean drinking water and also the complexity and scale of water transport in living organisms.

However, water is essential for many other biological activities, including:

- metabolism (e.g. all metabolic reactions occur in an aqueous solution);
- photosynthesis;
- lubrication (e.g. mucus in the mammalian gut; mucus to aid the movement of snails and earthworms; synovial fluid in the joints of vertebrates);
- transport of substances (e.g. movement of glucose, amino acids, nutrients and hormones in blood plasma; movement of nutrients in plants);
- transport of heat (e.g. from the core to the outside of the body in warm-blooded animals, and from the outside to the core of the body in cold-blooded animals);
- support (e.g. support of the hydrostatic skeleton in earthworms; turgidity in plant cells);
- temperature control (e.g. evaporation of water (sweating) helps to regulate body temperature).

Humans use water for a huge range of other activities, including power generation, irrigation, washing, sewage treatment, recreation, and in manufacturing industries. The scale of water use is staggering, as illustrated by the data presented in Table 7.3.

7.2 Types of Water

Environmental water may be divided into three categories, as follows:

- surface water;
- groundwater;
- precipitation.

Table 7.3   Water use in the world by human activity (adapted from Gleick, P.H. (Ed.), (1993). *Water in Crisis: A Guide to the World's Fresh Water Resources.* Oxford University Press, Oxford)

Water users[a]	1900 (km^3year^{-1})	1950 (km^3year^{-1})	1975 (km^3year^{-1})	2000 (km^3year^{-1})[b]	2000 (%)[b]
Agriculture					
Withdrawal	525	1 130	2 050	3 250	62.6
Consumption	409	859	1 570	2 500	86.2
Industry					
Withdrawal	37.2	178	612	1 280	24.7
Consumption	3.5	14.5	47.2	117	4.0
Municipal supply					
Withdrawal	16.1	52.0	161	441	8.5
Consumption	4.0	14	34.3	64.5	2.2
Reservoirs					
Withdrawal	0.3	6.5	103	220	4.2
Consumption	0.3	6.5	103	220	7.6
Total (rounded off)					
Withdrawal	579	1 360	2 930	5 190	100
Consumption	417	894	1 760	2 900	100

[a]Total water withdrawal is shown in the first line of each category; consumptive use (irretrievable water loss) is shown in the second line.
[b]Estimated (amounts).

We will briefly look at each category before seeing how they are linked together in the water cycle.

7.2.1   Surface water

Surface water may be divided into standing (oceans, lakes and reservoirs) and running (rivers and streams) water. It has been estimated that 1.34 billion km^3 of water are present on Earth, with 96.5% contained in the oceans. Some of the physical characteristics associated with the oceans are astonishing, namely:

- Oceans and marginal seas account for approximately 70% of the Earth's surface.

- The average depth of the global ocean is more than 3 660 meters.

- No sunlight penetrates below 1 000 meters of ocean depth.

- The oceans' current physico-chemical characteristics (i.e. salinity, density, etc.) were probably formed more than 1.5 billion years ago.

- The surface waters of the ocean to a depth of 2.5 meters hold as much heat as the entire atmosphere.

The oceans provide us with a rich supply of food (fish, shellfish, seaweed, salt, etc.) raw materials (sand, gravel, oil, coal, natural gas, sulfur, diamonds, metals, bromine, etc.) and fresh water (from desalination plants – see Section 7.6). The estimated abundance of selected elements in seawater (Table 7.4) and the estimated concentrations of the seven most abundant compounds in seawater (Table 7.5) provides us with some idea of the enormous capacity of the oceans.

Table 7.4 Estimated abundance of selected elements in seawater (near the surface) (adapted from Lide, D.R. (ed.), (1998). *CRC Handbook of Chemistry and Physics* (79th edn) CRC Press, Boca Raton H, FL, USA)

Element	Estimated percentage by mass	Estimated abundance $(mg\ l^{-1})$
Oxygen	85.4	8.57×10^5
Hydrogen	10.7	1.08×10^5
Chlorine	1.85	1.94×10^4
Sodium	1.03	1.08×10^4
Magnesium	0.127	1.29×10^3
Sulfur	0.087	9.05×10^2
Calcium	0.040	4.12×10^2
Potassium	0.038	3.99×10^2
Bromine	0.0065	6.73×10^1
Carbon	0.0027	2.8×10^1
Nitrogen	0.0016	5×10^{-1}
Strontium	0.00079	7.9
Boron	0.00043	4.44
Silicon	0.00028	2.2
Fluorine	0.00013	1.3

Table 7.5 Estimated concentrations of the seven most abundant compounds in seawater

Compound	Concentration $(tonnes\ km^{-3})$
Sodium chloride	27 500 000
Magnesium chloride	6 750 000
Magnesium sulfate	5 625 000
Calcium sulfate	1 800 000
Potassium chloride	750 000
Calcium carbonate	111 250
Potassium bromide	102 500

Because the oceans are so massive, their average composition varies little from place to place. However, the concentrations of natural dissolved substances found in surface waters other than oceans are extremely variable (see Section 7.3.1).

Lakes and reservoirs contain just 0.013% of the world's total water and supply 0.26% of freshwater. Nevertheless, many lakes are huge; Lake Tanganyika in Africa has a surface area of 32 900 km^2 and a volume of 18 900 km^3. Only a tiny fraction of the Earth's total water is present as running water (i.e. rivers and streams) – approximately 0.0002%. Nevertheless, running water is important for domestic and industrial use, agriculture, transportation and power generation. It is also the most important erosional factor modifying the Earth's surface (see Chapter 4). The surface flow of streams from the land to the oceans is known as run-off. Concern about the quality and management of urban run-off has increased significantly over the last 30 years as research has highlighted the presence of potentially harmful chemicals at elevated concentrations.

7.2.2 Groundwater

Groundwater is the name given to fresh water stored in open spaces within underground rocks and unconsolidated material. It arises from precipitation that seeps into the ground and water infiltration from lakes, streams and ponds. Groundwater passes through the zone of aeration and into the zone of saturation (see Figure 7.1). In the zone of aeration, the pores between the soil and rocks contain both air and water, while in the zone of saturation, all the pores are filled with groundwater.

The top of the zone of saturation is called the *water table*; where the water table occurs above the soil, lakes and streams are formed; where the water table occurs at the soil's surface, swamps are formed. Permeable rock bounded by an impervious layer (e.g. of clay), that has potential to hold groundwater, is called an *aquifer*. The latter have been likened to underground reservoirs.

Groundwater is traditionally regarded as very pure because organic matter and disease-causing microorganisms are filtered out as the groundwater seeps through soil, sediment and rocks. Fresh groundwater represents just 0.76% of the world's total water supply but around 30% of the world's supply of freshwater. In countries such as Denmark and Italy, almost all public drinking water is extracted from aquifers; in the UK, the figure is closer to 30%. In the USA, most groundwater is used for irrigation (approximately 65%), with industrial and domestic use being second and third, respectively.

Once groundwater is contaminated (by, for example, chemicals from industry, landfill or agricultural waste), it can take many years to clean up. Groundwater is particularly vulnerable to pollution by nitrates, predominantly from agricultural activities such as ploughing and the use of fertilizers. Protecting groundwater is

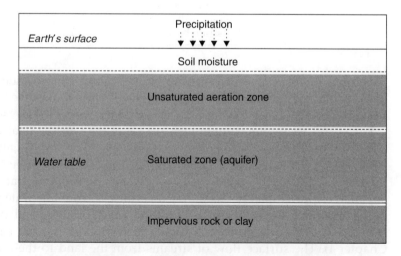

Figure 7.1 Location of groundwater in soil

complex because there are so many scientific and environmental factors to consider, including the nature and usage of the overlying soil, the presence and nature of unconsolidated deposits which overlay solid rock formations, the nature of the rock strata, and the depth down to the water table. Most countries have introduced legislation making it an offence to pollute groundwater.

The large-scale extraction of water from aquifers has led to problems such as ground subsidence, landslides, damage to building foundations, roads and bridges, etc., groundwater quality deterioration, flooding, saltwater contamination and severe depletion of groundwater in many areas of the world. For example, land subsidence due to groundwater extraction in Mexico City varies from 5–10 cm annually and reaches up to 35 cm in other regions. In the San Joaquin Valley of California, groundwater withdrawal between 1925 and 1975 caused subsidence of almost 9 m in certain areas. In Italy, the tilt of the Leaning Tower of Pisa is caused by subsidence due to excessive groundwater removal. Other areas severely affected by groundwater extraction include China (Shanghai and Hangu), England (London), Italy (Ravenna and Venice), Japan (Osaka and Tokyo), Taiwan (Yun-Lin), Thailand (Bangkok), Turkey (Sazlica) and the USA (Houston, Las Vegas, Los Angeles, New Jersey, New Orleans and the Santa Clara Valley).

7.2.3 Precipitation

Precipitation is one of the most variable elements of weather. It can take the form of rain, dew, fog, hail, sleet or snow, and may vary widely in its frequency, duration, intensity and spatial pattern. It is mainly derived from the oceans and plays a vital role in the cycling of water (see Section 7.3).

Rainfall frequency varies widely throughout the world, ranging from less than one rain day per year in arid regions to over 180 rain days per year in humid coastal areas. It is subject to large seasonal variations that are greatest in areas dominated by monsoons. Most rainfall events last up to a few hours but can last for days. Rainfall intensities vary from practically zero to over 100 mm h^{-1}. The intensity of rainfall is inversely related to its duration; the highest intensities are recorded over the smallest intervals. For example, the most intense rainfall ever recorded (in Maryland, USA in 1956) generated 1.23 inches of rainwater in just one minute. The most intense 24-hour rainfall ever recorded occurred on Ile de la Reunion (in the Indian Ocean) during March 1952, when 74 inches of rain fell.

The different physical states of precipitation mean that it is difficult to give general figures for its composition. Insoluble solids such as wind-blown dust and sand are often found in precipitation large distances from the source of the solids; Saharan sand may be transported by wind and deposited in precipitation many thousands of miles from the desert. However, it is soluble material that plays the major role in the chemistry of precipitation. The major soluble ions in precipitation and the water-soluble component of atmospheric particles are:

- SO_4^{2-} and NO_3^- (mainly from the oxidation of acid gases – see Section 7.10);

- H^+ (from atmospheric acids) and NH_4^+ (from ammonia neutralization of these acids);

- K^+ and Ca^{2+} (predominantly from wind-blown soil);

- Na^+, Mg^{2+} and Cl^- (mainly from sea spray).

Ions from seawater are present in significant amounts in the marine atmosphere. Indeed, the influence of the sea on rainwater composition is emphasized by Dean's Rule, which states that artificial rain may be made by diluting $1.5\ cm^3$ of seawater to a litre of distilled water. The composition of rain in coastal regions is dominated by the presence of sodium chloride; its concentration generally decreases exponentially with distance from the coast, levelling off at about 100 km. Gases (mainly acidic) and trace metal ions (from dust) are also dissolved in rainwater. Predicting rainfall composition accurately is almost impossible even though its composition is often affected by the origin of the air mass. This is because atmospheric processes, such as photochemical activity, and meteorological factors, such as rainfall rate, can affect the concentrations of dissolved species.

Snow and dew may have a very different composition from rainfall. Snow, for example, often has a higher ionic concentration than rain because it scavenges solutes more effectively from the atmosphere due to the relatively high surface area of its flakes. Dew may also accumulate ions from the surfaces of plants as it condenses.

Acid precipitation has been one of the most important causes of environmental damage this century and is discussed in more detail in Section 7.10.1.

7.3 The Water Cycle

In the water (or hydrologic) cycle (see Figure 7.2), water is continuously moving through the biosphere via the processes of evaporation, condensation, precipitation and transpiration. This is possible because water changes phases easily under Earth surface conditions. Radiant energy from the sun provides the driving force for the water cycle. The sun's radiation heats the water on the Earth's surface, causing it to evaporate into the atmosphere. Approximately 85% of all water that enters the atmosphere is derived from the oceans, an amount corresponding to a layer about 1 metre thick from all the oceans annually. The remaining 15% is obtained from evaporation of water from land.

Water vapour typically remains in 'atmospheric storage' for about 10 days, although this varies with latitude (longer at high latitudes and shorter at middle latitudes). It is important to realize that the atmosphere as a whole is not saturated with water vapour, although localized saturation is possible. At a given temperature, the ratio of the measured water vapour pressure to the saturation vapour pressure is expressed as the relative humidity. Atmospheric water may also be present in the solid and liquid phases.

Water changes from the aqueous to the liquid state through the process of condensation, forming clouds, before returning to Earth as precipitation. As evaporation is an endothermic process (requires energy), the evaporation of large volumes of water reduces the temperature of the air/water interface. Condensation on the other hand, is an exothermic process leading to an increase in temperature at the

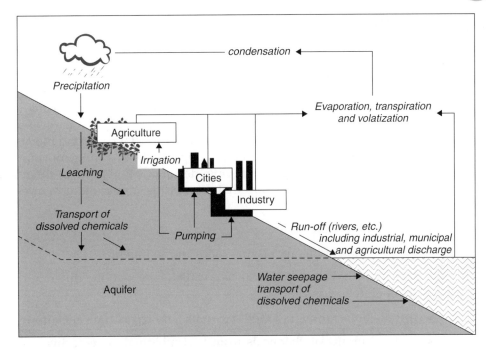

Figure 7.2   Schematic diagram of the water cycle

air/water interface. The overall effect of these processes is that the cycling of water plays an important role in the control of climatic conditions.

About 80% of precipitation falls directly into oceans, with 20% falling on land. In temperate regions, rainfall may be intercepted by vegetation and returned directly to the atmosphere by evaporation (rainfall can also evaporate as it falls). Rainfall reaching the ground may infiltrate into the ground or may collect to form surface run-off. Liquid water infiltrating the ground will either:

- percolate through the unsaturated layers of soil to reach the regions where the soil becomes saturated (the water table), or;

- be taken up by the roots of vegetation and transferred to the leaves via the xylem. The water is then lost to the atmosphere by evaporation from the leaves through pores called stomata – a process known as transpiration. The rate of transpiration depends upon factors such as relative humidity, wind speed, ambient temperature, light intensity and the structure of the plant itself.

Surface run-off collects in surface streams and rivers that eventually flow back to the ocean.

In Polar Regions, or in high mountains, snow may accumulate in snowfields or consolidate into glaciers, and some precipitation may be stored temporarily in lakes or may seep into groundwater. This water is effectively removed from the hydrological system and may be stored naturally for very long periods – up to thousands of years. However, snow and glaciers melt, and lakes and groundwater feed into streams and oceans, thus making the water available again for circulation.

7.3.1 Impurities in water

Naturally occurring water consists of 99.76% by weight of $^1H_2^{16}O$. The remainder consists of various isotopes, e.g. ^2H and ^{18}O. The commonest of these isotopes is deuterium (^2H), often given the symbol D. Deuterium is found in 'heavy water' (HDO and D_2O), which has a deleterious effect on living organisms.

However, all natural waters are impure due to the presence of dissolved and suspended substances. The various classes of substances found in natural water are shown in Table 7.6. These substances originate from a range of sources, including the atmosphere, chemical weathering, decomposition of plant litter, soil leaching, etc. The environmental factors that control the inputs of substances into natural waters include:

- distance from the ocean;

- composition of the underlying rocks and sediment;

- climate – the chemistry of surface waters may be quite different in tropical and temperate regions;

- production and degradation of aquatic and terrestrial vegetation;

- natural inputs of chemicals from hydrothermal vents, active volcanoes, landslides and weathering processes (see Chapter 4);

- anthropogenic inputs of chemicals.

Clearly, seawater contains more dissolved salts than freshwater and hence the land phases of the water cycle are particularly important–salty seawater is transferred into fresh precipitation that collects on land as freshwater. We have already seen that the concentrations of natural dissolved substances found in surface waters (other than oceans) are extremely variable. Some water found on land is unsuitable for drinking due to the presence of impurities from:

- anthropogenic pollution (e.g. oil, acids, industrial and sewage effluent);

- natural suspended solids;

- excess dissolved salts (brackish water).

However, a useful classification of the relative abundance of dissolved solids in potable (drinking) water is shown in Table 7.7. The labels of bottled drinking water detail the actual concentrations of dissolved materials in tables. We could not survive without appropriate amounts of these dissolved materials in our water (see Chapter 6).

Table 7.6 Classes of substances found in natural water

Class	Example
Suspended solids	Sand, silt, mud, silt, organics
Dissolved salts	Cl^-, SO_4^{2-}, HCO_3^-, Na, K, Ca, Mg, Fe, Al
Dissolved gases	O_2, N_2, CO_2, NO_2, NH_3
Dissolved organics	Decomposition products of organisms

Table 7.7 Relative abundance of dissolved solids in potable water (from Davis, S. and DeWiest, R. (1966). *Hydrogeology*. John Wiley & Sons, Chichester, UK)

Major constituents (1.0 to 1000 $mg\,l^{-1}$)	Secondary constituents (0.01 to 10.0 $mg\,l^{-1}$)	Minor constituents (0.0001 to 0.1 $mg\,l^{-1}$)	Trace constituents ($< 0.001 mg\,l^{-1}$)
Sodium	Iron	Antimony[a]	Beryllium
Calcium	Strontium	Aluminium	Bismuth
Magnesium	Potassium	Arsenic	Cerium[a]
Bicarbonate	Carbonate	Barium	Caesium
Sulfate	Nitrate	Bromide	Gallium
Chloride	Fluoride	Cadmium[a]	Gold
Silica	Boron	Chromium[a]	Indium
		Cobalt	Lanthanum
		Copper	Niobium
		Germanium[a]	Platinum
		Iodide	Radium
		Lead	Ruthenium[a]
		Lithium	Scandium[a]
		Manganese	Silver
		Molybdenum	Thallium[a]
		Nickel	Thorium[a]
		Phosphate	Tin
		Rubidium[a]	Tungsten[a]
		Selenium	Ytterbium
		Titanium[a]	Yttrium
		Uranium	Zirconium[a]
		Vanadium	
		Zinc	

[a]These elements occupy an uncertain position in the list.

7.4 Properties of Water

Water is the most common chemical on Earth and yet its properties are unique. To understand this statement, we need to look in more detail at water's physical and chemical properties.

7.4.1 Solvent ability

Water dissolves a wide variety of compounds and is known as the 'universal solvent'. This is clearly an exaggeration, but it does dissolve a huge number of ionic and molecular species. The excellent solvent ability of water is due to its high *dielectric constant* (80.4). The dielectric constant is defined by the following equation:

$$F = (q_1 q_2)/(\varepsilon r^2)$$

where F is the force of attraction, q_1 and q_2 are the electric charges (of the ions), *r* is the distance between the charges, and ε is the dielectric constant.

The greater the dielectric constant of the solvent, then the smaller the force is between two charged ions, and hence the more easily they separate i.e. dissolve. It is the excellent solvent properties of water that enables nutrients to be transferred to plants, and chemicals to be moved inside and out of organisms, and which also plays an important part in the weathering of rocks (see Chapter 4).

7.4.2 Hydrogen bonding

Water has relatively high melting and boiling temperatures compared to substances of similar molecular weight and structure. This suggests that the O−H bonds in water are strong. Water has a high dipole moment: the O−H bonds are highly polar (see Chapter 4) and dipoles tend to align with each other, thus increasing the cohesive energy. However, water boils at a higher temperature than hydrogen fluoride (HF) which has a greater dipole moment – why is this?

The high boiling and melting temperatures of water are due to the presence of hydrogen bonds (H-bonds). These bonds are formed because of the electrostatic interactions that exist between molecules, thus leading to the association of molecules shown in Figure 7.3. The oxygen atom in water is more electronegative than the two hydrogen atoms; it therefore attracts the bonding electrons away

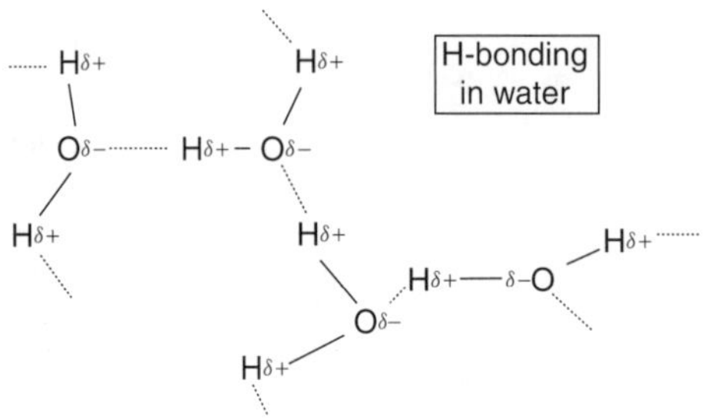

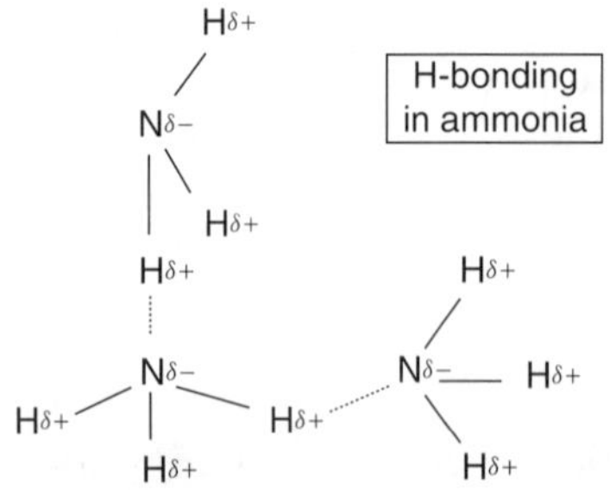

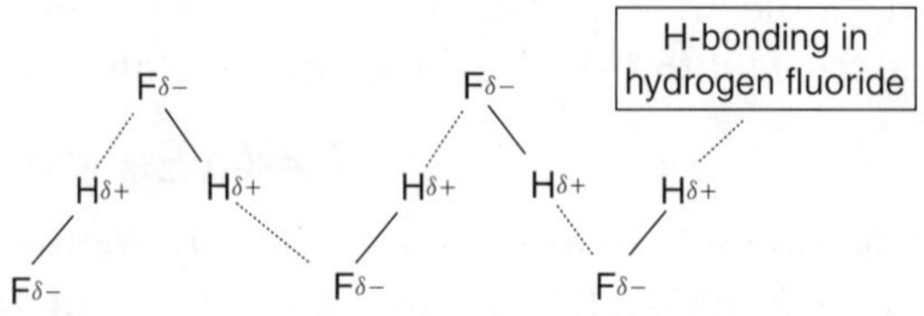

Figure 7.3 Hydrogen bonding in (a) water, (b) ammonia and (c) hydrogen fluoride

from the hydrogen atoms. The oxygen atom has a small negative charge ($\delta-$) and each hydrogen atom a small positive charge ($\delta+$). The positively charged hydrogen atom will be attracted to a lone pair of electrons on a neighbouring oxygen atom, giving rise to a *hydrogen bond* (H-bond). Consequently, H-bonds are intermolecular, permanent dipole – permanent dipole attractions. The requirements for a H-bond are:

- a hydrogen atom attached to a highly electronegative atom;

- a unshared pair on the electronegative atom.

Water contains two O—H bonds and two unshared electron pairs per molecule; ammonia (NH_3) contains three N—H bonds and one unshared electron pair per molecule; both can form multiple, networked H-bonds (see Figure 7.3). Hydrogen fluoride also contains H-bonds, but they are linear rather than networked, and hence are weaker than the H-bonds in water. H-bonding occurs most commonly with small, highly electronegative atoms such as O, F and N, but is not common with larger electronegative atoms such as S and Cl.

The hydrogen bonds present in the structures shown in Figure 7.3 are weak bonds, approximately 5–10% of an average covalent bond. Nevertheless, they are strong enough to affect certain properties of the substances in which they are present. For example, H-bonding also contributes to some of water's other unusual properties. Most substances contract upon freezing because the molecules take up more space in the chaotic liquid state than they do as a solid. The liquid expands as it is heated due to the increased motion of the molecules. However, when water freezes it expands by 9% – hence pipes burst in winter – and the density of the solid phase (ice) is lower than that of the liquid. These properties exist because of H-bonding in ice (see Figure 7.4).

The bonds in water are inclined approximately tetrahedrally. H-bonded networks exist in both water and ice, but the H-bonds fluctuate very rapidly in the liquid, thus allowing the water molecules to be mobile and close together. The arrangement of molecules is similar in ice but the regularity extends throughout the whole structure, so causing greater spacing of molecules. Hence, when ice melts, the crystal structure collapses upon itself and the density increases.

The consequences of H-bonding in water are important.

- Any substance that can H-bond with water can dissolve in it – hence its excellent solvent properties.

- Water has strong interactions with many common environmental substances (e.g. wood, soil and rock) because the molecules in their surfaces form H-bonds.

- Because ice is less dense than water, it floats, so causing ponds and lakes to freeze from the surface downwards. In harsh weather conditions, floating ice protects aquatic life from the inhospitable climate above, thus keeping it alive in winter. Fish and plants can survive under frozen ice for months. If ice was denser than water, than lakes and rivers would freeze from the bottom up, creating Arctic conditions.

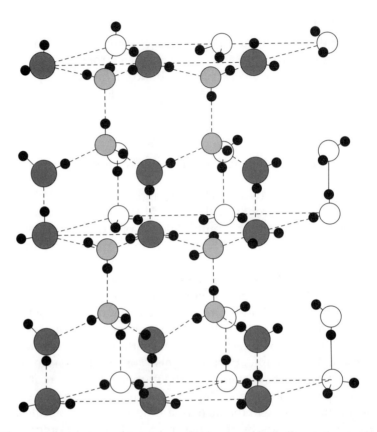

Figure 7.4 The crystal structure of ice, where the larger and smaller circles represent oxygen and hydrogen atoms, respectively; oxygens with the same colour shading indicate that they lie in the same plane. Chemistry of the environment by Spiro, © 1996. Reprinted by permission of Prentice-Hall, Inc., Upper Saddle River, NJ, USA

7.4.3 Surface tension

The surface of water (or indeed any liquid) is smooth. This is because the intermolecular (van der Waals) forces tend to pull the water molecules together and inward. The net inward pull of a liquid is called *surface tension*. The latter for water is approximately three times higher than that of most common liquids because of its strong H-bonds. Surface tension accounts for the spherical shapes of free-falling droplets of liquid – a sphere is the most compact shape.

The surface tension of water accounts for a number of everyday observations. Waxed jackets 'repel' water because the attractive forces between the water molecules are greater than the forces between water and wax. Consequently, a droplet of water will rest on the waxed surface rather than wetting it. Duck's feathers are coated with natural oils that repel water in a similar fashion. However, water will rise up narrow glass tubes – a process called *capillary action*. This is because the attractive forces between the water molecules and the glass surface are stronger than the cohesive forces between the water molecules. Thus, the meniscus of water in a glass capillary is curved upward at the edges (forming a ∪ shape) as the water tends to spread over the greatest possible area of glass.

7.4.4 Self-ionization of water

Water undergoes dissociation (self- or auto-ionization) to a small extent to produce H^+ and OH^- ions in equal amounts:

$$H_2O(l) \rightleftharpoons H^+(aq) + OH^-(aq)$$

or more precisely:

$$2H_2O(l) \rightleftharpoons H_3O^+(aq) + OH^-(aq)$$

| water molecule | oxonium ion | hydroxide (hydroxyl) ion |

This phenomenon means that the electrical conductivity of even the purest water never falls to exactly zero.

7.5 Properties of Solutions

We have already defined the key terms involved in solution chemistry (see Section 1.10) and discussed the environmental importance of aqueous solutions. In order to understand more fully the processes that occur in aqueous solutions, we need to study the factors that affect the solubilities of solutes in water.

7.5.1 Dependence of solubility on temperature

Figure 7.5 shows a solubility curve, a graph of solubility against temperature. Solubility curves demonstrate the effect of temperature on solubility and enable us to predict the mass of solute that will crystallize when a solution is cooled (or in some cases, heated). For example, a solution at point 'A' contains m_2 g of solute in 100 g of solvent at temperature t_3. It is clearly an unsaturated solution and will not deposit solute (crystallize) until it is cooled to temperature t_2. At this temperature solute is deposited, thus decreasing the solubility of the solution. The solubility of the solution will continue to decrease along the curve 'BC' as the temperature of the solution is reduced. The mass of solute that has crystallized between t_2 and t_1 is $(m_2 - m_1)$ g.

The solubility of most salts increases with temperature, e.g. $KClO_3$, KNO_3 and KCl. A number of salts show only a slight increase in solubility on heating. In some cases, the solubility actually decreases with the temperature. At typical environmental conditions for European and North American waters (10–25°C):

- all the common salts of sodium, potassium and ammonium (NH_4^+) are soluble;

- all nitrates are soluble;

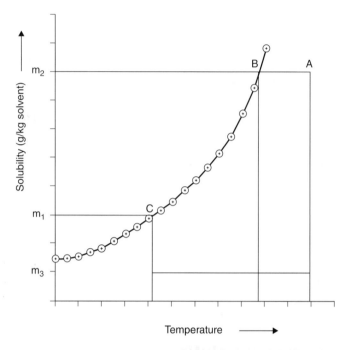

Figure 7.5 A typical solubility curve – a graph of temperature against solubility

- all common chlorides, except silver, lead (II) and mercury (I), are soluble;
- all common sulfates, except those of lead (II), barium and calcium, are soluble;
- all carbonates, except those of the alkali metals and ammonium, are insoluble.

The dependence of solubility on temperature can be exploited to provide us with valuable minerals. For example, the oceans contain about 2.7% by mass of dissolved NaCl. Enormous quantities of this salt (commonly used as table salt) are extracted by solar evaporation of seawater.

7.5.2 Dependence of solubility on the nature of a solvent

There is a simple rule relating to solubility stating that 'like dissolves like'. In other words, substances dissolve in solvents that are chemically similar to them.

- Non-polar organic solvents remove non-polar fat stains, but polar water does not.
- Ionic compounds are soluble in polar solvents such as water, but insoluble in non-polar solvents such as benzene.
- Metals are insoluble in both polar and non-polar solvents, but soluble in liquid metals.
- Giant molecular compounds are insoluble in all solvents.

When ionic compounds dissolve in polar solvents, the ionic lattice breaks up and the ions become solvated by the solvent molecules – the polar solvent becomes

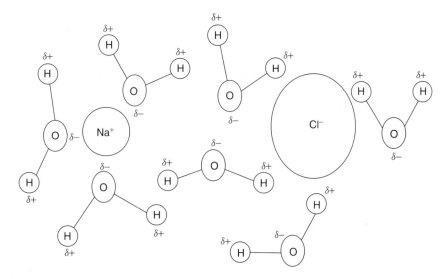

Figure 7.6 Illustration of solvation (hydration) of ionic compounds in water

oriented around the ions. This process, known as hydration when the solvent is water, is illustrated in Figure 7.6. It is usually, but not always, accompanied by a change in temperature. The energy required to break up the ionic lattice (lattice energy) is recovered by the energy released when polar solute ions interact with polar solvent molecules (heat of hydration). Both energy phenomena occur simultaneously, and whichever is the greater will determine whether the solution process is exothermic or endothermic.

All natural waters contain a variety of dissolved substances. Seawater is a solution of dissolved salts and gases, as well as some organic substances. Eleven ionic species make up 99.9% of the dissolved salts in seawater, although seawater probably contains all of the naturally occurring elements. The concentration of the minor components varies with location and depth, while the concentration of the major components is fairly constant.

7.6 Colligative Properties of Solutions

The presence of solutes affects certain properties of water. These include the following:

- a lowering of the vapour pressure;
- a depression of the freezing point;
- an elevation of the boiling point;
- osmotic pressure.

These properties are known as *colligative properties* – they depend upon the number of particles of the solute but not on the nature of the solute.

7.6.1 Lowering of vapour pressure

In a pure solvent, the solvent particles are free to escape from any position on the surface of the liquid. However, what if the solvent contains dissolved solute particles? Non-volatile solute particles will inhibit the escape of solvent particles from the liquid surface over time (the evaporation rate), by physically blocking the route of the solvent particles into the atmosphere. However, the returning (condensing) solvent particles can stick to any part of the solution surface – even a solute particle–and hence the solute particles have no effect on the rate of condensation. At equilibrium, the rate of evaporation equals the rate of condensation. Consequently, a solution will have a lower escape rate for solvent particles than the pure solvent. This is matched by a lower return rate and hence a lower rate of impact on the solution surface, i.e. the vapour pressure of the solution is lower than that of the pure solvent. Examples of this behaviour can be seen in many arid areas of the world. Salt lakes evaporate more slowly than freshwater lakes in desert conditions because the presence of the salt reduces their vapour pressure.

A hypothetical vapour pressure–composition diagram is shown in Figure 7.7. This graph represents the extreme case of a solution containing a non-volatile solute. In these circumstances, the vapour pressure of the solution is almost entirely due to the solvent particles. In reality, all solids exert a vapour pressure, although it may be extremely small.

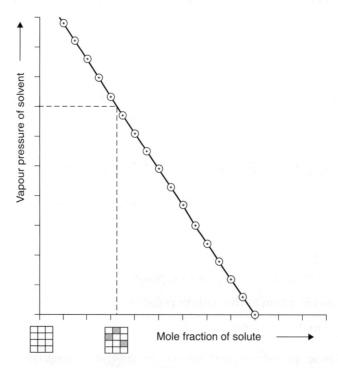

Figure 7.7 Vapour pressure–composition diagram for a solution containing a non-volatile solute

Composition is represented on the x-axis by the mole fraction, x. The mole fraction is the ratio of the number of moles of species to the total number of moles of all of the species present in a mixture. For solute molecules in a non-electrolyte solution:

$$x_{solute} = \frac{\text{number of solute molecules}}{\text{total number of solute and solvent molecules}} \qquad (7.1)$$

$$x_{solvent} = \frac{\text{number of solvent molecules}}{\text{total number of solute and solvent molecules}} \qquad (7.2)$$

For a two component mixture:

$$x_{solute} = \frac{n_{solute}}{n_{solute} + n_{solvent}} \qquad (7.3)$$

$$x_{solvent} = \frac{n_{solvent}}{n_{solute} + n_{solvent}} \qquad (7.4)$$

where n_x is the total number of moles of species 'x'. The mole fractions of a solvent and all the solute species always add up to one:

$$x_{solute} + x_{solvent} = 1 \qquad (7.5)$$

Example 7.1

Calculate the mole fractions of the non-electrolyte ethanol (CH_3CH_2OH) and water in a solution containing 10 g of ethanol dissolved in 20 g of water.

First, we calculate the number of moles of solute and solvent present. Since the molar mass of $CH_3CH_2OH = 46.068$ g mol^{-1} and the molar mass of $H_2O = 18.02$ g mol^{-1}:

$$\text{Moles of } CH_3CH_2OH = \frac{10}{46.068} = 0.2171$$

$$\text{Moles of } H_2O = \frac{20}{18.02} = 1.1099$$

We then determine the total number of moles in solution as $0.2171 + 1.1099 = 1.3270$

The mole fractions can then be calculated by using Equations 7.3 and 7.4:

$$x_{CH_3CH_2OH} = \frac{0.2171}{1.3270} = 0.1636$$

$$x_{H_2O} = \frac{1.1099}{1.3270} = 0.8364$$

The calculation of mole fractions is slightly different for dilute electrolyte solutions, e.g. solutions of ionic salts such as sodium and potassium chloride. Remember that colligative properties depend on the number of solute particles and not on their nature. Thus, if we were to completely dissolve one mole of sodium chloride in a solvent, the solution would contain one mole of sodium ions and one mole of chloride ions, i.e. two moles of ions. Similarly, a dilute magnesium chloride solution would contain one mole of magnesium ions and two moles of chloride ions, i.e. three moles of ions. Therefore, for dilute electrolyte solutions:

$$x_{cations} = \frac{n_{cation}}{n_{cation} + n_{anion} + n_{solute}} \tag{7.6}$$

$$x_{anions} = \frac{n_{anion}}{n_{cation} + n_{anion} + n_{solute}} \tag{7.7}$$

Example 7.2

Calculate the mole fractions of the species present in a solution containing 25.00 g of calcium chloride dissolved in 250 g of water.

First, we calculate the number of moles of anion, cation and solvent present. Since the molar mass of $CaCl_2 = 110.98$ g mol^{-1} and the molar mass of $H_2O = 18.02$ g mol^{-1}:

$$\text{Moles of solute} = \frac{25.00}{110.98} = 0.2253$$

Thus, the moles of Ca^{2+} present $= 0.2253 \times 1 = 0.2253$, while the moles of Cl^- present $= 0.2253 \times 2 = 0.4506$.

$$x_{H_2O} = \frac{250}{18.02} = 13.8735$$

Now we can calculate the mole fractions of each species:

$$x_{cations} = \frac{0.2253}{0.2253 + 0.4506 + 13.8735} = 0.0155$$

$$x_{anions} = \frac{0.4506}{0.2253 + 0.4506 + 13.8735} = 0.0310$$

$$x_{solvent} = \frac{13.8735}{0.2253 + 0.4506 + 13.8735} = 0.9535$$

Note that $0.0155 + 0.0310 + 0.9535 = 1$.

7.6.2 Depression of freezing point

In a pure solvent, the rate at which its molecules form a solid upon freezing and leave it to return to the liquid are equal. When a solute is added to a pure solvent, the solute particles block the route of the solvent molecules to the solid, thus slowing the freezing rate. However, because the solid is pure solvent, the rate at which the molecules leave the solid is unchanged. These circumstances result in a net flow of molecules away from the solid (i.e. the solid melts) – only lowering the temperature of the solution can stop this flow of molecules. Consequently, the freezing point of a solution is lower than that of the pure solvent. For example, aqueous solutions freeze at lower temperatures than water does – this can be very useful.

- The use of anti-freeze (an aqueous solution of glycol), in car engines, delays the freezing of water in the radiator.

- The use of salt on roads in cold weather delays the formation of ice because salt and water mixtures freeze at temperatures below 0°C. This phenomenon means that seawater freezes at about −1°C.

7.6.3 Elevation of boiling point

A solution that contains a non-volatile solute is less volatile than the pure solvent. Consequently, a higher temperature is required to raise the vapour pressure up to atmospheric pressure (when boiling begins) (see Figure 7.8). Aqueous solutions have a higher boiling point than water and thus solutions can prevent the boiling of water on hot summer days, i.e. anti-freeze in the engines of motor vehicles also acts as an anti-boil material.

7.6.4 Osmotic pressure

Osmosis is a process in which a substance passes through a membrane from an area of lower concentration to an area of higher concentration of a given solution. A membrane that allows the solvent particles but not the solute particles to pass through is called a *semi-permeable membrane*. Solvent molecules can move in both directions through a semi-permeable membrane.

We can demonstrate osmosis by setting up the experiment shown in Figure 7.9. A semi-permeable membrane is placed over the end of a glass tube that is filled with a sugar solution. This tube is then placed in a beaker of water. The water will pass

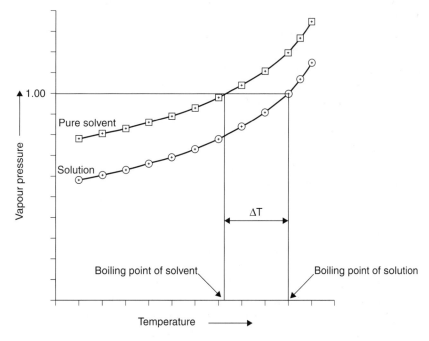

Figure 7.8 Illustration of the elevation of boiling point

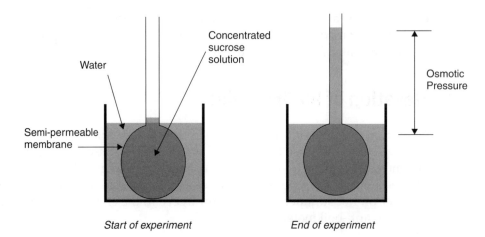

Figure 7.9   Experiment to demonstrate osmosis

through the membrane into the sugar solution. The pressure needed to stop the flow of solvent is called the osmotic pressure (π). Osmotic pressure is a colligative property since it depends on the number but not the nature of the solute particles.

When two solutions have the same osmotic potential, they are said to be *isotonic*. Isotonic drinks have become very popular 'sports drinks' in recent years, as such drink have the same concentration as body fluid. Consequently, these drinks are ideal for replacing fluid (sweat) lost during exercise

When one solution has a greater osmotic potential (i.e. is more concentrated) than another, it is said to be *hypertonic* to it. Hypertonic drinks are more concentrated than body fluids and are designed to replace and maintain energy levels during exercise of at least one hour. They are absorbed slowly into the body and consequently are not appropriate for fluid replacement. In fact, using these drinks as fluid replacers could increase dehydration as the body will release water into the intestines in order to dilute the drink enough for it to be absorbed. *Hypotonic* drinks are less concentrated than body fluid and thus help the body to speed up water absorption. They are best used when the body needs urgent fluid replacement (after exercise), rather than for energy replacement.

Osmosis plays an important role in many biological systems. For example, red blood cells contain salt solutions surrounded by a plasma membrane. When they are put into water, they swell and burst due to osmosis; conversely, when they are put into concentrated salt solutions, they shrink. Similarly, the uptake of water by a plant cell is controlled by the concentrations of ions present within the cell. This intake of water helps to keep the cell turgid (swollen), which is necessary for continued growth. If the water supply is cut off (e.g. by a drought), cells lose their turgidity, become dehydrated, and the plant wilts.

In *reverse osmosis* (RO), the solvent's direction of movement is opposite to that experienced during osmosis, i.e. the solvent molecules move through a semi-permeable membrane from a more concentrated to a more dilute solution. This is achieved by applying a pressure greater than the osmotic pressure to the solution side of the semi-permeable membrane. This technique is used in desalination plants to remove salts from seawater to produce fresh water for drinking and irrigation.

The energy requirements of RO desalination plants depend on the concentration of salts in the feed-water. To desalinate brackish water using RO requires about 15 kJ dm^{-3}, whereas to desalinate seawater requires about 90 kJ dm^{-3}. Most of the RO plants in the world are located in the US, where over 1350 plants have capacities greater than 20 000 m^3 (per day). However, most desalination occurs in the Middle East, where they tend to use fossil-fuelled distillation processes.

7.7 Solution Equilibria

Aquatic equilibria are important in industrial, analytical, biological and environmental processes. For example, we have studied the carbonate–bicarbonate system in Chapter 6 and we will study it further in this present Chapter. However, we need to study solution equilibria in order to understand water's solvent properties in more detail. Let us consider the general case of a reversible reaction:

$$aA + bB \rightleftharpoons cC + dD$$

At equilibrium, it can be shown experimentally that:

$$K_C = \left(\frac{[C]^c[D]^d}{[A]^a[B]^b} \right)_{eq} \tag{7.8}$$

where K_C is the equilibrium constant, and [X] is the concentration of component X.

This is a mathematical statement of the *Equilibrium Law* , which states that for any equilibrium at a constant temperature, the product of the concentrations of the products (raised to the appropriate powers) divided by the product of the concentrations of the reactants (raised to the appropriate power), has a constant value. The appropriate power is the coefficient of that substance in the stoichiometric equation for the reaction. By convention, the concentrations of the products are divided by the concentrations of the reactants. We should really write this equation in terms of the *activities* of the chemical species. The activity, a_i, of a chemical species in solution is a formal thermodynamic representation of concentration (its 'effective' concentration). However, as activity and concentration are almost identical in dilute solutions, we will confine ourselves to using concentration terms (mol dm^{-3}).

The constant in the Equilibrium Law is called the *equilibrium constant*, K_C, and applies to reactions in solution. The dimensions of K_C are concentration$^{(c+d-a-b)}$, and the units vary from one equilibrium to another. The equilibrium constant may or may not have units depending upon the form of the equilibrium expression. The right hand side of Equation 7.8 is called the equilibrium quotient and it contains only concentration terms. It should not contain terms for pure solids or liquids, or any solvent, since the activity (and thus the concentration) of a pure substance is defined as unity. The magnitude of K_C provides a useful indication of the extent of a chemical reaction. A large value for K_C indicates a high proportion of products to reactants (i.e. an almost complete reaction), and vice versa. Note that K_C gives us no information about the rate of reaction.

Example 7.3

Write an expression for the equilibrium constant for the following reaction:

$$Sn^{2+}(aq) + 2Fe^{3+}(aq) \rightleftharpoons Sn^{4+}(aq) + 2Fe^{2+}(aq)$$

What are the units of this equilibrium constant?
Using the definition above, the equilibrium constant is:

$$K_C = \left(\frac{[Sn^{4+}][Fe^{2+}]^2}{[Sn^{2+}][Fe^{3+}]^2} \right)_{eq}$$

The units are therefore:

$$\frac{(mol\ dm^{-3})(mol\ dm^{-3})^2}{(mol\ dm^{-3})(mol\ dm^{-3})^2}$$

All the units cancel out; consequently, K_C for this reaction has no units.
The French scientist, Henry Le Chatelier, studied the influence of temperature, pressure and concentration on equilibria for many years. His famous principle (*Le Chatelier's Principle*) has been controversial since its inception because it does not always provide the correct answer. However, it is useful for studying shifts in equilibrium in a qualitative manner. A simplified statement of Le Chatelier's Principle is 'when a system in equilibrium is subjected to change, the system will alter in such a way as to lessen the effect of that change'. For example, adding product 'C' to the system described in Equation 7.8 will make the reaction shift to the left to produce more 'A' and 'B' and consume 'C' and 'D' until the equilibrium constant is satisfied, i.e. the position of the equilibrium changes; the equilibrium constant remains unchanged (at constant temperature). This is known as the *mass action effect* and is the driving force behind many environmental phenomena. For reactions involving gases, the equilibrium constant is more conveniently expressed in terms of the partial pressures of the gas rather than their concentrations. The equilibrium constant is then given the symbol K_P.
Some examples of solution equilibria are outlined in Examples 7.4–7.7.

Example 7.4 Dissociation of Water and the pH Scale

As we have previously seen, water can self-ionize:

$$2H_2O(l) \rightleftharpoons H_3O^+(aq) + OH^-(aq)$$

$$K_W = \left(\frac{[H_3O^+][OH^-]}{[H_2O]^2} \right)_{eq}$$

Water is a pure liquid and thus its concentration is defined as unity. Hence:

$$K_W = ([H_3O^+][OH^-])_{eq}$$

Experimental determination shows that $K_W = 1.0 \times 10^{-14}$ mol^{2}dm^{-6} at 25°C (this gives a value of p$K_W = 14$, where p$K_W = -\log K_W$). Thus, it can be assumed that:

$$[H_3O^+][OH] = 1.0 \times 10^{-14} \text{ mol}^2 \text{ dm}^{-6} \text{ and } [H_3O^+]$$

$$= [OH^-] = 1.0 \times 10^{-7} \text{ mol dm}^{-3}$$

Consequently, in one litre of water, this dissociation will produce $1/(10\,000\,000)$ mole of H_3O^+ ions and $1/(10\,000\,000)$ mole of OH^- ions. The concentration of H_3O^+ ions in solution can be assigned a number from 0 to 14. This scale, called the *pH scale*, was originally introduced by the Danish biochemist Soren Peter Lauritz Sorensen while he was working on methods to improve the quality control of beer. The pH of a solution is the negative logarithm to the base ten of the numerical value of the H_3O^+ ion concentration.

$$pH = -\log[H_3O^+] \tag{7.9}$$

The pH scale is shown in Table 7.8. If $[H_3O^+] > [OH^-]$ the solution is acidic, while if $[H_3O^+] < [OH^-]$, the solution is alkaline. The concentration of OH^- ions in a solution can be expressed in terms of pOH.

$$pOH = -\log[OH^-] \tag{7.10}$$

Thus, at equilibrium during the self-ionization of water:

$$pH = -\log [1 \times 10^7] = pOH = -\log[1 \times 10^7] = 7$$

Therefore, neutral solutions by definition have a pH of 7. We can show that K_W links pH and pOH as follows:

$$K_W = [H_3O^+][OH^-]$$

$$\log K_W = \log[H_3O^+] + \log[OH^-] \Rightarrow -pK_W = -pH - pOH$$

$$\Rightarrow pH + pOH = pK_W$$

Table 7.8 The pH scale

Example	pH	$[H_3O^+]$	$[OH^-]$	
	0	10^0	10^{-14}	
Stomach juices (1.6–1.8)	1	10^{-1}	10^{-13}	Increasing
Orange juice (2.6–4.4)	2	10^{-2}	10^{-12}	acidity
Vinegar	3	10^{-3}	10^{-11}	
	4	10^{-4}	10^{-10}	
Natural rain (5.6)	5	10^{-5}	10^{-9}	
Saliva (6.35–6.85)	6	10^{-6}	10^{-8}	
Human blood (7.35–7.45)	7	10^{-7}	10^{-7}	Neutrality
Sea water	8	10^{-8}	10^{-6}	
Soap	9	10^{-9}	10^{-5}	
	10	10^{-10}	10^{-4}	
Milk of magnesia	11	10^{-11}	10^{-3}	
	12	10^{-12}	10^{-2}	Increasing
	13	10^{-13}	10^{-1}	alkalinity
	14	10^{-14}	10^0	

Thus:

$$pH + pOH = 14 \tag{7.11}$$

The sum of pH and pOH must be 14 in aqueous solutions. Equation 7.11 allows the pH value for bases to be calculated directly, and consequently pOH values are rarely quoted. The pH of a solution is usually measured potentiometrically by using a pH meter.

Example 7.5 Dissociation of Aqueous Ethanoic Acid

The Equilibrium Law can be applied to aqueous solutions of acids, e.g. the dissociation of ethanoic acid in water:

$$CH_3COOH(aq) + H_2O(l) \rightleftharpoons CH_3COO^-(aq) + H_3O^+(aq)$$

The equilibrium constant is given by:

$$K_a = \left(\frac{[H_3O^+][CH_3COO^-]}{[CH_3COOH]} \right)_{eq}$$

The units of K_a in this equilibrium are mol dm^{-3} (you should check this). In this case, the equilibrium constant is called the *acid dissociation constant* (K_a), which is a measure of the strength of an acid. An acid such as hydrochloric acid is virtually fully dissociated in aqueous solution and has a large K_a. It is often more convenient to compare the strengths of acids using pK_a values, where:

$$pK_a = -\log K_a$$

For most acids, this gives a range of values from 1 to 14, with strong acids having low pK_a values and weak acids having large values. A list of pK_a values for common industrial, laboratory and environmental acids is given in Table 7.9. Acids containing more than one replaceable hydrogen atom (e.g. H_2SO_4, and H_3PO_4) are called polyprotic acids and have more than one dissociation constant.

Example 7.6 Dissociation of Aqueous Ammonia

The Equilibrium Law can also be applied to aqueous solutions of bases, e.g. the dissociation of aqueous ammonia in water:

$$NH_3(aq) + H_2O(aq) \rightleftharpoons NH_4^+(aq) + OH^-(aq)$$

The *base dissociation constant*, K_b, for this equilibrium is given by:

$$K_b = \left(\frac{[NH_4^+][OH^-]}{[NH_3]} \right)_{eq}$$

As with acids, the strength of bases can be compared on the pK_b scale, where:

$$pK_b = -\log K_b$$

Table 7.9 Dissociation constants of acids and bases[a] in water at 25°C (adapted from Lide, D.R. (ed.), (1998). *CRC Handbook of Chemistry and Physics* (79th edn). CRC Press, Boca Raton, FL, USA)

Species	Equilibrium in aqueous solution	Step	pK_a
Aluminium (III) ion	$Al(H_2O)_6^{3+}(aq) + H_2O(l) \rightleftharpoons H_3O^+(aq) + Al(H_2O)_5(OH)^{2+}(aq)$		5.0
Ammonium ion	$NH_4^+(aq) + H_2O(l) \rightleftharpoons H_3O^+(aq) + NH_3(aq)$		9.25
Boric acid[b]	$H_3BO_4(aq) + H_2O(l) \rightleftharpoons H_3O^+(aq) + H_2BO_3^-(aq)$	1	9.27
	$H_2BO_3^-(aq) + H_2O(l) \rightleftharpoons H_3O^+(aq) + HBO_32-(aq)$	2	>14
Carbonic acid	$CO_2(aq) + H_2O(l) \rightleftharpoons H_3O^+(aq) + HCO_3^-(aq)$	1	6.35
	$HCO_3^-(aq) + H_2O(l) \rightleftharpoons H_3O^+(aq) + CO_3^{2-}(aq)$	2	10.33
Ethanoic acid	$CH_3CO_2H(aq) + H_2O(l) \rightleftharpoons H_3O^+(aq) + CH_3CO_2^-(aq)$		4.76
Formic acid[b]	$HCO_2H(aq) + H_2O(l) \rightleftharpoons H_3O^+(aq) + HCO_2^-(aq)$		3.75
Nitric acid	$HNO_3(aq) + H_2O(l) \rightleftharpoons H_3O^+(aq) + NO_3^-(aq)$		−1.4
Nitrous acid	$HNO_2(aq) + H_2O(l) \rightleftharpoons H_3O^+(aq) + NO_2^-(aq)$		3.25
Phosphoric acid	$H_3PO_4(aq) + H_2O(l) \rightleftharpoons H_3O^+(aq) + H_2PO_4^-(aq)$	1	2.16
	$H_2PO_4^-(aq) + H_2O(l) \rightleftharpoons H_3O^+(aq) + HPO_4^{2-}(aq)$	2	7.21
	$HPO_4^{2-}(aq) + H_2O(l) \rightleftharpoons H_3O^+(aq) + PO_4^{3-}(aq)$	3	12.32
Silicic acid[c]	$H_4SiO_4(aq) + H_2O(l) \rightleftharpoons H_3O^+(aq) + H_3SiO_4^-(aq)$	1	9.9
	$H_3SiO_4^-(aq) + H_2O(l) \rightleftharpoons H_3O^+(aq) + H_2SiO_4^{2-}(aq)$	2	11.8
	$H_2SiO_4^{2-}(aq) + H_2O(l) \rightleftharpoons H_3O^+(aq) + HSiO_4^{3-}(aq)$	3	12
	$HSiO_4^{3-}(aq) + H_2O(l) \rightleftharpoons H_3O^+(aq) + SiO_4^{4-}(aq)$	4	12
Sulfuric acid	$H_2SO_4(aq) + H_2O(l) \rightleftharpoons H_3O^+(aq) + HSO_4^-(aq)$		
Sulfurous acid	$H_2SO_3(aq) + H_2O(l) \rightleftharpoons H_3O^+(aq) + HSO_3^-(aq)$	1	1.85
	$HSO_3^-(aq) + H_2O(l) \rightleftharpoons H_3O^+(aq) + SO_3^{2-}(aq)$	2	7.2
Water	$H_2O(l) + H_2O(l) \rightleftharpoons H_3O^+(aq) + OH^-(aq)$		13.995

[a]In the case of bases, the entry in the table is for the conjugate acid, e.g. ammonium ion for ammonia. The K_b value may be calculated form the equation, $K_b = K_w/K_a$, where $K_w = 1 \times 10^{-14}$ at 25°C.

Strong bases have low pK_b values, and while bases have large values. The pK_b values may be calculated from the pK_a values (see Table 7.9).

Example 7.7 Formation and Dissociation of Complex Ions

The formation of a complex can remove an ion from solution and disturb the solubility equilibrium until more solid dissolves, for example:

$$Cu^{2+}(aq) + 4NH_3(aq) \rightleftharpoons [Cu(NH_3)_4]^{2+}(aq)$$

$$K_f = \left(\frac{[Cu(NH_3)_4^{2+}]}{[Cu^{2+}][NH_3]^4} \right)_{eq} = 6.8 \times 10^{12} \text{ mol}^{-4} \text{ dm}^{12}$$

where K_f is known as the formation constant.

The very large formation constant in this case shows that the reaction goes completely to the right – there is hardly any free Cu^{2+} ion left. The ability of metal ions in solution to form complexes can be utilized in analytical determinations (see Experiment 7.1).

7.8 Solubility Product

Ionic compounds dissolve in water, establishing ionic equilibria. When increasing quantities of a sparingly soluble ionic solid are added to water, a saturated solution is eventually formed. Ions in the saturated solution are in dynamic equilibrium with the excess undissolved solute:

$$MX(s) \rightleftharpoons M^+(aq) + X^-(aq)$$

For example, when equilibrium between pure silver chromate, Ag_2CrO_4, and its solution is reached, we have:

$$Ag_2CrO_4(s) \rightleftharpoons 2Ag^+(aq) + CrO_4{}^{2-}(aq)$$

The equilibrium constant expression is as follows:

$$K_f = \left(\frac{[Ag^+]^2[CrO_4{}^{2-}]}{[Ag_2CrO_4]} \right)_{eq}$$

However, $[Ag_2CrO_4(s)]$, which represents the concentration of a pure solid, is constant. So,

$$K_c[Ag_2CrO_4(s)] = [Ag^+(aq)]^2[CrO_4{}^{2-}(aq)] = \text{a new constant}$$

This new constant is called the *solubility product* and is given the symbol K_{sp}. Values for K_{sp} are normally quoted only for electrolytes that are sparingly soluble in water. An electrolyte is generally considered to be sparingly soluble if its solubility is less than 0.01 mol dm^{-3} of water. For concentrations greater than this, the value of K_{sp} is no longer constant. Consequently, the K_{sp} concept is not suitable for use with soluble compounds such as halite (NaCl), and K_{sp} values are always very small, rarely exceeding 10^{-4} (see Table 7.10).

Example 7.8

Given that the molar solubility s, of silver chromate is 6.5×10^{-5} mol dm^{-3}, determine the value of K_{sp}.

At equilibrium:

$$Ag_2CrO_4(s) \rightleftharpoons 2Ag^+(aq) + CrO_4{}^{2-}(aq)$$

$$\Rightarrow \quad 1 \text{ mol } Ag_2CrO_4 = 2 \text{ mol } Ag^{2+} \text{ and } 1 \text{ mol } Ag_2CrO_4 = CrO_4{}^{2-}$$

$$\therefore \quad [Ag^+] = 2s \text{ and } [CrO_4{}^{2-}] = s$$

We already know that $K_{sp}[Ag_2CrO_4(s)] = [Ag^+(aq)]^2[CrO_4{}^{2-}(aq)]$

$$\therefore \quad K_{sp}[Ag_2CrO_4(s)] = (2s)^2(s) = (2 \times 6.5 \times 10^{-5})^2(6.5 \times 10^{-5})$$

$$= 1.1 \times 10^{-12} \text{ mol}^3 \text{ dm}^{-9}$$

Alternatively, $K_{sp}[Ag_2CrO_4(s)] = (2s)^2(s) = 4s^3 = 1.1 \times 10^{-12} \text{mol}^3 \text{ dm}^{-9}$

Table 7.10 Solubility product constants (K_{sp}) for selected sparingly soluble salts in water at 25°C (adapted from Lide, D.R. (ed.) (1998). *CRC Handbook of Chemistry and Physics* (79th edn). CRC Press, Boca Raton, FL, USA)

Compound	Formula	K_{sp}
Aluminium phosphate	$AlPO_4$	9.84×10^{-21}
Barium nitrate	$Ba(NO_3)_2$	4.64×10^{-3}
Cadmium carbonate	$CdCO_3$	1.0×10^{-12}
Calcium carbonate	$CaCO_3$	3.36×10^{-9}
Calcium fluoride	CaF_2	3.45×10^{-11}
Calcium hydroxide	$Ca(OH)_2$	5.02×10^{-6}
Calcium phosphate	$Ca_3(PO_4)_2$	2.07×10^{-33}
Calcium sulfate	$CaSO_4$	4.93×10^{-5}
Copper (I) bromide	$CuBr$	6.27×10^{-9}
Copper (I) chloride	$CuCl$	1.72×10^{-7}
Iron (II) carbonate	$FeCO_3$	3.13×10^{-11}
Iron (II) hydroxide	$Fe(OH)_2$	4.87×10^{-17}
Iron (III) hydroxide	$Fe(OH)_3$	2.79×10^{-39}
Lead (II) bromide	$PbBr_2$	6.60×10^{-6}
Lead (II) carbonate	$PbCO_3$	7.40×10^{-14}
Lead (II) chloride	$PbCl_2$	1.70×10^{-5}
Magnesium carbonate	$MgCO_3$	6.82×10^{-6}
Magnesium hydroxide	$Mg(OH)_2$	5.61×10^{-12}
Nickel (II) carbonate	$NiCO_3$	1.42×10^{-7}
Silver (I) bromide	$AgBr$	5.35×10^{-13}
Silver (I) carbonate	Ag_2CO_3	8.46×10^{-12}
Silver (I) chloride	$AgCl$	1.77×10^{-10}
Zinc carbonate	$ZnCO_3$	1.46×10^{-10}
Zinc hydroxide	$Zn(OH)_2$	3.10×10^{-17}

As with other equilibrium constants, the K_{sp} of a salt is temperature dependent. However, the concentrations of the individual ions may vary over a wide range. We can illustrate this by comparing the solubility of $BaSO_4$ in water and in 0.1 mol dm^{-3} of sodium sulfate solution (Examples 7.9 and 7.10).

Example 7.9

The molar solubility of a salt may be obtained from its solubility product by using the following calculation; $K_{sp}(BaSO_4) = 1.08 \times 10^{-10}$ mol^2 dm^{-6}.

Let the solubility of $BaSO_4$ in water $= s$ mol dm^{-3}

At equilibrium:

$$BaSO_4(s) \rightleftharpoons Ba^{2+}(aq) + SO_4{}^{2-}(aq)$$

$$K_{sp}(BaSO_4) = (Ba^{2+})(SO_4{}^{2-}) = 1.08 \times 10^{-10}$$

$$\Rightarrow \quad K_{sp}(BaSO_4) = s \times s = s^2 = 1.08 \times 10^{-10}$$

$$\Rightarrow \quad s = \sqrt{(1.08 \times 10^{-10})} = 1.04 \times 10^{-5}$$

Therefore the solubility of $BaSO_4$ in water at 25°C $= 1.04 \times 10^{-5}$ mol dm^{-3}.

Example 7.10

Let the solubility of $BaSO_4$ in 0.1 mol dm^{-3} $Na_2SO_4 = b$ mol dm^{-3}.
At equilibrium:

$$BaSO_4(s) \rightleftharpoons Ba^{2+}(aq) + SO_4^{2-}(aq)$$

$$Na_2SO_4(s) \rightleftharpoons 2Na^{2+}(aq) + SO_4^{2-}(aq)$$

In this situation:

$$[Ba^{2+}] = b \, mol \, dm^{-3}$$

$$[SO_4^{2-}] = (b + 0.1) \, mol \, dm^{-3} \text{(since sulfate is the common ion)}$$

$$K_{sp}(BaSO_4) = (Ba^{2+})(SO_4^{2-}) = 1.08 \times 10^{-10}$$

$$\Rightarrow \quad K_{sp}(BaSO_4) = b(b + 0.1) = 1.08 \times 10^{-10}$$

Since $b \leqslant 0.1$, we can estimate that $(b + 0.1) \approx 0.1$ (alternatively we could expand the equation and solve for b):

$$\Rightarrow \quad K_{sp}(BaSO_4) = 0.1b = 1.08 \times 10^{-10}$$

$$\Rightarrow \quad b = 1.08 \times 10^{-9}$$

Therefore the solubility of $BaSO_4$ in 0.1 mol dm^{-3} Na_2SO_4 at $25°C = 1.08 \times 10^{-9}$ mol dm^{-3}.

   If we compare the solubility values calculated in Examples 7.9 and 7.10, we can see that the solubility of an electrolyte in aqueous solution containing a common ion (e.g. $BaSO_4$ in Na_2SO_4) is less than its solubility in water. This generalization – the solubility of MA is reduced by the addition of $M^{n+}(aq)$ or $A^{n-}(aq)$ – is known as the *common ion effect*.

   An application of solubility products is that they enable us to predict the maximum concentrations of ions in solution. It is thus possible to tell whether or not precipitation will occur. For example, if we mix a dilute solution of Ba^{2+} ions with a dilute solution of SO_4^{2-} ions at $25°C$, will precipitation of $BaSO_4(s)$ occur? We can use $K_{sp}(BaSO_4)$ to predict precipitation. If $(Ba^{2+})(SO_4^{2-}) > K_{sp}(BaSO_4)$, then precipitation occurs until $[Ba^{2+}][SO_4^{2-}] = K_{sp}(BaSO_4)$. Precipitation will not occur if $[Ba^{2+}][SO_4^{2-}] < K_{sp}(BaSO_4)$. The term $[Ba^{2+}][SO_4^{2-}]$ is called the *ionic product*. Similarly, where $[H_3O^+][OH^-] = K_W$, the latter is known as *the ionic product of water*.

   From an environmental point of view, precipitation from aqueous solution is important. The formation of stalagmites and stalactites in caves is due to the precipitation of calcium carbonate from saturated solution, i.e. where $[Ca^{2+}][CO_3^{2-}] > K_{sp}(CaCO_3)$. Coral reefs and shells of marine crustacea (e.g. oysters) grow by precipitation in a similar fashion.

7.9 Acids and Bases

As chemistry has developed, various definitions have been proposed to define acidic and basic properties. In the nineteenth century, the Swedish scientist Svante Arrhenius suggested that acids were substances containing hydrogen that dissociated in water to produce oxonium ions, for example:

$$HCl(aq) + H_2O(l) \rightleftharpoons H_3O^+(aq) + Cl^-(aq) \qquad (7.12)$$

He proposed that strong acids such as HCl were completely dissociated in water, while weak acids were only partially dissociated, with a high proportion of the acid remaining in solution, for example:

$$CH_3COOH(aq) + H_2O(l) \rightleftharpoons CH_3COO^-(aq) + H_3O^+(aq)$$

Most acids that occur naturally in the environment are weak acids (e.g. carbonic, boric and phosphoric acids).

Arrhenius also suggested that bases were substances that reacted with hydrogen ions to produce water and that soluble bases dissociated in water to produce hydroxide ions, for example:

$$CuO(s) + 2H^+(aq) \longrightarrow Cu^{2+}(aq) + H_2O(l)$$

$$NH_3(aq) + H_2O(l) \longrightarrow NH_4^+(aq) + OH^-(aq)$$

However, Arrhenius' definitions apply only to aqueous solutions and a restricted number of bases. Consequently, new definitions were required for non-aqueous systems and a larger number of bases. In 1923, the Danish chemist Johannes Brønsted and the English chemist Thomas Lowry independently suggested new definitions.

- An acid is a proton donor.

- A base is a proton acceptor.

The theory based on these definitions is called the Brønsted–Lowry theory of acids and bases. Thus, in the dissociation of HCl (Equation 7.12), hydrochloric acid is an acid because it donates an H^+ ion to form a Cl^- ion. Water is a base because it accepts an H^+ ion to form an H_3O^+ ion. In the reverse reaction, oxonium ions lose H^+ ions in forming water and thus H_3O^+ is acting as an acid. Chloride ions act as a base since they accept H^+ ions to form HCl. In the case of a general acid, HA:

$$\underset{\text{acid}}{HA(aq)} + \underset{\text{base}}{H_2O(l)} \rightleftharpoons \underset{\text{conjugate acid}}{H_3O^+(aq)} + \underset{\text{conjugate base}}{A^-(aq)}$$

HA and A^- are said to be conjugate and to form a conjugate acid–base pair. The Brønsted–Lowry theory includes:

- all substances we commonly regard as acids, including acid salts and the ammonium ion;

- all substances we commonly regard as bases, including anions, ammonia, oxide and hydroxide ions.

Water is *amphiprotic*, meaning it can act as both a proton donor and acceptor:

$$H_2O(l) + H_2O(l) \rightleftharpoons H_3O^+(aq) + OH^-(aq)$$

$$\text{base} \qquad \text{acid} \qquad\qquad \text{acid} \qquad\quad \text{base}$$

The American chemist, Gilbert Newton Lewis, extended the Brønsted–Lowry theory even further. In 1938, he defined an acid (Lewis acid) as a substance that can form a covalent bond by accepting an electron pair from a base, and a base (Lewis base) as a substance that has an unshared electron pair that can form a covalent bond with an atom, molecule or ion. These definitions include acids and bases that would not normally be recognized by using the other definitions, e.g. those that do not involve protons. For example, boron trifluoride (a Lewis acid) reacts with trimethylamine (a Lewis base) to form a solid salt:

$$
\begin{array}{ccccccc}
\text{F} & \text{CH}_3 & & \text{F} & & & \text{CH}_3 \\
| & | & & | & & & | \\
\text{F--B} & + \quad \text{:N}-\text{CH}_3 & \longrightarrow & \text{F}-\text{B} & \leftarrow & \text{N}-\text{CH}_3 \\
| & | & & | & & & | \\
\text{F} & \text{CH}_3 & & \text{F} & & & \text{CH}_3
\end{array}
$$

Lewis acid   Lewis base   Lewis acid–base complex

(electron pair acceptor)   (electron pair donor)

In Chapter 4, we studied the environmental importance of acids and bases in chemical weathering and ion-exchange reactions. Later in this present chapter, we will study the environmental damage caused by acid precipitation.

7.10 Buffer Solutions

Buffer solutions can maintain a given pH value on the addition of a small amount of acid or base. Strong acid and base buffers are often used industrially in aqueous solutions with low (<4) and high pH (>10) values, respectively. Weak acid and alkaline buffer solutions are more important in environmental chemistry since extremely low or high pH values occur rarely in nature.

A weak acidic buffer solution (pH 4–7) is prepared by mixing together definite amounts of a weak acid (to supply protons to a strong base) and its conjugate base (to accept protons from a strong acid), e.g. ethanoic acid and sodium carbonate, and nitrous acid and sodium nitrite. A weak alkaline buffer (pH 7–10) is prepared by mixing a weak base (to accept protons from a strong acid) with its conjugate acid (to transfer protons to a strong base), e.g. ammonia solution and ammonium chloride. We will explain the action of weak buffer solutions by using two examples.

Example 7.11

Consider a weak acidic buffer solution containing equal concentrations of ethanoic acid and sodium ethanoate at 25°C.

(i) $CH_3COOH(aq) + H_2O(l) \rightleftharpoons CH_3COO^-(aq) + H_3O^+(aq)$

(ii) $CH_3COONa(s) \longrightarrow CH_3COO^-(aq) + Na^+(aq)$

The resulting mixture will clearly contain a relatively large concentration of ethanoate ions ($CH_3COO^-(aq)$). Most of the ethanoate ions are derived from the sodium ethanoate since the acid dissociation constant for ethanoic acid is quite small (see Table 7.9). If a strong base is added to this buffer (e.g. $OH^-(aq)$, they will remove the H_3O^+ ions from the solution:

$$OH^-(aq) + H_3O^+(aq) \longrightarrow 2H_2O(l)$$

This will disturb the equilibrium of the first equation shown above (i), moving it to the right to produce more $H_3O^+(aq)$. Thus, the pH of the solution is restored. If H_3O^+ ions are added to the buffer, the equilibrium in (i) is disturbed and moves to the left (in accordance with Le Chatelier's Principle). Consequently, the concentration of $H_3O^+(aq)$ in solution remains unchanged.

We saw in Example 7.5 that the acid dissociation constant (K_a) for the dissociation of ethanoic acid is given by:

$$K_a = \left(\frac{[H_3O^+][CH_3COO^-]}{[CH_3COOH]} \right)_{eq}$$

If we take logarithms to the base ten and rearrange the equation, we obtain:

$$pH = pK_a + \log \left(\frac{[CH_3COO^-]}{[CH_3COOH]} \right) \text{ (You should check this for yourself.)}$$

However, the value of $[CH_3COO^-]$ is almost entirely from the sodium ethanoate, namely:

$$[CH_3COO^-] \approx [CH_3COONa] = [A^-]$$

We have already seen that ethanoic acid is largely undissociated in the buffer solution and so the concentration of the acid in the equilibrium mixture is approximately the same as the initial concentration of the acid, namely:

$$[CH_3COOH] = [HA]$$

Substituting these values into the equation above gives us:

$$pH = pK_a + \log \left(\frac{[A^-]}{[HA]} \right) \tag{7.13}$$

This is known as the *Henderson–Hasselbach Equation* for a buffer solution consisting of a weak acid and its salt. In the above example, the concentration of the acid and the salt are equal and therefore the pH of the buffer solution may be calculated.

$$pH = 4.76 + \log 1 = 4.76$$

Thus, a solution of equal concentrations of ethanoic acid and sodium ethanoate at 25°C will act as a buffer close to pH $= 4.76$.

If a solution is prepared from a weak base B and its conjugate acid (BH^+), the Henderson–Hasselbach Equation becomes:

$$pH = pK_a + log \left(\frac{[B]}{[BH^+]} \right) \tag{7.14}$$

where pK_a is the acid dissociation constant of the weak acid BH^+.

Example 7.12

Consider one litre of a weak alkaline buffer containing 0.023 M aqueous ammonia and 0.030 M ammonium chloride.

$$NH_3(aq) + H_2O(l) \rightleftharpoons NH_4^+(aq) + OH^-(aq)$$

$$NH_4Cl(s) \longrightarrow NH_4^+(aq) + Cl^-(aq)$$

The buffer contains a relatively large concentration of $NH_4^+(aq)$ ions, mainly from the complete ionization of ammonium chloride. If H_3O^+ ions are added to the buffer, they remove $OH^-(aq)$ ions from solution and the equilibrium moves to the right, so producing more $OH^-(aq)$ and returning the original pH. If $OH^-(aq)$ ions are added to the buffer, the equilibrium moves to the left. This reduces the concentration of $OH^-(aq)$, leaving the pH of the solute unchanged.

In order to calculate the pH of this buffer solution at 25°C, we use Equation 7.14. You should be able to work out which substances are acting as the base and acid by using the Brønsted-Lowry definitions.

$$pH = pK_a + log \left(\frac{[0.023]}{[0.030]} \right) = 9.13$$

i.e. this solution will act as a buffer close to pH $= 9.13$.

To demonstrate buffer activity let us consider an additional example.

Example 7.13

Calculate the pH of the resulting solution when an acid gas (0.01 M HCl(g)) dissolves in 300 cm^3 of a weak acid buffer solution consisting of 0.080 M ethanoic acid and 0.080 M sodium ethanoate at 25°C. What is the change in pH?

The number of moles of acid in the buffer solution is initially:

0.3 (volume of the buffer solution in dm^3) $\times$ 0.080 (molarity in mol dm^{-3}) $=$ 0.024 mol

The addition of 0.01 mol HCl(g) will increase the number of moles of acid from 0.024 to $(0.024 + 0.01) = 0.025$ mol. The molarity of the acid is thus increased to:

$$\text{Molarity of } CH_3COOH = \frac{0.025}{0.3} = 0.083 \text{ mol dm}^{-3}$$

The number of moles of base in the buffer solution is initially:

$$0.3 \times 0.080 = 0.024 \text{ mol}$$

The additional acid will react with the ethanoate, reducing the number of moles from 0.024 to $(0.024-0.01) = 0.023$. The molarity of the base is thus reduced to:

$$\text{Molarity of } CH_3COO^- = \frac{0.023}{0.3} = 0.077 \text{ mol dm}^{-3}$$

We can now estimate the new pH for the solution by using the Henderson–Hasselbach Equation (Equation 7.13).

$$pH = 4.75 + \log\left(\frac{[0.077]}{[0.083]}\right) = 4.71$$

We know from Example 7.11 that a solution of equal concentrations of ethanoic acid and sodium ethanoate at 25°C will act as a buffer close to $pH = 4.76$. Consequently, the pH has reduced from 4.76 to 4.71, i.e. a reduction of 0.05.

Natural buffer solutions are ubiquitous. The ability of natural waters to absorb large quantities of acid or alkali is known as its *buffer capacity*. The oceans are maintained at approximately $pH = 8.4$ by a complex buffering process that depends upon the presence of hydrogen carbonates and silicates. The poor buffer capacity of many Scandinavian inland lakes has led to disastrous environmental consequences (see Section 7.10.1). Many biological systems depend on buffer action to preserve a constant pH. A mixture of phosphate, bicarbonate and proteins maintains the pH of blood between 7.35–7.45; the pH of tears is maintained at 7.4 by protein buffers. Buffer solutions also play an important part in a large number of industrial processes, including the manufacture of dyes and photographic materials, and in electroplating.

7.11 Water Pollution

Water pollution may be defined as the degradation of water quality by the introduction of chemical, physical or biological parameters into rivers, lakes, streams, oceans, etc. Water pollutants include the following:

- waste (e.g. industrial effluents, domestic and agricultural sewage, radioactive discharge and thermal pollution);
- urban run-off;
- agricultural run-off (including fertilizers and pesticides);
- oil spillages (see Chapter 5);
- acid precipitation;
- suspended solids (e.g. litter and rubbish).

Globally, the scale of water pollution is huge. Billions of metric tons of silt, sewage, industrial waste, and chemical residues are discharged into the world's oceans; rivers alone pour an estimated 9.3 billion metric tons of silt and other wastes into coastal waters each year.

All water pollution incidents may be categorized on the basis of their severity. For example, the UK Environment Agency defines three categories of water pollution incident (Environment Agency, 1998). *Category 1* incidents, the most serious, may involve one or more of the following:

- closure of a source of water abstraction;
- an extensive fish kill;
- a potential or actual persistent effect on water quality or aquatic life;
- a major effect on the amenity value of the receiving water;
- extensive subsequent remedial measures.

Category 2 incidents are significant but less severe, and may:

- involve the necessity to notify downstream abstractors;
- result in a significant fish kill;
- render water unfit for livestock;
- have a measurable effect on animal life in the water;
- contaminate the bed of the river or canal;
- reduce the amenity value of the water to their owners or to the general public.

Category 3 incidents are relatively minor and have no significant or lasting effect on the receiving water. Other countries use similar, although not identical, classifications of water pollution.

7.11.1   Acid precipitation

Acid precipitation (sometimes referred to as acid rain) is not a recent discovery. Robert Angus Smith, Britain's first Air Pollution Inspector, first used the phrase in the late nineteenth century. Smith was the first person to make the link between sulfur pollution and acidic rainfall, realizing that it was attacking vegetation, stone and iron in his home town of Manchester. He published his findings in *Acid and Rain: The Beginnings of Chemical Climatology* in 1872.

However, Smith's work was largely forgotten until the 1960s when Scandinavian scientists began to link pollution blown across the sea from the UK to their own acidified fishless lakes. A political row between the UK and Norway (and also between the USA and Canada) broke out which lasted for well over 15 years. The recognition that air pollution from one country can be transported for 1000s of kilometres before being deposited upon another has had an important impact on international pollution control legislation. There is evidence of acid damage throughout the world – from Australia to Mexico to China – although Europe has been the worst affected.

Acidity is measured by using the pH scale, and as we have already seen, a substance with a pH value of less than 7 is acidic. The pH scale is logarithmic, and hence a substance with a pH of 6 is 10 times more acidic than another with a pH of 7. Generally, the pH of 5.6 has been used as the baseline in identifying acid rain because this is the pH value of carbon dioxide in equilibrium with distilled water. Hence, *acid rain* is defined as any rainfall that has an acidity level beyond which is expected in non-polluted rainfall. In essence, any precipitation – rain, snow, hail, etc. – that has a pH value (expressed as [H⁺]) of less than 5.6 is considered to be acid precipitation. The term *acid deposition* is used to describe the total deposition of acid (H⁺) or acid-forming species (e.g. $SO_2(g)$) by both *wet and dry deposition*.

The main cause of acid precipitation is the presence of strong mineral acids, mainly sulfuric (H_2SO_4) and nitric (HNO_3) acids, derived from the atmospheric oxidation of sulfur dioxide (SO_2) and nitrogen oxides (NO_x). These gases are emitted from both natural and anthropogenic sources, with the combustion of fossil fuels being the most important anthropogenic source. Sulfur dioxide and nitrogen oxides are converted into acids by a number of complex atmospheric processes involving several chemical reactions. The emission and deposition of pollutants leading to the detrimental environmental effects of acid precipitation is shown in Figure 7.10. It is important to consider both solution-and gas-phase chemistries in the formation of acid rain (see Figure 7.11), although we will only study the aqueous-phase reactions.

Sulfur dioxide is partially soluble in water. Consequently, only a fraction of atmospheric SO_2 exists in the dissolved aqueous form when there is a cloud or mist content in the air. In the aqueous phase, sulfur dioxide ($SO_2(aq)$) exists in equilibrium with sulfite ($SO_3^{2-}(aq)$) and bisulfite ($HSO_3^-(aq)$) ions. The dissociation of gaseous sulfur dioxide in water occurs by a threefold process, as follows:

1) $SO_2(g) + H_2O(l) \rightleftharpoons SO_2(aq)$

2) $SO_2(aq) + 2H_2O(l) \rightleftharpoons H_3O^+(aq) + HSO_3^-(aq)$

3) $HSO_3^-(aq) + H_2O(l) \rightleftharpoons H_3O^+(aq) + SO_3^{2-}(aq)$

These equilibria are critically dependent upon the pH of the precipitation (since hydrogen ions are involved in reactions (2 and (3)) and droplet size. The sulfite and bisulfite ions may be oxidized by a number of atmospheric mechanisms to sulfuric acid.

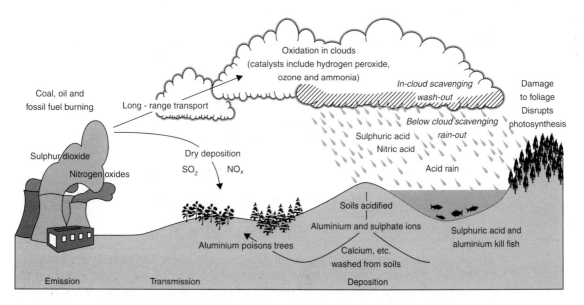

Figure 7.10 The emission and deposition of atmospheric pollutants

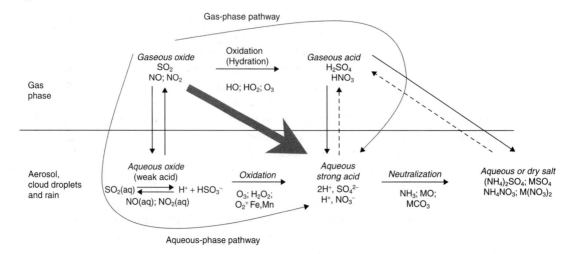

Figure 7.11 Schematic representation of pathways for atmospheric formation of sulfuric and nitric acids and their salts. (Adapted from Schwartz, S.E. (1987), in *The Chemistry of Acid Rain: Sources and Atmospheric Processes*, American Chemical Society, Washington, DC, USA.) p. 95

The oxidation of aqueous SO_2 by molecular oxygen relies on a metal catalyst such as Fe^{3+} or Mn^{2+}, or a combination of the two. Oxidation by ozone, however, is a more appreciable process because it does not require a catalyst. The dominant oxidation process occurs by the action of hydrogen peroxide (formed in the gas phase from free radicals). The reaction involves the formation of an intermediate (A^-), possibly HSO_4^- (aq), and may proceed as follows:

1) $HSO_3^-(aq) + H_2O_2(g) \longrightarrow HSO_4^-(aq) + H_2O(l)$

2) $HSO_4^-(aq) + H^+(aq) \longrightarrow H_2SO_4(aq)$

Gaseous sulfur dioxide will also dissolve in water to form sulfurous acid ($H2SO_3$):

$$SO_2(g) + H_2O(l) \rightleftharpoons H_2SO_3(aq)$$

The relative concentrations of gaseous SO_2 and H_2SO_3(aq) are related by the equilibrium constant for this reaction.

There are three equilibria to consider in the aqueous oxidation of NO_x:

1) $2NO_2(g) + H_2O(l) \longrightarrow 2H^+(aq) + NO_3^-(aq) + NO_2^-(aq)$

2) $NO(g) + NO_2(g) + H_2O(l) \longrightarrow 2H^+(aq) + 2NO_2^-(aq)$

3) $3NO_2(g) + H_2O(l) \longrightarrow 2H^+(aq) + 2NO_3^-(aq) + NO(g)$

These reactions are limited by their dependence upon the partial pressures of NO_x present in the atmosphere, and the low solubility of these oxide. Potential for increase in reaction rate exists with the use of metal catalysts, similar to those used in the aqueous oxidation of SO_2.

7.11.2 The effects of acid precipitation

Acid precipitation has been a major environmental concern for many decades now. Many studies have been made to determine its effects, including those on:

- lakes and aquatic ecosystems;
- trees and soils;
- the atmosphere;
- architecture;
- materials;
- humans.

Effects on lakes and aquatic ecosystems

Acid precipitation can have a direct effect on lakes and the ecosystems within them. This effect can be very dramatic – between 1961 and 1971, Lumsden Lake in the Killarney region of Ontario, Canada, went from a pH reading of 6.8 to 4.4 – a two-hundredfold increase in acidity. Most acidified lakes occur at high elevations, since these lakes are usually small and located in watersheds where the rock and soil have little ability to neutralize the additional acid.

There are several routes through which acidic chemicals can enter lakes.

- They may enter water directly – as dry particles falling through the air or as wet particles falling as precipitation, including rain, snow, sleet, hail dew or fog.
- Lakes can also be thought of as the 'sinks' of the earth, whereby precipitation falling on to land is drained into them via surface run-off and groundwater. Acid precipitation falling on to the Earth washes nutrients out of soil and carries toxic metals that have been released from soil into lakes.

A particularly harmful example of the latter method by which acids can enter lakes is *spring acid shock*. When snow melts rapidly in spring due to a sudden temperature change, any acids and chemicals in the snow are released into the soils. The melted snow then runs off into streams and rivers and gradually makes its way into lakes. The introduction of these acids and chemicals causes a sudden dramatic change in the pH of the lakes – hence, the term 'acid shock'.

One effect of spring acid shock is that the aquatic ecosystem does not have sufficient time to adjust to the sudden change. To make matters worse, springtime is an especially vulnerable time for many aquatic species since this is when amphibians, fish and insects reproduce. Many of these species lay their eggs in the water to hatch. The sudden pH change is dangerous because the acids can cause serious deformities in the immature creatures or even annihilate whole species since the young of many such species spend a significant part of their life cycle in water.

Acids in water can affect the fish in lakes in two ways.

Direct effects

Acid directly interferes with the fish's ability to take in the oxygen, salt and nutrients needed to stay alive. For freshwater fish, maintaining osmoregulation is key to their survival. *Osmoregulation* is the process of maintaining the delicate balance of salts and minerals in their tissues. Acid molecules in water irritate the gills of fish, causing mucus to form that prevents the fish absorbing oxygen efficiently. If the buildup of mucus increases, the fish cannot absorb sufficient oxygen and effectively suffocate.

In addition, the low pH associated with acidified lakes affects the balance of salts in fish tissue. Salts containing essential ions such as Ca^{2+} migrate away from acid waters, depleting the calcium present in fish tissue. Decreased Ca^{2+} levels result in weak spines and deformities – for example, crayfish need Ca^{2+} to maintain a healthy exoskeleton, with low Ca^{2+} levels resulting in a weak skeleton. Low calcium also results in poor reproduction – the fish eggs produced are either too brittle or too weak to produce healthy young.

Indirect effects

Indirectly, acids cause heavy metals present in soils to be dissociated and released. For example, aluminium is one of the most abundant metals found in soils (see Chapter 4), but acid precipitation causes Al^{3+} to be released into the soils and gradually (through leaching and percolation) into lakes. Like acid molecules, Al^{3+} ions irritate the gills of fish, causing mucus to form, and accumulates in their organs to toxic levels. Both these effects can result in fish kills, which can be economically as well as ecologically significant if acidification takes place in fisheries.

In general, as acidification begins and the pH of a lake decreases, crustaceans start to die out, mainly because of reproduction problems. At a pH of 5.6, algal growth in lakes becomes hindered and some species die. As the pH falls, larger fish start to die through suffocation and reproductive problems. At pH 5, any surviving fish tend to be thin, deformed and unable to reproduce. These effects have been observed in a number of countries for many years, particularly in Scandinavia and Canada.

Limestone and/or calcium hydroxide has been added to lakes in severely affected areas of Canada and Scandinavia in an attempt to neutralize the affects of acid precipitation. However, it is an expensive process which needs to be repeated every few years in order maintain an acceptable lake pH.

Effects on trees and soils

One of the most serious impacts of acid precipitation is on forests and soils. Minerals present in soils can be washed away by percolating acidic waters (see the discussion of weathering processes in Chapter 4). Hydrogen ions from the acidified water replace important metal ions (e.g. potassium, calcium, sodium and magnesium) in soil minerals and these metal ions are transferred to soil solutions and removed as run-off or in groundwater. If this process continues for some time, the silicate structure of the soil is gradually destroyed. If soils become very acidic, aluminium from clay minerals is freed (often as $Al(H_2O)_6^{3+}$ – see Chapter 4) and this element can be absorbed by the roots of trees. Trees present in these acidified soils will gradually become starved of their vital nutrients (calcium and magnesium) and poisoned by the aluminium.

Trees can also be affected severely by gaseous sulfur dioxide and acidic aerosols, which can affect the stomata in leaves and hinder photosynthesis. Research has been carried out on red spruce seedlings by spraying them with different combinations of sulfuric and nitric acid, with pH values ranging from 2.5 to 4.5. The needles of these seedlings were observed to develop brown lesions and, eventually, the needles fell off. It was also found that new needles grew more slowly at higher concentrations of the acid used. Because the rate at which the needles were falling was greater than the rate at which they were being replenished, photosynthesis was greatly affected. The actual way in which these needles were killed is still not yet fully understood. However, studies have shown that calcium and magnesium nutrients are washed away from their binding sites when sulfuric acid enters the system. Hydrogen atoms replace the ions and this inhibits photosynthesis.

Climate and elevation impact upon the severity of forest damage. Tree damage is more severe at high elevations, in trees exposed on ridge crests or at the edge of cleared forests where stress from weather exposure aggravates the acid-induced damage. Forest damage in northern Sweden and western Britain seems to reflect severe winters, while forest damage in Germany increased rapidly after the dry summers of 1982 and 1983. When sulfur dioxide, ammonia, ozone and acid aerosols are present in the air, the frost-hardiness of trees is reduced, thus weakening them further. Ammonia from ground-level biological decay processes reacts with droplets of atmospheric sulfuric acid to form ammonium sulfate $((NH_4)_2SO_4)$, which can form on the surfaces of trees:

$$H_2SO_4(aq) + 2NH_3(g) \rightleftharpoons (NH_4)_2SO_4(aq)$$

When ammonium sulfate reaches the soils, it reacts to form both sulfuric and nitric acids, so further increasing soil acidification. These conditions also stimulate the growth of fungi and pests because trees under stress release chemicals such as terpenes, with the latter attracting pests such as the ambrosia beetle.

Nitrogen oxide and nitric oxide, also components of acid rain, can force trees to grow even though they do not have the necessary nutrients. Such trees are sometimes forced to grow well into late autumn when it is actually time for them to prepare for severe frosts in the winter.

Atmospheric effects

Some of the constituents of acid pollution – atmospheric sulfates, nitrates and acids – can contribute to haze. Atmospheric haze can reduce visibility and interfere with atmospheric energy processes by absorbing sunlight. In the Arctic, this process limits the growth of lichens, thus reducing its availability as a foodstuff for caribou and reindeer.

Effects on architecture

Acid particles and precipitation deposited on to stone buildings, monuments and statues causes corrosion and stone damage. Many historic buildings have been disfigured or damaged irreparably by acid deposition, including St Paul's Cathedral in London and the Acropolis in Athens.

Many ancient buildings and monuments are composed of limestone, sandstone and marble. Limestone ($CaCO_3$) and marble turn to a crumbling substance called gypsum (calcium sulfate) upon contact with the acid, which explains the corrosion of buildings and statues. Gypsum is readily soluble and will be washed off the stone surface by the action of rain, thus leaving a fresh surface of limestone exposed to further attack.

$$CaCO_3(s) + 2H^+(aq) \rightleftharpoons Ca^{2+}(aq) + CO_2(g) + H_2O(l)$$

In addition, building materials such as paint, plastics and steel may be damaged by acid attack. Any iron that is exposed to acid in the presence of oxygen will rapidly dissolve and wash away, so weakening the building structure and exposing fresh iron to further corrosion.

$$2Fe(s) + O_2(g) + 4H^+(aq) \rightleftharpoons 2Fe^{2+}(aq) + 2H_2O(l)$$

Consequently, metal bridges, trains, cars and aeroplanes will corrode at a faster rate in an acid atmosphere than in a 'clean' atmosphere. This is economically taxing – costly protective coatings have to be replaced with greater frequency – as well as a safety hazard to the general public; in 1967, the bridge over the Ohio River in the USA collapsed, killing 46 people, because of corrosion caused by acid rain.

Effects on materials

Acid rain may damage materials such as fabrics – there are many examples of flags being 'eaten away' by the acidic chemicals in the precipitation. Even materials inside buildings are not immune to such effects – many old books and works of art have deteriorated because the ventilation systems of the libraries and museums that hold them do not prevent acidic particles from entering the building.

Nowadays, metal water pipes are predominantly made of copper, although lead pipes still exist in older buildings. Acidic water corrodes water pipes, so leading to more frequent pipe replacement. Copper solubility increases sharply below pH 5.0 and also with increasing temperature. Metals leached from pipe walls can be consumed and bio-accumulated by humans, thus leading to detrimental health effects.

Effects on humans

There are no known direct effects of acid rain upon humans. However, occasional exposure to acidic aerosols or mists can produce detrimental effects such as dry coughs, headaches, and eye, nose and throat irritations, as well as exacerbating existing respiratory and pulmonary conditions. Small droplet sizes produce the most harmful effects.

It has been suggested that acid precipitation can have an indirect effect on the human health of populations exposed to contaminated acid water. Metals may be leached into public water supplies from pipes and soils under acidic conditions. Long-term exposure to elevated concentrations of metals may lead to bioaccumulation in tissues and subsequent detrimental health effects (e.g. kidney damage).

7.12 Water Quality

Environmental and climatic conditions and anthropogenic activities predominantly determine the quality of our water. Indicators of water quality used by environmental agencies may be subjective as well as objective and include:

- turbidity (cloudiness);
- colour;
- temperature;
- taste (for drinking water);
- odour;
- pH;
- conductivity;
- hardness;
- dissolved oxygen content (which is dependent upon temperature);
- dissolved inorganic anions (e.g. CO_3^{2-}, HCO_3^-, Cl^-, SO_4^{2-} and NO_3^-) and cations (e.g. Na^+, K^+, Mg^{2+}, Ca^{2+}, Al^{3+} and NH_4^+);
- dissolved organic substances (e.g. organic acids and phenols);
- presence or absence of microbiological organisms (e.g. bacteria);
- presence or absence of animals, algae, flora and fauna;
- presence or absence of litter and rubbish, sewage, oil, etc.

We will study the analytical determination of some of these water quality indicators in Experiments 7.1–7.4.

7.12.1 Inland waters

In many countries, the chemical and biological quality of rivers and canals is monitored regularly by appointed government agencies. Each agency uses a different approach depending upon its regulatory authority. Consequently, it is almost impossible to describe a generic approach which is suitable for every country.

In the UK, the chemical quality of river and canal water is assessed by using a classification scheme called the General Quality Assessment (GQA) Scheme which is divided into chemical, biological and aesthetic quality assessments.

Assessments using the GQA Chemical Scheme use samples of water taken on a monthly basis at intervals of 5–8 km, for approximately 40 000 km of UK rivers and canals. This chemical scheme is based on dissolved oxygen, biochemical oxygen demand, and ammonia. Stretches of water are classified into six bands: Very Good (Class A), Good (Class B), Fairly Good (Class C), Fair (Class D), Poor (Class E) and Bad (Class F), depending on their chemical water quality. The class of a specific stretch is calculated in a standard fashion throughout the UK by using the combined results of three consecutive years in order to ensure that there are sufficient data to provide a reliable assessment.

Assessments using the GQA Biological Scheme are currently based on samples of river macro-invertebrates (small animals) taken twice during the year (spring and autumn). The biological scheme is based on the number of different macro-invertebrate families (taxa) found at a sampling site, which are given weighted scores according to their tolerance, or sensitivity to pollution. Taxa that are sensitive to pollution (i.e. more likely to be affected) score more highly than insensitive taxa. The number of taxa found and the average weighted score at a site, are compared with the values expected at the site, taking into account the location, and natural chemical and physical characteristics of the river. Where the observed macro-invertebrate scores for a river are similar to those expected, the river is classified as Very Good (Class A) or Good (Class B). Where the observed macro-invertebrate scores for a river are slightly lower than those expected, the river is classified as Fairly Good (Class C) or Fair (Class D). Rivers classified as Poor (Class E) or Bad (Class F) are those where the observed macro-invertebrate scores are much less than expected.

The General Quality Assessment (GQA) for Aesthetics Scheme is used to classify the aesthetic quality of rivers. It considers the amount of litter, colour, odour and stagnation, as well as the presence of oil and foam on riverbanks. The scheme is mainly used at sites with public access and with known problems from intermittent sewerage system discharges, although 'clean' sites are often included for reference purposes.

In the USA, individual states (and certain Indian Tribes) establish water quality standards (WQSs) for inland surface waters, under the United States Federal Clean Water Act. Such WQSs define the water quality goals of a water body by designating the use or uses to be made of the water, by setting criteria necessary to protect the uses, and by preventing degradation of water quality through anti-degradation provisions. The US Environmental Protection Agency (EPA), which oversees the process, provides guidance and technical assistance to ensure that federal requirements are satisfied.

7.12.2   Coastal and marine water quality

In many countries, the quality of coastal and marine waters is vital to the tourist and fishing industries, and consequently to both local and national economies. Water quality may be controlled by limiting discharges into tidal and coastal waters to a fixed distance out to sea from certain fixed baselines on the shore. Marine pollutants include sewage (treated and untreated), marine litter, petroleum, synthetic organic compounds, metals, radioactive materials and thermal pollution. The major US sources of ocean pollutants include industrial (10%), municipal (56%), and run-off (34%).

International agreements to limit inputs of contaminants to the sea are listed in Box 7.1. Historically, these have tended to focus on accidental and intentional ship-based marine pollution, primarily involving petroleum, which can affect many countries. Under the auspices of the International Maritime Organisation (IMO) and the United Nations Environmental Program (UNEP), several marine pollution agreements have established multinational mechanisms to reduce marine pollution. However, international and regional agreements have evolved to include all possible emissions sources and special approaches to certain areas, including the North, Caribbean and Mediterranean Seas.

Box 7.1 International Agreements

International agreements to limit the inputs of contaminants to the sea include:

Global conventions

- International Convention Relating to Intervention on the High Seas in Cases of Oil Pollution Casualties (Intervention Convention), Brussels, 1969.
- International Convention on the Establishment of an International Fund for Compensation for Oil Pollution Damage 1971 (1971 Fund Convention), Brussels, 1971.
- Convention on the Prevention of Marine Pollution by Dumping of Wastes and Other Matter (London Convention 1972), London, 1972.
- International Convention for the Prevention of Pollution from Ships, 1973, as modified by the Protocol of 1978 relating thereto (MARPOL 73/78), London, 1973 and 1978.
- International Convention on Civil Liability for Oil Pollution Damage 1969 (1969 CLC), Brussels, 1969, 1976 and 1984.
- United Nations Convention on the Law of the Sea (UNCLOS), Montego Bay, 1982.
- International Convention on Oil Pollution Preparedness, Response, and Co-operation (OPRC), London, 1990.
- Convention on Liability and Compensation for Damage in Connection with the Carriage of Hazardous and Noxious Substances by Sea (HNS), London, 1996.

Regional conventions

- Convention for the Prevention of Marine Pollution by Dumping from Ships and Aircraft (Oslo Convention), Oslo, 1972.
- Convention on the Protection of the Marine Environment of the Baltic Sea Area (1974 Helsinki Convention), Helsinki, 1974.
- Convention for the Prevention of Marine Pollution from Land-based Sources (Paris Convention), Paris, 1974.
- Convention for the Protection of the Marine Environment of the North East Atlantic (OSPAR Convention), Paris, 1992.
- Convention on the Protection of the Marine Environment of the Baltic Sea Area (1992 Helsinki Convention), Helsinki, 1992.

Conventions within the UNEP Regional seas programme

- Convention for the Protection and Development of the Marine Environment and Coastal Region of the Mediterranean Sea (Barcelona Convention), Barcelona, 1976.

- Kuwait Regional Convention for Co-operation on the Protection of the Marine Environment from Pollution, Kuwait, 1978.

- Convention for the Protection of the Marine Environment and Coastal Area of the South-East Pacific, Lima, 1981.

- Regional Convention for the Conservation of the Red Sea and the Gulf of Aden Environment, Jeddah, 1982.

- Convention for the Protection and Development of the Marine Environment of the Wider Caribbean Region, Cartagena de Indias, 1983.

- Convention for the Protection, Management, and Development of the Marine and Coastal Environment of the Eastern African Region, Nairobi, 1985.

- Convention for the Protection of the Natural Resources and Environment of the South Pacific Region, Noumea, 1986.

- Convention on the Protection of the Black Sea against Pollution, Bucharest, 1992.

The quality of coastal bathing waters is regulated in Europe by the EC Bathing Water Directive (76/160/EEC). Monitoring of bathing water is undertaken in a defined bathing season and water sampling usually begins prior to the start of the season. At each monitoring site, about 20 samples are taken at regular time periods (often weekly) throughout the season. Samples are taken at predetermined points on the beach at a depth of approximately 30 cm below the surface. Additional surface samples may be taken to test for the presence of mineral oils. The samples are then analysed and assessed for compliance against the standards of the EC Directive.

The results of the various bathing water analyses are used to assist in the giving of 'awards' to bathing beaches, such as the *European 'Blue' Flag Scheme*. In this Blue Flags are awarded to beaches that fulfil specific criteria in three areas. These include environmental education and information, beach area management and safety, and the quality of the bathing water, which must be in accordance with the EC Directive. The bathing water quality in the previous year needs to have complied with the *guideline standards* of the EC Directive for indicators of sewage pollution such as total coliforms, faecal coliforms and faecal streptococci. These require that:

- At least 80% of the samples (i.e. 16 or more out of 20) must meet the standards of not more than 500 total coliforms and 100 faecal coliforms per 100 ml of water.

- At least 90% of the samples (18 or more out of 20) must meet the standard of not more than 100 faecal streptococci per 100 ml of water.

In the USA, coastal water quality is controlled primarily through state-run regulatory programme administered by the US EPA under the Clean Water Act (CWA).

7.13 Summary

In this chapter, we have studied the physical and chemical properties of water, as well as many of the essential chemical principles necessary for an understanding of its behaviour. We have discussed our potential for causing water pollution and described some of the detrimental effects of contaminated water. In addition, we have identified indicators of water quality that may be used to help us to maintain Earth's most valuable resource for future generations. Your practical skills will be further developed by attempting Experiments 7.1–7.4, in which selected indicators of water quality can be determined.

Experiment 7.1 Determination of Hardness in Water

Introduction

Environmental chemists use the *hardness index* to measure the concentration of Ca^{2+} and Mg^{2+} ions in water. Chemically, the hardness index is defined as the sum of the concentrations of Ca^{2+} and Mg^{2+}, i.e. $[Ca^{2+}] + [Mg^{2+}]$. Hardness is traditionally expressed as the mass (in milligrams) per litre of water of calcium carbonate that contains the same number of dipositive (2+) ions.

The most common type of hardness in water is *temporary hardness*, caused by the presence of dissolved calcium and/or magnesium compounds. The main natural source of these salts is from rocks containing calcium or magnesium carbonates, such as limestone ($CaCO_3$) and dolomite ($CaCO_3 \cdot MgCO_3$).

Calcium carbonate is insoluble in water but reacts with carbon dioxide (which is easily soluble in water and is in plentiful supply from the atmosphere) to form calcium hydrogen carbonate (or bicarbonate), which is soluble in water. The chemical reaction that takes place is:

$$CaCO_3(s) + CO_2(g) + H_2O(l) \rightleftharpoons Ca(HCO_3)_2(aq)$$

These dissolved salts cause the hardness of water. The reaction is reversible; on heating water that contains bicarbonates of calcium and magnesium, the bicarbonates are precipitated as insoluble carbonates as the water evaporates:

$$Ca(HCO_3)_2(aq) \xrightarrow{\text{heat}} CaCO_3(s) + CO_2(g) + H_2O(g)$$

A similar reaction occurs if bicarbonates of magnesium are present in the water. Consequently, temporary hardness may be removed simply by boiling. This process is responsible for the deposition of scale (insoluble carbonates) in boilers, water pipes and kettles during the heating of water.

A less common type of hardness in fresh waters is *permanent hardness*, due to the presence of calcium and magnesium sulfates and chlorides. These salts are not precipitated when water is boiled. Hard water can be softened by various methods, including the use of ion exchangers.

As calcium and magnesium ions react with ethylenediamminetetraacetic acid (EDTA) to form strong complexes, their concentration in water can be determined by titrating a known volume of water with a standard EDTA solution using Eriochrome Black T (EBT) indicator to show the endpoint of the reaction. The EDTA structure is:

$$^-O_2C\text{-}H_2C \diagdown \qquad\qquad\qquad CH_2\text{-}CO_2^-$$
$$N\text{-}CH_2\text{-}CH_2\text{-}N$$
$$^-O_2C\text{-}H_2C \diagup \qquad\qquad\qquad CH_2\text{-}CO_2^-$$

At the start of the reaction, the EBT indicator forms complexes with the cations (Ca^{2+} and Mg^{2+}) and creates a red coloured solution. As EDTA is added, this forms new complexes with Ca^{2+} and Mg^{2+} because the affinity between EDTA and the cations is bigger than that between EBT and the cations. When all of the cations are linked to the EDTA, the indicator changes from red to blue. Consequently, the hardness of water can be calculated by measuring the amount of EDTA consumed by a water sample and relating this to the concentrations of dissolved calcium and magnesium ions.

Time required

Two–three hours, depending upon the number of water samples determined.

Equipment

250 cm^3 conical flasks
100 cm^3 pipettes and pipette filler
50 cm^3 burette
10 cm^3 measuring cylinders
Boss, clamp, retort stand and white tile
Bunsen burner, tripod stand and metal gauze
Tongs

Reagents

Ammonium hydroxide/ammonium chloride (NH_4OH/NH_4Cl) buffer at
    pH 10
0.01 M ethylenediamminetetraacetic acid (EDTA)
Eriochrome Black T indicator
1.00 M hydrochloric acid (HCl)
Suitable water samples (e.g. tap, lake or river waters)

Significant experimental hazards

- Students should be aware of the hazards associated with the use of all glassware (cuts), Bunsen burners (burns and fire), tongs to move hot objects (burns and dropping hazards) and fume cupboards.

- Hydrochloric acid is harmful by ingestion and may cause severe burns to eyes and skin.

- Eriochrome Black T indicator may be harmful if ingested in quantity and may irritate eyes and skin. It may also stain clothing and skin.

Procedure

(1) Pipette 100 cm^3 of tap water into a 250 cm^3 conical flask.

(2) In a fume cupboard, slowly and carefully add 2 cm^3 of dilute hydrochloric acid to the tap water sample. Boil the solution by using a Bunsen burner on a tripod stand and gauze for 2–3 minutes. This will remove any dissolved carbon dioxide.

(3) Cool the solution to room temperature by holding the flask with tongs under running cold water (indicator composition becomes a problem if the solution is hot).

(4) Add 5 cm^3 of NH_4Cl/NH_4OH buffer and four drops of Eriochrome Black T indicator (you should add two drops per 50 cm^3 to be titrated).

(5) Titrate the solution with 0.01 M EDTA until the indicator colour changes from red to pure blue.

(6) Repeat this rough titration at least twice more, or until two concordant results are obtained.

(7) Carry out the same determination by using lake water and river water instead of tap water. (You may, of course, substitute other appropriate waters.)

Calculation

EDTA reacts with calcium and magnesium in a ratio of 1:1.

$$1 \text{ mole EDTA} = 1 \text{ mole Ca}^{2+} = 1 \text{ mole Mg}^{2+}$$

The total hardness, which is due to the dissolved calcium and magnesium ions, is normally reported as the concentration of calcium carbonate in water, $[CaCO_3]$ in mg dm^{-3}. (For example, a water sample (RIV1) that contains 0.0006 moles of Ca^{2+} and 0.0004 moles of Mg^{2+}, has a total of 0.0010 moles of Ca^{2+} plus Mg^{2+}. Since the molar mass $(CaCO_3) = 100$ g mol^{-1}, 0.0010 moles of calcium carbonate weigh 0.1 g, or 100 mg. RIV1 therefore has a hardness value of 100 mg dm^{-3}.)

Thus:

$$1 \text{ cm}^3 \text{ 0.01 M EDTA} = 1 \times 10^{-5} \text{ mole} = 1 \times 10^{-5} \times 100 \text{ g CaCO}_3$$

$$= 1 \text{ mg CaCO}_3$$

Use this relationship to determine the hardness in each of your water samples and comment on the results obtained. The verbal descriptions of water hardness usually associated with values of $CaCO_3$ in mg dm^{-3} are listed in Table 7.11.

Issues to consider for your practical report

- What are the potential sources of error in this experiment? How could they be overcome?

- Are there alternative methods for determining the hardness of water? If so, how do they compare to this method?

- What is the source of the unknown river and lake water samples?

- What are typical water hardness values in rivers, streams, lakes and drinking water in your local area? How do your data compare with these values?

- How can permanent hardness be removed for domestic and industrial purposes?

- What are the legal limits (if any) of hardness in drinking water? Do your values exceed such legal limits?

- What are the potential human health and environmental effects (if any) of excess hardness in potable water?

Useful references

Greenberg, A.E., Clesceri, L.S. and Eaton, A.D. (Eds), (1992). Standard Methods for the Examination of Water and Wastewater. American Public Health Association, Washington, DC, USA. ISBN 0-87553-207-1.

Betz, J.D. and Noll, C.A. (1950). 'Total hardness determination by direct colorimetric determination'. *Journal of the American Water Works Association*, **42**, 49.

Experiment 7.2 Investigation of the Carbonate–Bicarbonate System (Alkalinity of Aqueous Systems)

Introduction

An aqueous solution of carbon dioxide produces a mixture of carbonate and bicarbonate ions. Determining the carbonate and bicarbonate ions in each other's presence is often important in environmental chemistry.

(a) $CO_2(g) + H_2O(l) \rightleftharpoons H_2CO_3(aq)$

(b) $H_2CO_3(aq) \rightleftharpoons HCO_3^-(aq) + H^+(aq)$

(c) $HCO_3^-(aq) \rightleftharpoons H^+(aq) + CO_3^{2-}(aq)$

The *alkalinity* of water is its capacity to act as a base by reacting with protons. In practice, it may be determined by measuring the number of moles of H^+ required to titrate one litre of a water sample to the end point. Alkalinity is therefore a useful measure of the capacity of water to resist acidification from acid precipitation. The presence of carbonate, bicarbonate and hydroxide ions imparts the alkalinity of natural or treated waters. In natural waters, the hydroxide ions are generally much less important than the other two ions and they will be neglected in the calculations associated with this experiment.

Initially, your water sample will contain both carbonate and bicarbonate ions and will be alkaline. The addition of acid to the water sample will convert the carbonate to bicarbonate (reaction(c)) until no more carbonate remains. The addition of further acid will convert the bicarbonate to carbonic acid until no more bicarbonate remains (reaction(b)). The carbonate and carbonic acid equivalence points may be determined either by titration using indicators or by pH titration. Both methods will be employed in this experiment.

Phenolphthalein, an indicator that produces a colour change in the pH range 8.3–10, enables the measurement of the alkalinity fraction contributed by the carbonate ion. The end point determined with this indicator represents the *completion* (equivalence point or stoichiometric end point) of the following reaction:

$$CO_3^{2-}(aq) + H^+(aq) \longrightarrow HCO_3^-(aq)$$

i.e. half of the carbonate has been neutralized by the acid-forming bicarbonate ions. This is sometimes referred to as the *phenolphthalein alkalinity*.

$$\text{phenolphthalein alkalinity} = [CO_3^{2-}]$$

Methyl orange indicator responds in the pH range 3.2–4.5, and may therefore be used to assess the alkalinity due to carbonate and bicarbonate. At this pH, all of the bicarbonate ions initially present in the water sample, together with all of those produced from the half-reaction of the carbonate ions, will be neutralized. The resulting alkalinity is known as the *total alkalinity*.

$$HCO_3^-(aq) + H^+(aq) \longrightarrow H_2O(l) + CO_2(g)$$

$$\text{(Estimated) total alkalinity} = 2[CO_3^{2-}] + [HCO_3^-]$$

The importance of the carbonate/bicarbonate system in natural waters stems from their ability to act as buffers in natural waters. The oceans are described as being buffered since relatively large quantities of acid or base can be added to seawater without causing much change to its pH. However, many freshwater lakes do not have a large buffer capacity and consequently a small addition of acid (e.g. from acid precipitation or industrial effluent) can cause large changes in pH without warning. The phenolphthalein alkalinity and the total alkalinity are useful for the calculation of chemical dosages required in the treatment of natural water supplies.

Time required

Two–three hours for each part ((a) and (b)) of the experiment.

Equipment

Part (a) Titration with indicators
100 cm^3 conical flasks
25 cm^3 pipettes and pipette filler
50 cm^3 burette
Boss, clamp, retort stand and white tile

Part (b) pH Titration
250 cm^3 beakers
25 cm^3 pipettes
50 cm^3 burette
Boss, clamp and retort stand
pH meter
Magnetic stirrer

Reagents

Part (a) Titration with Indicators
   Carbonate–bicarbonate mixture, either a natural water sample or a synthetic mixture. (An effective synthetic mixture can be made by dissolving 4.110 ± 0.001 g of sodium carbonate in a small amount of distilled water and then making up to 1 l in a volumetric flask. Similarly, dissolve 2.250 ± 0.001 g of anhydrous sodium bicarbonate in distilled water and make up to 1 l. Mixing these two solutions gives 2 l of a carbonate–bicarbonate mixture.)

Phenolphthalein indicator (2%)
Methyl orange indicator
0.100 M hydrochloric acid (HCl)

Part (b) pH Titration
Carbonate–bicarbonate mixture (as in part (a))
0.100 M hydrochloric acid (HCl)
Buffer solution (pH 9.2)
Double-distilled, deionized water

Significant experimental hazards

Part (a) Titration with indicators
- Students should be aware of the hazards associated with the use of all glassware (cuts).
- Hydrochloric acid is harmful by ingestion and may cause severe burns to eyes and skin.
- Phenolphthalein indicator may be harmful if ingested in quantity as it has a strong purgative effect. It may irritate eyes and skin and cause stains to clothing and skin.
- Methyl orange indicator may be harmful if ingested in quantity and may irritate eyes and skin. It may also stain clothing and skin.

Part (b) pH Titration
- Students should be aware of the hazards associated with the use of all glassware (cuts) and electrical equipment (shock and burns).
- Hydrochloric acid is harmful by ingestion and may cause severe burns to eyes and skin.
- Alkaline buffer solution (pH 9.2) may be mildly irritant to eyes and skin.

Procedure

Part (a) Titration with indicators
1) Pipette 25.0 cm^3 of the carbonate–bicarbonate mixture into a 100 cm^3 conical flask.
2) Add a few drops of phenolphthalein indicator.
3) Titrate the solution with 0.100 M hydrochloric acid until the indicator colour changes from pink to colourless. The volume of acid used (z cm^3) corresponds to half of the carbonate present.
4) Repeat this rough titration at least twice more, or until two concordant results are obtained.
5) Pipette 25.0 cm^3 of the carbonate–bicarbonate mixture into a 100 cm^3 conical flask.
6) Add a few drops of methyl orange indicator.
7) Titrate the solution with 0.100 M hydrochloric acid until the indicator colour changes from yellow to red. The volume of acid used (y cm^3) corresponds to the sum of the carbonate and bicarbonate present

in the aqueous solution. The bicarbonate is therefore equivalent to $(y - 2z)$ cm^3 of acid.

8) Repeat this rough titration at least twice more, or until two concordant results are obtained.

Part (b) pH Titration

(1) Calibrate the pH meter using a buffer solution (pH 9.2).

(2) Pipette 25.0 cm^3 of the carbonate–bicarbonate mixture into a 250 cm^3 beaker.

(3) Place the beaker on the base of a magnetic stirrer and drop a stirrer bar carefully into the beaker.

(4) Insert the electrodes of the pH meter into the beaker.

(5) Add distilled water, if necessary, to ensure complete coverage of the electrodes. It is essential that adequate clearance is achieved between the electrodes and the magnetic stirrer or the stirrer bar will not rotate.

(6) Place a burette (previously filled with 0.100 M hydrochloric acid) over the apparatus so that acid can be run slowly into the beaker. Ensure that you have sufficient room to turn the tap of the burette freely.

(7) Record the burette reading and start the magnetic stirrer.

(8) Add approximately 1.0 cm^3 of acid from the burette while the mixture is stirred.

(9) Measure and record the pH of the solution after addition of each 1.0 cm^3 of acid. You should record the following information in your logbook at each stage:

- pH;
- volume of acid added (cm^3);
- change in pH after addition of acid (δpH);
- total volume of acid added (cm^3).

(10) The shape of the pH curve we would expect has been discussed previously in Chapter 6. Two end points should appear on the curve, representing the bicarbonate and total alkalinity respectively. You should see an increase in pH change per volume of acid added as you approach an end point. When this happens, take measurements at more frequent intervals (0.3, 0.2, 0.1 cm^3 etc.).

(11) For an accurate determination of the end points, plot the change in pH (δpH) divided by the change in volume (δvol) (i.e. (δpH)/(δvol)), on the y-axis, against the volume of 0.100 M HCl added, on the x-axis.

(12) From your graph, determine the amount of carbonate and bicarbonate present.

(13) Compare your results for part (b) with those obtained for part (a) of the experiment.

Calculation

Carbonate reacts with hydrochloric acid in a ratio of 1:2.

$$CO_3{}^{2-}(aq) + 2HCl(aq) \longrightarrow H_2O(l) + CO_2(g) + 2Cl^-(aq)$$

1 mole carbonate = 2 moles hydrochloric acid

We can use this relationship to determine the carbonate levels in the water sample, $[CO_3{}^{2-}]$ in mg dm^{-3}.

1 cm^3 of 0.1 M HCl = 1×10^{-4} mole HCl = $1/2 \times 10^{-4}$ mole$CO_3{}^{2-}$

Relative molecular mass of $CO_3{}^{2-} = 60$

$$1/2 \times 10^{-4} \times 60 = 3 \times 10^{-3} \text{ g } CO_3{}^{2-} = 3 \text{ mg } CO_3{}^{2-}$$

Consequently, z cm^3 of 0.1 M HCl = $3z$ mg $CO_3{}^{2-}$

Bicarbonate reacts with hydrochloric acid in a ratio of 1 to 1:

$$HCO_3{}^-(aq) + HCl(aq) \longrightarrow H_2O(l) + CO_2(g) + Cl^-(aq)$$

1 mole bicarbonate = 1 mole hydrochloric acid

We can use this relationship to determine the bicarbonate levels in the water sample, $[HCO_3{}^-]$ in mg dm^{-3}.

1 cm^3 of 0.1 M HCl = 1×10^{-4} mole HCl = 1×10^{-4} mole HCO$_3{}^-$

Relative molecular mass of $HCO_3{}^- = 61$

$$1 \times 10^{-4} \times 61 = 6.1 \times 10^{-3} \text{ g } HCO_3{}^- = 6.1 \text{ mg } HCO_3{}^-$$

Consequently, $(y - 2z)$ cm^3 of 0.1 M HCl = $6.1(y - 2z)$ mg HCO$_3{}^-$

Issues to consider for your practical report

- Identify other ions that could make a minor contribution to the alkalinity of natural waters.

- What are the potential sources of error in this experiment? How could they be overcome?

- How do the two methods you have used to determine the alkalinity of water compare?

- Are there other methods for determining the alkalinity of water?

- What are typical phenolphthalein and total alkalinity values for rivers, streams, lakes and drinking water in your local area? How do your data compare with these values?

- What are the legal limits (if any) of alkalinity in drinking water? Do your values exceed such legal limits?

- Explain, in detail, the importance of the buffering capacity of natural waters. Make sure you refer to the scientific literature (journals and periodicals) as well as textbooks.

- How can the pH of natural waters be restored once their buffering capacity has been exceeded and they have become acidified?

Useful References

Greenberg, A.E., Clesceri, L.S. and Eaton, A.D. (Eds), (1992). *Standard Methods for the Examination of Water and Wastewater*. American Public Health Association, Washington, DC, USA. ISBN 0-87553-207-1.

Thomas, J.F.J. and Lynch, J.J. (1960). Determination of carbonate alkalinity in natural waters. *Journal of the American Works Association*, **52**, 259.

Experiment 7.3 Cation Concentration of River Water by Ion Exchange

Introduction

Ion exchange resins are complex organic molecules, insoluble in water, which possess ionizable atoms or groups. Those possessing replaceable cations, such as sodium or hydrogen, are called cation exchangers, whereas those possessing replaceable anions are called anion exchangers.

When solutions of electrolytes are passed through a column of ion exchanger, the following reactions take place:

Cation exchanger

$$R-SO_3^-H^+ + M^+ \longrightarrow RSO_3^- + H^+ (R = \text{organic part of the molecule})$$

Anion exchanger

$$2R-(CH_3)_3N^+Cl^- + SO_4^{2-} \longrightarrow R - [(CH_3)_3N^+]_2SO_4 + 2Cl^-$$

The above equations show that the cation exchanger releases hydrogen ions and the anion exchanger releases chloride ions. Ion exchangers can be used for the determination of the salt concentration of solutions and commercially for the desalination of water.

In this experiment, the total cation concentration of river water will be determined by passing a sample through a cation exchanger in the hydrogen form. During this procedure, all of the cations will be replaced by hydrogen ions, i.e. the salts dissolved in the water sample will be converted to the corresponding mineral acids. The effluent from the ion-exchange column is titrated with dilute sodium hydroxide in order to estimate the total cation content of the sample.

Silicate minerals and soils also show ion-exchange properties, which is usually expressed as their cation-exchange capacity in milliequivalents (meq) per 100 g of material (see Chapter 4).

Time Required

Three hours per water sample, although the time taken depends upon whether you repeat your determination using a fresh ion exchanger or you regenerate the resin between samples. You will probably require an extra hour to properly regenerate the ion exchanger.

Equipment

50 and 100 cm^3 beakers
50 cm^3 pipettes and pipette filler
50 cm^3 burette
250 cm^3 conical flasks
Boss, clamp, retort stand and white tile
Cation-exchange resin in glass column, held by using a retort stand and clamp
Glass funnel
Wristwatch or stopwatch

Reagents

Suitable water sample(s)
0.02 M sodium hydroxide (NaOH)
Double-distilled, deionized water
Methyl orange indicator
Silver nitrate solution (as indicator)
Suitable acidic solution (if regeneration of the ion-exchange column is performed)

Significant experimental hazards

- Students should be aware of the hazards associated with the use of all glassware (cuts) and ion exchange columns (tipping).

- Acidic cation-exchange resin and effluent is harmful by ingestion and may cause irritation and burns to eyes and skin.

- Sodium hydroxide solution is harmful by ingestion and may cause irritation and burns to eyes and skin.

- Methyl orange indicator may be harmful if ingested in quantity and may irritate eyes and skin. It may also stain clothing and skin.
- Silver nitrate solution is harmful by ingestion and corrosive to eyes and skin. Prolonged skin contact may cause blackening of the skin and argyria.

Procedure

(1) Your apparatus will consist of an ion-exchange resin packed into a glass column held upright by a clamp attached to a retort stand. It is vital that the ion-exchange resin does not dry out – ensure the water level is always above the upper surface of the resin in the column and that the tap is not leaking.

(2) Wash the resin with double-distilled, deionized water and collect the effluent from the bottom of the column in a 50 cm^3 beaker.

(3) Test the effluent for chloride ions by adding some silver nitrate to the beaker. If the effluent turns cloudy, chloride ions are present and the ion-exchange resin requires further washing. Continue this process until the effluent is free of chloride, i.e. the tested effluent remains clear.

(4) Pipette 50.0 cm^3 of river water into a 100 cm^3 beaker. (You may, of course, substitute other appropriate waters.)

(5) Pour some of the water sample into the top of the column via a glass funnel. Make sure that the column does not overflow or you will lose some of your sample.

(6) Place a clean 250 cm^3 conical flask under the tap of the column.

(7) Adjust the tap to allow the water sample to flow through the column at a rate of about 1–2 cm^3 per minute (you will need a wristwatch or stopwatch to do this). This will clearly be accomplished by trial and error, but it is better to start with a slow flow rate and gradually speed up than to start too quickly. If the flow rate is too fast, the ions in the water sample may not have enough time to exchange properly with the resin.

(8) Add your water sample to the top of the column bit-by-bit, ensuring that the resin never dries out, until all of your water sample is used. Follow this by passing 20–30 cm^3 of double-distilled, deionized water through the column at the same flow rate. Collect the water in the same conical flask.

(9) Titrate the contents of the conical flask with 0.02 M sodium hydroxide by using methyl orange as an indicator. Take great care as you get close to the end point (red $\rightarrow$ yellow) – exchanging the ions properly with the resin is very time consuming and you probably won't have time to perform three determinations. At the end point, the volume of 0.02 M sodium hydroxide used $= z$ cm^3.

(10) Repeat the determination with another 50.0 cm^3 of river water by using a fresh ion exchanger. (Alternatively, you can regenerate the old column by flushing with an appropriate acidic solution. This is obviously more time consuming, but it will help you to improve your practical skills.)

(11) Calculate the total cation concentration of the river water, expressing your results as mg CaCO$_3$ dm^{-3} of water.

Calculation

By convention, we will be expressing our results as mg CaCO$_3$ per litre of water. Each Ca^{2+} ion from the water sample will liberate two H$^+$ ions from the cation-exchange column (in order to balance the charge). Thus:

$$2 \text{ moles H}^+ = 1 \text{ mole Ca}^{2+}$$
$$1 \text{ mole H}^+ = 1/2 \text{ mole Ca}^{2+}$$

The hydrogen ions liberated from the cation exchanger and collected in the conical flask react with sodium hydroxide in a ratio of 1 : 1.

$$\text{H}^+(\text{aq}) + \text{NaOH(aq)} \longrightarrow \text{H}_2\text{O(l)} + \text{Na}^+(\text{aq})$$
$$1 \text{ mole H}^+ = 1 \text{ mole NaOH}$$
$$\therefore \quad 1 \text{ mole H}^+ = 1 \text{ mole NaOH} = 1/2 \text{ mole Ca}^{2+}$$

We can use this relationship to determine the total cation concentration of the water sample, [CaCO$_3$] in mg dm^{-3}.

$$1 \text{ cm}^3 \, 0.02 \text{ M NaOH contains } 2 \times 10^{-5} \text{ moles}$$
$$\therefore 2 \times 10^{-5} \text{ moles NaOH} = 1 \times 10^{-5} \text{ moles CaCO}_3$$

Molar mass (CaCO$_3$) = 100 g mol^{-1}

$$\therefore 1 \times 10^{-5} \text{ moles} \times 100 \text{ g} = 1 \times 10^3 \text{ g CaCO}_3 = 1 \text{ mg CaCO}_3$$
$$\Rightarrow 1 \text{ cm}^3 \, 0.02 \text{ M NaOH} = 1 \text{ mg CaCO}_3$$
$$\Rightarrow z \text{ cm}^3 \, 0.02 \text{ M NaOH} = z \text{ mg CaCO}_3 (\text{per 50 cm}^3 \text{ of water sample})$$

Thus, the total cation concentration of the water sample $= 20\,z$ mg CaCO$_3$ dm^{-3}.

Issues to consider for your practical report

- What are the potential sources of error in this experiment? How could they be overcome?

- Do you think that this experiment could be automated? If your answer is yes, explain how you would do this, outlining the main technical difficulties.

- Are there any other methods for determining the total cation concentration of natural waters?

- What are typical total cation concentration values for rivers, streams, lakes and drinking water in your local area? How do your data compare with these values?

- How does your total cation concentration compare with the hardness of water value you obtained?

- Identify the main cations present in natural freshwaters. What are their typical concentration ranges in freshwaters? Make sure you refer to the scientific literature (journals and periodicals) as well as textbooks.

- What are the legal limits (if any) of cations in drinking water? Do your values exceed such legal limits?

- What are the potential human health and environmental effects (if any) of excess cations in potable and natural waters?

Experiment 7.4 Methods of Chloride Determination

Introduction

Chloride (Cl^-) is one of the major inorganic ions in water and wastewaters. In potable water, the saltiness is variable and dependent upon both the concentration of chloride ions and the presence of other ions. Waters containing 250 mg Cl^- dm^{-3} can taste salty if sodium ions are also present, and yet waters containing 1000 mg Cl^- dm^{-3} may not taste salty if the predominant cations present are calcium and magnesium.

There are many potential sources for the chloride in natural and drinking waters and consequently its determination is important. We will study two widely used methods, i.e. Mohr's Titration and the use of ion-selective electrodes.

In Mohr's Titration, chloride is titrated with silver nitrate in neutral or slightly alkaline solution using potassium chromate as an end-point indicator. Silver chloride is quantitatively precipitated before red silver chromate is formed. However, the reaction is potentially subject to interference that may limit its accuracy. Interference from bromide, iodide and cyanide is not important as they register as equivalent chloride concentrations. Sulfide, thiosulfate and sulfite ions interfere with the reaction but may be removed. Sulfite can be removed by treatment with hydrogen peroxide (H_2O_2) in a neutral solution, while sulfide and thiosulfate can be removed by treatment with H_2O_2 in alkaline solution. The presence of selected other ions should be recognized as limiting the accuracy of the determination. Orthophosphate in excess of 25 mg dm^{-3} interferes by precipitation as silver phosphate; iron in excess of 10 mg dm^{-3} will conceal the end point.

Ion-selective electrodes respond to specific anions and cations and are a development of the long established pH glass electrode. The chloride electrode is usually a solid-state membrane electrode whose sensing element is a single crystal of pure silver chloride, although other types are available. The solid-state electrode gives a very good response towards Cl^- ions down to the level of solubility of AgCl, which is 10^{-5} M.

Time Required

Three hours for each part ((a) and (b)) of the experiment.

Equipment

Part (a) Mohr's titration
50 cm^3 conical flasks
25 cm^3 pipettes and pipette filler
50 cm^3 burette
Boss, clamp, retort stand and white tile

Part (b) Ion-Selective electrodes
Chloride-ion-selective electrode
50 cm^3 volumetric flasks
Pipettes of various sizes (e.g. 1–25 cm^3) and pipette filler
25 cm^3 plastic beakers and a rinsing beaker
Magnetic stirrer
Five-cycle semi-logarithmic graph paper

Reagents

Part (a) Mohr's titration
0.0141 M sodium chloride (NaCl) solution
0.0141 M silver nitrate ($AgNO_3$) solution
10% potassium chromate ($K_2Cr_2O_7$) solution (as indicator) in a dispenser
Suitable water sample(s)

Part (b) Ion-Selective electrodes
1.0 M sodium chloride (NaCl) solution
Double-distilled, deionized water
Suitable water sample(s)

Significant experimental hazards

- Students should be aware of the hazards associated with the use of all glassware (cuts) and electrical equipment (shocks and burns).

- Hydrochloric acid is harmful by ingestion and may cause severe burns to eyes and skin.

- Potassium chromate indicator may be harmful if ingested in quantity and may irritate eyes and skin.

Procedure

Part (a) Mohr's titration

(1) Pipette 25.0 cm^3 of the water sample into a 50 cm^3 conical flask.

(2) Add 0.25 cm^3 of potassium chromate indicator from a dispenser. You must take great care to add the correct amount of indicator. If you add too much, silver nitrate will precipitate before the end point is reached; if you add too little, formation of silver chloride is less distinct, thus making it difficult to see the end point.

(3) Titrate the water sample and indicator with 0.0141 M silver nitrate solution. You should add the silver nitrate very slowly from the burette until the red colour formed by each addition begins to disappear more slowly. This indicates that most of the chloride has been precipitated. Continue the titration until a faint but distinct colour change occurs (z cm^3). This faint reddish-brown colour should persist after brisk shaking.

(4) Repeat this rough titration at least twice more, or until two concordant results are obtained.

Part (b) Ion-Selective electrodes

(1) Prepare a series of standard solutions of concentration 10^{-1} M, 10^{-2} M, 10^{-3} M, 10^{-4} M and 10^{-5} M (blank) by using the 1.0 M sodium chloride solution provided. (The 1.0 M solution contains 35.45 g dm^{-3} of chloride.)

(2) Starting with the most dilute solution (why?), place 10 cm^3 of the standard solution in a plastic beaker containing a magnetic stirrer bar. Immerse both the chloride and reference electrodes in the beaker, ensuring that there is sufficient clearance between the stirrer bar and the bottom of the electrodes for the stirrer to rotate unimpeded.

(3) Allow about one minute for the reading to stabilize before recording your measurement. The reading will stabilize more slowly for the more concentrated solutions.

(4) Obtain measurements for all five standard solutions and plot a calibration graph of millivolt reading against chloride concentration by using the five-cycle semi-logarithmic graph paper. Between measurements you should thoroughly rinse both electrodes with double-distilled, deionized water.

(5) Once you have prepared an adequate calibration curve, you should obtain a reading for each of the unknown water samples provided and calculate the chloride concentration by using the calibration

curve. Each unknown sample should be measured several times to test the standard deviation of the method. The calibration curve is only accurate over the concentration range for which it has been prepared and therefore any unknown samples that give readings outside this range should be diluted by a known factor. If the reading is lower than the calibrated range, the chloride concentration in the sample is below the detection limit of the instrument.

Calculation

Part (a) Mohr's titration

The precipitation reaction is described by the equation:

$$AgNO_3(aq) + NaCl(aq) \longrightarrow AgCl(s) + NaNO_3(aq)$$

1 mole silver nitrate = 1 mole sodium chloride

We can use this relationship to determine the chloride levels in the water sample, $[Cl^-]$ in mg dm^{-3}. If the concentration of $AgNO_3$ is exactly 0.0141 M, then:

$$1 \text{ cm}^3 \text{ of } 0.0141 \text{ M } AgNO_3 \text{ contains } (1 \times 0.0141)/1000 \text{ moles}$$

At the end point, this would also be the amount of NaCl present.
Relative molecular mass of NaCl = 58.45
If a = the mass of NaCl present in the water sample (in g), then:

$$\frac{a}{58.45} = \frac{1 \times 0.0141}{1000}$$

and therefore:

$$a = 0.824 \times 10^{-3} \text{ g} = 0.824 \text{ mg}$$

Therefore, the mass of chloride present in the sample $= 0.824 \times \dfrac{35.45}{58.45} = 0.500$ mg

Consequently, 1 cm^3 of 0.0141 M $AgNO_3$ = 0.500 mg Cl^-

$\Rightarrow z$ cm^3 0.0141 M $AgNO_3$ = 0.500 z mg Cl^- (per 25 cm^3 of water sample).

Thus, the chloride concentration of the water sample = 20z mg Cl^- dm^{-3}.

Part (b)

The chloride concentration should be estimated by using the calibration graph (see Chapter 3)

Issues to consider for your practical report

- What are the potential sources of error in each experiment? How could they be overcome?

- Identify alternative methods for determining chloride concentrations in water samples. What are their advantages/disadvantages/detection limits?

- Identify the advantages and disadvantages of using ion-selective electrodes for analytical determinations.

- What are typical chloride concentrations for rivers, streams, lakes and drinking water in your local area? How do your data compare with these values?

- Identify the main sources of chloride in natural waters.

- Explain why chloride concentrations in surface waters are subject to seasonal fluctuations.

- What are the legal limits (if any) of chloride in drinking water? Do your values exceed such legal limits?

- What are the potential human health and environmental effects (if any) of excess chloride in potable and natural waters?

Useful references

Greenberg, A.E., Clesceri, L.S. and Eaton, A.D. (Eds), (1992). Standard Methods for the Examination of Water and Wastewater. American Public Health Association Washington, DC, USA. ISBN 0-87553-207-1.

ISO Compendia on the Environment: Water Quality. Volume 2 Chemical Methods (1992). 'Water quality – Determination of chloride – Silver nitrate titration with chromate indicator (Mohr's Method), ISO9297:1989(E). International Organization for Standardization (ISO), Geneva, Switzerland. ISBN 92-67-10193-5.

Self-Study Exercises

Introduction to Water

7.1   What is the unique chemical feature of water?

7.2   Explain why water is essential for many biological purposes.

7.3   List the main uses of water.

Types of Water

7.4   Define the following terms:

(a)   groundwater;

(b)   zone of aeration;

(c) subsidence;

(d) urban run-off;

(e) aquifer;

(f) water table.

7.5 Outline the main differences between surface water and groundwater.

7.6 Identify the environmental problems caused by the large-scale extraction of ground-water.

Water cycle

7.7 Draw a schematic diagram of the water cycle.

7.8 Outline the energy processes involved in the water cycle.

7.9 Explain how water can be temporarily removed from the water cycle.

Properties of water

7.10 Explain why water is known as the 'universal solvent'.

7.11 Which would be the better solvent, H_2O or CH_2Cl_2 (dichloromethane), for each of the following:

(a) NaCl;

(b) NH_3;

(c) C_6H_6 (benzene);

(d) CH_3OH (methanol);

(e) HCl;

(f) I_2?

7.12 Decide whether the following statements are true or false, giving reason(s) for your answers.

(a) Water has a low dielectric constant.

(b) The larger the force between two ions, then the smaller the dielectric constant.

(c) Water has a high dipole moment because the O–H bonds are very polar.

(d) H-bonds are relatively weak.

(e) Water contracts on freezing.

(f) When ice melts, its density increases.

(g) H-bonds contribute to the high surface tension of water.

(h) Water will rise up the walls of a glass tube.

(i) Pure water will not conduct electricity.

(j) Water has acidic but not basic properties.

Properties of solutions

7.13   Decide whether the following statements are true or false, giving reason(s) for your answers.

   (a)   The solubility of all aqueous solutions increases with temperature.

   (b)   Sodium carbonate is insoluble in water.

   (c)   The lattice energy is the energy released when solute ions interact with solvent molecules.

   (d)   Colligative properties depend upon the nature of the solute.

   (e)   A solution will have a lower escape rate for solvent particles than the pure solvent.

   (f)   Solute particles in solution slow the freezing rate.

   (g)   An aqueous salt solution boils at a lower temperature than pure water.

   (h)   Hypertonic drinks are less concentrated than body fluids.

   (i)   In reverse osmosis, solvent molecules pass through a semi-permeable membrane from a more concentrated to a more dilute solution.

7.14   Calculate the mole fractions of each component in the following solutions:

   (a)   20.0 g of water and 40.0 g of methanol, CH_3OH.

   (b)   A sucrose solution that is 0.025 M $C_{12}H_{22}O_{11}$(aq).

   (c)   35.2 g of benzene, C_6H_6 and 78.7 g of toluene, $C_6H_5CH_3$.

   (d)   125.0 g of benzene, 25.0 g of dichloromethane, CH_2Cl_2, and 220.0 g of naphthalene, $C_{10}H_8$.

7.15   Calculate the mole fractions of the anions, cations and water in the following:

   (a)   0.25 M KCl(aq);

   (b)   0.15 M $CaCO_3$(aq);

   (c)   7% by mass $MgSO_4$(aq);

   (d)   0.0125 M $Na_2Cr_2O_7$(aq).

Solution equilibria

7.16   Express the Equilibrium Law in words rather than mathematically.

7.17   Write expressions for the equilibrium constants (in terms of concentration) for each of the following reactions:

   (a) SCN^-(aq) $+ Fe^{3+}$(aq) $\rightleftharpoons$ $[Fe(SCN)]^{2+}$(aq);

   (b) Ag^+(aq) $+ 2CN^-$(aq) $\rightleftharpoons$ $Ag(CN)_2$(aq);

   (c) Hg^{2+}(aq) $+ 4Cl^-$(aq) $\rightleftharpoons$ $HgCl_2{}^{2+}$(aq);

(d) $Ni^{2+}(aq) + 6NH(aq) \rightleftharpoons Ni(NH_3)_6^{2+}(aq)$;

(e) $NH_3(aq) + H_2O(l) \rightleftharpoons NH_4^+(aq) + OH^-(aq)$.

7.18   For each of the reactions outlined in Exercise 7.17, determine the units of the equilibrium constant.

7.19   Explain the following terms, using examples to highlight your answers.

(a)   Le Chatelier's Principle;

(b)   polyprotic acid;

(c)   the mass action effect;

(d)   pH;

(e)   pOH.

Solubility product

7.20   Write expressions for the solubility products of the following salts:

(a) Bi_2S_3;

(b) $BaCO_3$;

(c) $AgCl$;

(d) BaI_2;

(e) Sb_2S_3;

(f) Ag_3PO_4.

7.21   Given that the molar solubility of lead iodide, PbI_2, is 1.5×10^{-3} mol dm^{-1}, determine the value of K_{sp}.

7.22   Given that the molar solubility of Hg_2Cl_2, is 5.2×10^{-7} mol dm^{-1}, determine the value of K_{sp}.

7.23   Estimate the molar solubilities of the following salts, using the appropriate K_{sp} values from Table 7.10:

(a) $AlPO_4$;

(b) $Ba(NO_3)_2$;

(c) $NiCO_3$;

(d) Ag_2CO_3;

(e) $Ca_3(PO_4)_2$.

7.24   Use your knowledge of the solubility product concept to explain the common-ion effect, giving an example to illustrate your answer.

7.25   The solubility product for calcium sulfate $(CaSO_4)$ in water at 25°C is 2×10^{-5} mol^2 dm^{-6}. Calculate the solubility of $CaSO_4$ in the following:

(a) pure water;

(b) 0.100 M $Na_2SO_4(aq)$;

(c) 0.200 M $Ca(NO_3)_2(aq)$.

7.26 The solubility product for lead sulfate ($PbSO_4$) in water at 25°C is 1.6×10^{-8} mol^2 dm^{-6}. Calculate the solubility of $PbSO_4$ in the following:

(a) pure water;

(b) 0.100 M $Pb(NO_3)_2(aq)$;

(c) 0.010 M $Na_2SO_4(aq)$.

Acids and bases

7.27 Explain the following terms, using examples to highlight your answers.

(a) a Lewis acid;

(b) a Conjugate acid;

(c) amphiprotic;

(d) a Brønsted–Lowry base;

(e) acid dissociation constant.

Buffer solutions

7.28 Explain why a mixture of ammonium chloride and ammonia in solution has a buffering action.

7.29 Calculate the pH of the following buffer solutions at 25°C:

(a) 0.30 M $HNO_2(aq)$ and 0.40 M $NaNO_2(aq)$;

(b) equal concentrations of $Na_2HPO_4(aq)$ and $KH_2PO_4(aq)$;

(c) 100 cm^3 of 0.10 M $CH_3COOH(aq)$ and 0.10 M $NaCH_3CO_2(aq)$.

Water pollution

7.30 Explain why water is so easily polluted.

7.31 List common types of water pollutants and specify two examples of each.

7.32 Explain why the pH of acid rain is generally regarded as being <5.6.

7.33 Identify the main human health and environmental effects of acid precipitation.

Water quality

7.34 Identify common indicators of water quality.

7.35 Explain how the water quality of river, canal and marine waters can be compared against national standards.

Challenging Exercises

7.36 What experimental evidence do we have that ions exist in solution?

7.37 Explain the variation in the elemental compositions of seawater, river water, acid mine drainage and fog.

7.38 Explain how you would experimentally determine the exact value of the depression of the freezing point of water by using de-icing salt.

7.39 If we mix two aqueous solutions, i.e. 27.0 cm^3 of 0.0001 M sodium chloride with 73.0 cm^3 of 0.0040 M silver nitrate at 25°C, will precipitation of AgCl(s) occur?

7.40 Is pH affected by temperature? Give a reason for your answer.

7.41 A 0.5 l sample of rainwater is titrated by using 0.010 M NaOH(aq). If 16.1 cm^3 of NaOH(aq) is required to reach the end point, what is the pH of the rainwater?

7.42 Calculate the pH of the resulting solution when an acid gas (0.005 M HCl(g)) dissolves in 120 cm^3 of a weak acid buffer solution that consists of 0.050 M ethanoic acid and 0.050 M sodium ethanoate at 25°C. What is the change in pH?

7.43 Explain, in detail, the difference between the total alkalinity and the phenolphthalein alkalinity.

7.44 Explain why the total alkalinity of a lake is a more useful indicator of its ability to resist acidification than its pH.

References

Schwartz, S.E. (1987). 'Aqueous phase reactions in clouds', in *The Chemistry of Acid Rain: Sources and Atmospheric Processes*, Johnson, R.W., Gordon, G.E., Calkin, W. and Elzerman, A.Z. (Eds), American Chemical Society, Washington, DC, USA. ISBN 0-8412-1414-X.

UK Environment Agency, (1998). The State of the Environment in England and Wales: Fresh Waters. HMSO, London, ISBN 0-1131 0148 1.

Further Sources of Information

Further reading

Franks, F. (2000). *Water: A Matrix of Life* (2nd Edn). The Royal Society of Chemistry, Cambridge, UK. ISBN 0-85404-583-X.

Franks, F. (1983). *Water*. The Royal Society of Chemistry, Cambridge, UK. ISBN 0-85186-473-2.

Belousova, A.P., Krainov, S.R. and Ryzhenko, B.N. (1999). Evolution of groundwater chemical composition under human activity in an oilfield. *Environmental Geology*, **38**(1), 34–46.

Hodge, V.F., Stetzenbach, K.J. and Johannesson, K.H. (1998). Similarities in the chemical composition of carbonate groundwaters and seawater. *Environmental Science and Technology*, **32**(17), 2481–2486.

Balarew, C. (1993). 'Solubilities in seawater-type systems – some technical and environmental friendly applications'. *Pure and Applied Chemistry*, **65**(2), 213–218.

Precoda, N. (1991). Requiem for the Aral Sea. *Ambio*, **20**(3–4), 109–114.

Sun, H., Grandstaff, D. and Shagam, R. (1999). 'Land subsidence due to groundwater withdrawal: potential damage of subsidence and sea level rise in southern New Jersey, USA.' *Environmental Geology*, **37**(4), 290–296.

Scholes, L., Shutes, R.B.E., Revitt, D.M., Forshaw, M. and Purchase, D. (1998). The treatment of metals in urban runoff by constructed wetlands. *The Science of the Total Environment*, **214**, 211–219.

Useful reference books

Van der Leeden, F., Troise, F.L. and Todd, D.K. (1990). The Water Encyclopedia (2nd Edn). Lewis Publishers, Chelsea, MI, USA. ISBN 0-87371-120-3.

Useful websites

Groundwater

Groundwater Foundation
http://www.groundwater.org

Acid precipitation

US EPA
http://www.epa.gov/docs/acidrain/ardhome.html

State of the Environment, Norway
http://www.grida.no/soeno95/acidrain/acidrain.htm

National Environment Technology Centre, UK
http://www.aeat.co.uk/netcen/airqual/reports/acidrain/clair.html
http://www.aeat.co.uk/netcen/airqual/reports/acidrain/fresh.html

Water Quality

US EPA
http://www.epa.gov/watrhome/you.html

UK Environment Agency
http://www.environment.detr.gov.uk/wqd/index.htm
http://www.environment.detr.gov.uk/clearnerseas/index.htm

Environment Canada
http://www.mb.ec.gc.ca/english/water/

Environment Australia
http://www.environment.gov.au/science/water/index.html
http://www.affa.gov.au/nwqms/

US National Safety Council
http://www.nsc.ogr/ehc/water.htm (great links page)

Glossary

You'd better watch out, you'd better beware, Albert said $E = mc^2$.
Landscape, from *Einstein-a-Go-Go*.

An **abiotic factor** is a non-living variable (such as temperature) that affects the life of organisms.

Accuracy refers to the closeness of agreement between the experimental results and the true result – the ultimately correct value.

Acid deposition is the term used to describe the total deposition of acid (H^+) or acid-forming species (e.g. $SO_{2(g)}$) by both wet and dry deposition.

Acid rain is defined as any rainfall that has an acidity level beyond what is expected in non-polluted rainfall i.e. having a $pH < 5.6$.

In modern chemistry, **acids** are defined as substances that are proton donors and accept electrons to form ionic bonds.

The **activation energy** is the energy which colliding molecules must possess before a collision will result in a reaction.

The **activity**, A_i, of a chemical species in solution is a formal thermodynamic representation of concentration (its effective concentration).

Adsorption may be defined as the adherence of one particle, ion or molecule to the surface of another.

An **aerobic** process can only take place in the presence of oxygen or air. For example, an aerobic organism requires oxygen for the efficient release of energy from food molecules.

In soil science, the term **aggregate** is used to describe a single mass or cluster of soil particles.

Algae are primitive plants that may be one or many-celled which are capable of synthesizing their cell material by photosynthesis. Algae are usually aquatic.

An **alkali** is a base that is soluble in water.

The **alkalinity** of water is its capacity to act as a base by reacting with protons.

Elements that can exist in several different physically different but chemically identical forms are said to exhibit **allotropy**.

Ammonification refers to the production of ammonia as a result of the biological decomposition of organic nitrogen compounds.

Amphiprotic substances can act as both a proton donor and a proton acceptor.

An **anaerobic** process can take place in the absence of oxygen or air. For example, an anaerobic organism does not require oxygen or air for the efficient release of energy from food molecules.

Atoms that attain a stable electronic configuration and achieve an overall negative charge (A^{ne^-}) by gaining an electron (or electrons) are called **anions**.

The **anode** is the site (electrode) of an electrochemical cell where oxidation occurs as the principal reaction. Electrons (e) flow away from the anode.

An **anoxic** organism, or environment, is one that is being deprived of oxygen.

Groundwater held in porous and permeable rock bounded by an impervious layer is called an **aquifer**.

The **atmosphere** is the envelope of gases that surrounds the Earth.

An **atom** is the smallest unit of matter that can take part in a chemical reaction. It is made up of protons and neutrons in a central nucleus surrounded by orbiting electrons and it cannot be broken down chemically into anything smaller.

Autotrophic bacteria are not dependent on organic material for their carbon.

Avogadro's number is the number of carbon atoms in 12 g of carbon-12 (6.022045×10^{23}). Note, however, that it is now the recommended practice to refer to the related physical quantity, the **Avogadro constant** (L, N_A), which is a fundamental constant and has a value of 6.0221367×10^{23} mol^{-1}.

The **biochemical oxygen demand** (BOD) is the amount of dissolved oxygen taken up by microorganisms in a sample of water.

The cyclical movements of elements throughout the Earth's biosphere are called **biogeochemical cycles**.

The **biosphere** (also known as the ecosphere) is the narrow zone that supports life on Earth. It is limited to a fraction of the Earth's crust (the lithosphere), the hydrosphere and the lower regions of the atmosphere.

A **biotic** factor is a living variable that affects the life of organisms.

A **blank** is a sample containing a negligible (or zero) amount of the analyte being determined.

The ability of natural waters to absorb large quantities of acid or alkali is known as its **buffer capacity**.

A **buffer solution** can maintain a given pH value on the addition of a small amount of acid or base.

The term **calibration** may be defined as the preparation of a scale that can be used to determine the values of unknown quantities. The scale (calibration curve) is prepared by using a calibration standard (a standard solution) in order to obtain a reading from an instrument.

The **carbon cycle** is the cyclic movement of carbon in different chemical forms from the environment to organisms, and then back to the environment.

Carcinogenic substances are chemicals, ionizing radiation and viruses that cause or promote the growth of cancer.

The **cathode** is the electrode of an electrochemical cell where reduction is the principal reaction. Electrons flow towards the cathode.

Cathodic protection is an electrochemical means of corrosion protection in which the oxidation reaction of a galvanic cell is concentrated at the anode and suppresses corrosion of the cathode in the same cell.

The **cation exchange capacity** (CEC) is the quantity of monovalent cations that can be exchanged by 100 g of dry soil.

Atoms that attain a stable electronic configuration and achieve an overall positive charge (M^{ne^+}) by losing an electron (or electrons) are called **cations**.

A **chelate** is a compound formed when metallic cations combine with organic acids, e.g. in the soil. A **chelating agent** is a particular class of complexing agent that possesses two or more sites at which it can bind a metal ion. **Chelation** in soil science is the process by which soil and rocks decompose to chelates as a result of the action of organic acids.

A **chemical bond** is the force that holds together two or more atoms, ions or molecules, or any combination of these. This force is the electrostatic force of attraction between positively charged nuclei and negatively charged orbiting electrons.

A **chemical formula** is a shorthand way to show the number of atoms or ions in the basic structural unit of a compound.

The **chemical oxygen demand** (COD) is the amount of oxygen consumed in the complete oxidation of carbonaceous matter in a wastewater sample.

Clay minerals are naturally occurring inorganic crystalline or amorphous materials in soils (or other earth deposits) of clay size – usually less than 0.002 mm in diameter.

Chemical structures in which the gaps between the atoms are kept to a minimum are called **close-packed structures**.

Coal is a mixture of organic compounds of high molecular weight and complex structure, containing a large percentage of carbon with small amounts of hydrogen, oxygen, nitrogen and sulfur.

A **colligative property** is one that depends upon the number of particles of a solute in solution, but not on the nature of the solute.

A **colloid** is a substance composed of minute particles of one material evenly distributed in another material.

A **complex ion** is formed when one or more distinct molecules or negatively charged ions become attached to a central atom.

A **complexing agent** is a chemical species that combines with a metal ion to form a complex.

The number of ligands attached to the central atom of a co-ordination compound is known as the **co-ordination number**. The latter for metals is the number of nearest neighbours of each atom.

The **core** is the innermost central part of the Earth below a depth of 2900 km that consists of a solid inner core surrounding a liquid inner core.

Corrosion is usually regarded as the chemical attack of a metal by the environment, although environmental corrosion of stone and other materials can also occur.

A **covalent bond** occurs when two electrons are shared between two adjacent atoms, with each atom contributing one electron. In some molecules and ions, both the shared electrons come from one atom, thus forming a **co-ordinate bond**.

The **crust** is the solid outer zone of the Earth that consists of the oceanic and continental crusts.

Crystals are substances with orderly three-dimensional arrangements of atoms or molecules and smooth external surfaces.

Deforestation is the removal of trees from a forested area without adequate replacement.

The **deionization** of water occurs when cations are exchanged for hydrogen ions (H^+) and anions are exchanged for hydroxyl ions (OH^-), producing pure, demineralized water.

The reduction of nitrates to nitrites, ammonia and free nitrogen, particularly in soil by soil organisms, is known as **denitrification**.

Desalination is the purification of brackish (slightly salty) or salty water by the removal of dissolved salts by processes such as distillation, reverse osmosis, deionization, electrodialysis or freezing.

The **detection limit** of an analytical method or instrument may be defined as the lowest mass or concentration of analyte that can meaningfully be detected (i.e. discriminated from zero).

The separation of charge in a polarized molecule is referred to as a **dipole**.

The amount of oxygen gas ($O_2(g)$) dissolved in a known volume of water at a particular pressure and temperature is known as the **dissolved oxygen** (DO) content.

Distillation is a process involving evaporation and recondensation that can be used for producing pure water, separating mixtures of hydrocarbons, or for separating polluting material from water.

The term **ecosystem** is used to describe a community of different species interacting with one another and with the physical and chemical factors making up the non-living environment.

An **electrode** is any terminal by which an electric current passes in and out of a conducting substance.

The term **electrolysis** is used to describe the production of chemical changes by passing an electric current through a solution or a molten salt, resulting in the migration of ions to the electrodes.

An **electron** is a stable, negatively charged, elementary particle. The electrons in an atom orbit the nucleus in groups called shells.

The **electronegativity** of an atom represents its ability to attract electrons.

Electroplating is the electrodeposition of an adherent metal coating upon an electrode to create a surface with different properties or dimensions from the original.

An **element** is a substance that cannot be split chemically into smaller substances. The atoms of an element all have the same number of protons in their nuclei.

An **Ellingham diagram** is a graph showing the variation of standard molar free energy of formation (ΔG_f°) with temperature (T).

An **endothermic** process requires an input of heat in order to proceed.

Enthalpy (H) is a measure of the total energy in a system at constant pressure.

Entropy (S) is a measure of the state of disorder of a system; the greater the disorder, then the greater the entropy.

The excessive enrichment of watercourses by nutrients, leading to the rapid growth of algae and bacteria, and the subsequent reduction of dissolved oxygen, is called **eutrophication**.

Evaporites are minerals formed from the direct evaporation of seawater.

Evapotranspiration is the process by which moisture is lost from the Earth's surface by means of direct evaporation allied to transpiration from vegetation.

Heat is given out during an **exothermic** process.

Fermentation is the term given to the breakdown of sugars by bacteria and yeast by anaerobic respiration.

A **fertilizer** is a material applied to soil in order to provide chemicals essential to plant life.

A **formula unit** is the ratio in which positive and negative ions are present in a crystal structure.

Fossil fuels are non-renewable sources of energy that include peat, coal, oil (petroleum) and natural gas, formed from the remains of plants that lived hundreds of millions of years ago.

The term **free radical** refers to any atom or group of atoms that has an odd number of electrons and is capable of independent existence.

The **Gibbs free energy** (G) is a measure of the available energy in a system at constant pressure.

The trapping of heat energy in the troposphere is called the **greenhouse effect**. The **enhanced greenhouse effect** is the extra absorption of re-emitted heat due to emissions of greenhouse gases.

Groundwater is the name given to fresh water stored in open spaces within underground rocks and unconsolidated material that arises from precipitation which seeps into the ground via water infiltration from lakes, streams and ponds.

A **herbicide** is a chemical used to destroy plants or impede their growth.

Heterotrophic bacteria are sulfate-reducing bacteria that are dependent on organic material for their carbon.

The term **humus** is used to describe the partially decomposed organic material or the slightly soluble residues of undigested material present in topsoil.

Hydrogen bonds are intermolecular, permanent dipole–permanent dipole attractions.

In the **hydrologic** (or water) **cycle** , water is continuously moving through the biosphere via the processes of evaporation, condensation, precipitation and transpiration.

Hydrolysis is a process in which water both reacts with and dissolves mineral constituents.

The **hydrosphere** is the water component of the Earth and includes the oceans, seas, rivers, streams, lakes, swamps, groundwater and atmospheric water vapour.

Igneous rocks result from the crystallization of magma or lava, or from the accumulation and consolidation of volcanic material such as ash.

The term **infiltration** is used to describe the downward movement of water through soil.

An **insecticide** is a chemical used to kill insects.

An **ionic bond** is the electrostatic force of attraction between opposite charges of cations and anions.

Atoms achieving their most stable electronic configuration by gaining or losing an electron (or electrons) are called **ions**.

Isomorphism is the replacement of one atom or ion in a crystal structure by another atom or ion of similar size in such a way that the geometrical arrangement of the crystal lattice remains intact.

Isotopes are forms of the same element, differing in the neutron content of their nuclei, and therefore in their masses.

Kerogen is a solid, waxy mixture of hydrocarbons found in oil shale rock.

Le Chatelier's Principle states that when a system in equilibrium is subjected to change, the system will alter in such a way as to lessen the effect of that change.

The term **leaching** is used to describe the process in which chemicals in the upper layers of soil are dissolved and transferred to lower layers, and in some cases, to groundwater.

A **ligand** is an ion or molecule that is firmly bonded to a central atom or ion. Ligands that can form one co-ordinate bond to the central atom are known as **monodentate ligands**. **Polydentate ligands** can simultaneously form more than one co-ordinate bond with the central atom.

The **lithosphere** is the outer shell of the Earth, composed of the crust, material found in the Earth's plates and the outermost section of the mantle.

A **lone pair** of electrons on an atom, molecule or ion does not participate in bonding.

The **mantle** is the intermediate zone of the Earth between the crust and the core. It may be further sub-divided into the upper and lower mantles.

Metamorphic rocks result from the alteration of other rocks by pressure, heat or chemically active fluids after their original formation.

Minerals are the natural inorganic, crystalline materials that make up the earth and most of its rock. **Primary minerals** are essential constituents of the rocks in which they occur (their presence is typically indicated by the rock name), whereas **secondary minerals** result from the decomposition of earlier minerals.

The **molar mass** is the mass of one mole of an element or compound (molecular or ionic).

The **molarity** of a solution is defined as the number of moles of solute in 1.00 litre of solution.

A **molecule** is the smallest part of an element or a compound that can exist independently under ordinary conditions.

The **morphology** of a soil refers to its physical constitution, including the texture, porosity, structure, consistence and colour of the various soil horizons, their thicknesses, and their arrangement in the soil profile.

Natural gas typically consists of 50–90% methane (depending on its source), together with varying proportions of ethane, propane, butane, nitrogen and carbon dioxide.

A **neutron** is an uncharged elementary particle that forms part of the nucleus of an atom.

The term **nitrification** refers to the formation of nitrates from ammonia, particularly in soils by soil organisms.

The **nitrogen cycle** is the cyclic movement of nitrogen in different chemical forms from the environment to organisms, and then back to the environment.

The process by which dinitrogen gas is converted to ammonia, and ammonium and other nitrogen compounds is called **nitrogen fixation**.

The **nucleus**, in chemistry, is the positively charged central part of an atom.

Any chemical element or compound an organism must take in to live, grow or reproduce is called a **nutrient**. Macro-nutrients are required in large quantities, whereas micro-nutrients are those required in small (trace) quantities.

An **ore** is a mineral from which a metal can be profitably extracted, although the term may be extended to non-metallic rocks such as coal.

Osmoregulation is the process of maintaining the delicate balance of salts and minerals in the tissues of living organisms.

Osmosis is a process in which a substance passes through a membrane from an area of lower concentration to an area of higher concentration of a given solution. In **reverse osmosis** , the solvent's direction of movement is opposite to that experienced during osmosis.

Oxidation may be defined as the loss of electron(s) by an element, compound or ion.

The **oxygen cycle** is the cyclic movement of oxygen in different chemical forms from the environment to organisms, and then back to the environment.

The process by which water moves through the soil in response to the force of gravity and the downward pull of soil pores is known as **percolation**.

Permafrost is permanently frozen material underlying the upper part of the soil profile or a perennially frozen soil horizon.

A **pesticide** is the common name given to any chemical used to combat pests. Pesticides include fungicides, herbicides and insecticides.

Petroleum is a biological product which contains a complex mixture of hydrocarbons-alkanes, cycloalkanes and aromatics-as well as oxygen, nitrogen and sulfur-containing organic and inorganic compounds and trace amounts of metallic substances.

The **pH** of a solution is the negative logarithm to the base ten of the numerical value of the H_3O^+ ion concentration.

The **phosphorus cycle** is the cyclic movement of phosphorus in different chemical forms from the environment to organisms, and then back to the environment.

Photosynthesis is the process by which green plants use energy from sunlight to convert carbon dioxide and water to carbohydrates and dioxygen gas.

Plankton are small, frequently microscopic, forms of animal and plant life that live in the surface layers of fresh and sea water.

Polymorphism refers to a mineral of fixed chemical composition existing in different crystal forms.

Precipitation is the collective name given to rain, dew, fog, hail, sleet and snow.

Precision refers to the closeness of agreement between replicate determinations.

A **proton** is a positively charged, elementary particle found in the nucleus of all atoms.

The ratio of the ionic radius of a cation (r^+) to the ionic radius of an anion (r^-) is called the **radius ratio**.

The **relative atomic mass** (A_R) is defined as the mass of an atom relative to one-twelfth of the mass of an atom of carbon-12. Relative atomic mass is often used now as the preferred term for atomic weight, which takes into account the different isotopes (with different atomic masses) of the element.

The **relative molecular mass** (M_R) of a compound is defined as the mass of one molecule of the compound relative to one-twelfth of the mass of an atom of carbon-12.

A **resonance structure** is a hybrid (blend) of the various possible distributions of electrons in a polyatomic ion. The different distributions of electrons are known as **limiting** or **canonical forms**.

A **resource** may be defined as a concentration of naturally occurring material (solid, liquid or gas) in or on the Earth's crust in such a form and amount that its extraction is currently or potentially feasible.

Aerobic **respiration** is the process in which organisms release energy by using dioxygen gas to break down carbohydrates. Aerobic respiration is the reverse of photosynthesis.

Reverse osmosis (*see* osmosis).

Salinization is the process whereby salts are drawn to the surface of the soil and deposited by evaporation.

Sedimentary rocks are formed by the accumulation and consolidation of rock deposits (mainly from the products of weathering processes), precipitation of matter from solution or compaction of plant and animal remains.

The **sensitivity** of an analytical method or instrument is a measure of its ability to distinguish between small differences in concentration of an analyte.

Soil erosion occurs when topsoil components are moved from one place to another.

A **soil horizon** is the layer of soil roughly parallel to the ground surface that differs from adjoining layers in its chemical, physical and biological properties.

A **soil's texture** gives an indication of the relative proportion of individually sized mineral particles in a soil.

Soils are complex mixtures of inorganic materials (clay, silt, gravel and sand), decaying organic matter, water, air and living organisms, formed by the weathering of rocks.

Solubility is usually expressed as the mass of solute required to saturate 100 g of solvent at a given temperature.

A **solution** is a homogeneous mixture of two or more substances with no definite composition. A solution is saturated when no more of the solute will dissolve in it. Standard solutions are solutions of known concentrations.

Spring acid shock is the term used to describe the rapid release of acids and chemicals from melting snow into soils and surface waters in spring due to a sudden temperature change, causing a dramatic change (decrease) in pH.

Stoichiometry is the branch of chemistry that deals with quantitative relationships between the amounts of reactants and products in chemical reactions.

The **stratosphere** is the atmosphere's second layer and extends from about 17–48 km above the Earth's surface.

The intermolecular force that pulls the liquid molecules on the surface of a liquid together and inward is called **surface tension**.

Surface water may be divided into standing (oceans, lakes and reservoirs) and running (rivers and streams) water.

Thermodynamics is the branch of physics that deals with the transformation of heat into and from other forms of energy.

A **titration** is an analytical technique in which a standard solution is added from a burette to an unknown sample or solution in a conical flask in order to determine its concentration.

Transpiration is the process by which water is lost to the atmosphere by evaporation from the leaves through pores called stomata.

The **troposphere** is the inner layer of the atmosphere that extends for about 17 km above sea level at the equator and 8 km over the poles. It contains about 75% of the mass of the Earth's air.

The **unit cell** of an ionic crystal is the basic building block that is repeated to create a crystal lattice.

The **water table** is the upper limit of the part of the soil (or underlying material) that is totally saturated with water.

Waterlogging is the state of being saturated with water.

The process of **weathering** converts chemically inactive, relatively hard non-porous rocks into soils which are soft, porous and chemically active. **Physical weathering** is essentially erosion (fragmentation) of rocks that increases the surface area exposed to chemical action. During **chemical weathering**, the chemical composition of surface rocks and fragments is changed by chemical reactions, producing more stable compounds or species.

Useful Addresses

American Institute of Biological Sciences, Inc., 730 11th St., NW, Washington, DC 20001, USA (www.aibs.org).

American Water Resources Association , 5410 Grosvenor Lane, Suite 220, Bethesda, MD 20814, USA.

Centre for Marine Conservation , 1725 DeSales St., NW, Suite 500, Washington, DC 20036, USA (www.cmc-ocean.org).

Centre for Plant Conservation , PO Box 299, St Loius, MO 63166, USA (www.rbg.ca/cbcn/en/orgs/op_cpc.html).

Centre for Science in the Public Interest , 1875 Connecticut Ave. NW, Suite 300, Washington, DC 20009, USA (www.cspinet.org).

Clean Water Action , 1320 18th St., NW, Washington, DC 20036, USA (www.cleanwateraction.org).

Coastal Society , 5410 Grosvenor Lane, Suite 110, Bethesda, MD 20814, USA (www.thecoastalsociety.org).

Earth First!, 731 State St., Madison, WI 53711, USA (www.earthfirst.org).

Environmental Law Institute, 1616 P St., NW, Suite 200, Washington, DC 20036, USA (www.eli.org).

Environmental Policy Institute, 218 D St., SE, Washington, DC 20003, USA.

Friends of the Earth (UK), 26–28 Underwood St., London, N1 7JQ. UK (www.foe.co.uk).

Friends of the Earth (US), 218 D St., SE, Washington, DC 20003, USA (www.foe.org).

The Geological Society, Keyworth, Nottingham, NG12 5GG, UK (www.bgs.ac.uk).

Greenpeace, Canada, 427 Bloor St., West Toronto, Ontario M5S 1X7, Canada (www.greenpeacecanada.org).

Greenpeace, UK, Canonbury Villas, London, N1 2PN, UK (www.greenpeace.org.uk).

Greenpeace, USA, Inc., 1436 U St., NW, Washington, DC 20009, USA (www.greenpeace.org).

Hong Kong Environmental Protection Department, 28/F, Southorn Centre, 130 Hennessy Road, Wan Chai, Hong Kong (www.info.gov.hk/epd/).

Ministry for Agriculture, Fisheries and Food, Nobel House, 17 Smith Square, London, SW1P 3JR, UK (www.maff.gov.uk).

National Centre for Atmospheric Research , 1850 Table Mesa Drive, Boulder, CO 80305. (www.ncar.ucar.edu).

National Centre for Urban Environmental Studies , 516 North Charles St., Suite 501, Baltimore, MD 21201, USA.

National Clean Air Coalition, 801 Pennsylvania Ave. SE, Washington, DC 20003, USA.

National Environmental Health Association, 720 S. Colorado Blvd, South Tower, Suite 970, Denver, CO 80222, USA (www.neha.org).

National Geographic Society, 1145 17th St., NW, Washington, DC 20036–4688, USA (www.nationalgeographic.com).

National Institute for Environmental Studies, 16–2 Onogawia, Tsububa-Shi, Ibaraki, 305-0053, Japan (www.nies.go.jp).

National Oceanic and Atmospheric Administration, 14th St., and Constitution Ave. NW, Room 6013, Washington, DC 20230, USA (www.noaa.gov).

National Resources Conservation Service, 14th and Independence Ave., Washington, DC 20250, USA (www.usda.gov).

National Society for Clean Air and Environmental Protection, 44 Grand Parade, Brighton, BN2 2QA, UK (www.nsca.org.uk).

Rachael Carson Council , 8940 Jones Mill Rd, Chevy Chase, MD 20815, USA (members.aol.com/rccouncil/ourpage/rcc_page.html).

Royal Society of Chemistry, Thomas Graham House, The Science Park, Cambridge, CB4 4WF, UK (www.rsc.org).

Soil and Water Conservation Society , 7515 NE Ankeny Rd, Ankeny, IA 50021, USA (www.swcs.org).

Soil Survey and Land Research Centre , Cranfield Rural Institute, Silsoe, Bedfordshire, MK45 4DT, UK (www.silsoe.cranfield.ac.uk/sslrc/).

Transport Research Laboratory, Old Wokingham Rd, Crowthorne, Berkshire, RG11 6AU, UK (www.trl.co.uk).

UK Department of the Environment, Transport and the Regions, Eland House, Bressenden Place, London, SW1E 5DU, UK (www.detr.gov.uk).

UK Environment Agency (South East), Swift House, Frimley Business Park, Camberley, Surrey, GU16 5SQ, UK (www.environment-agency.gov.uk).

UK National Environment Technology Centre, AEA Technology, Harwell, Didcot, Oxfordshire, OX11 0QS, UK (www.aeat.co.uk).

United Nations Environment Programme, Regional North American Office, United Nations Room DC2–0803, New York, NY 10017, and 1889 F St., NW, Washington, DC 20006, USA (www.unep.org).

University of Central Lancashire, Department of Environmental Management, Preston, PR1 2HE, UK (www.uclan.ac.uk).

Urban Pollution Research Centre, Middlesex University, Bounds Green Rd, Bounds Green, London, N11 2NQ, UK (www.mdx.ac.uk/www/uprc/uprc.html).

US Department of Agriculture, 14th St., and Jefferson Dr., SW, Washington, DC 20250, USA (www.usda.gov).

US Department of the Interior, 18th and C Sts NW, Washington, DC 20240, USA (www.doi.gov).

US Environmental Protection Agency, 401 M St., SW, Washington, DC 20460, USA (www.epa.gov).

US Geological Survey, 12201 Sunrise Valley Dr., Reston, VA 22092, USA (www.usgs.gov).

Water Pollution Control Federation, 601 Wythe St., Alexandria, VA 22314, USA.

Women's Environmental Network, Aberdeen Studio, 22 Highbury Grove, London, N5 2EA, UK (www.wen.org.uk).

World Resources Institute, 10 G St., NE, Suite 800, Washington, DC 20002, USA (www.wri.org).

World Wildlife Fund, 1250 24th St., NW, PO Box 97180, Washington, DC 20037, USA (www.wwf.org).

Worldwatch Institute, 1776 Massachusetts Ave. NW, Washington, DC 20036-1904, USA (www.worldwatch.org).

Index

Note: Bold italic numbers refer to figures; italic numbers refer to tables and numbers within [ ] refer to boxes.